Contents

Preface

Introduction to Satellite Communication is designed to meet the needs of working professionals and students. The first edition was a response to a request by many friends and associates for a basic and clear book that provides newcomers with an accessible way to gain knowledge and become productive. The second edition follows the same approach, updating the older and dated material and adding background in newer systems, particularly the global mobile personal communications systems now being introduced worldwide. Whether the reader is technically trained or not, the need exists for an authoritative guidebook to the construction and usage of satellite networks.

This book is designed to give you, the reader, an understanding that should permit you to begin work as a satellite professional or as a user of satellite communication. Sufficient technical information has been included to instill a feeling for how systems are designed and operate. However, it is not essential that every reader comprehend all the concepts presented herein. Readers seeking greater detail can use this book as a basis for further study of the engineering aspects, consulting some of the suggested references. Every effort has been made to simplify complex concepts so that a person without a technical degree or background can grasp them.

Many categories of professionals and students should profit from a significant portion of the material. A major group includes newly hired personnels working for satellite and earth station equipment manufacturers as well as operators of satellites and earth stations. The latter group includes TV and radio networks and their affiliated local stations, direct-to-home satellite service providers (e.g., DBS), and communications service companies such as telephone companies, Internet service providers, and cellular operators. Large corporations employ telecommunication and information services (information technology) professionals who work with terrestrial and satellite-based networks and therefore need to understand many of the principles outlined in this book.

The book's explanatory nature and broad coverage make it suitable as a textbook for university programs and internal training in communication systems design and planning. The first edition was used for an introductory satellite communication systems engineering course at UCLA Extension for many years, and students have remarked how helpful it was in learning the basic principles. It also can be found within easy reach of many industry professionals, who use it as a handy reference. That is because the book covers essentially all aspects of commercial satellite communication systems, from the ground and into space. Nontechnical professionals in associated business management, contracts, legal, and financial fields will find the book particularly helpful when they must deal with telecommunication projects and issues.

The information in the book has been drawn from the author's 30 years of satellite communication experience, which involved assignments in system design and implementation, program management, satellite and ground operations, and marketing of hardware and services. The companies that provided that experience include Hughes Electronics, Communications Satellite Corporation (COMSAT), and Western Union. In addition, the material has evolved through courses the author conducted over the years at UCLA Extension. The author is pleased to help other instructors at universities and technical schools who wish to use the book as a text.

The book is organized into 13 chapters (expanded from nine in the first edition) to correspond to the major areas of commercial satellite communication systems. The text emphasizes the commercial approach throughout because that is where the applications are (government and military systems and applications are outside the scope and content of this book).

Chapter 1, "Fundamentals of Satellite Systems," identifies the key features of satellite communication and reviews some of the more basic concepts in a nontechnical style. It is understandable to all readers, including high school students. Likewise, Chapter 2, "Evolution of Satellite Communication," provides an easy-to-understand history of the technology and its commercial applications. It begins with geostationary Earth orbit (GEO) systems, which are the foundation of the industry, and moves into the new era of non-GEO systems, which promise to double industry size. Another purpose of Chapter 2 is to capture in one place the background with which many newcomers otherwise would not be acquainted (e.g., how we got from Arthur C. Clarke's concept in 1945 to the present day). Chapter 3, "Satellite Network Architectures," covers the ways in which satellite links can be applied to practical communication problems. It gives the reader an appreciation for the variety of uses in which satellites have gained a stronghold. Technologists involved with spacecraft or communication systems will find that Chapter 3 explains many of the mysteries surrounding the business of using satellites. Chapters 1, 2, and 3 have been thoroughly revised and updated from the first edition.

The next two chapters begin the core technical material, focusing on the engineering and design of radio transmissions to and from satellites. Chapter 4, "Microwave Link Engineering," gives the reader a basic understanding of the physics of the radio link between the Earth station and the satellite and covers the factors that are under the designer's control as well as those that are not. It is assumed that the reader has little or no technical training, so the only form of mathematics used is arithmetic. Chapter 5, "Modulation and Multiple Access," rounds out the basic theory of communication as it relates to efficient satellite transmission. The chapter is fairly compact and may be helpful for interested readers to supplement their study with a basic textbook on communication engineering. Nontechnical readers can examine Chapters 4 and 5 but not delve deeply into the engineering details (which are more important to technical professionals, who need to understand features and trade-offs).

Chapters 6, 7, and 8 provide a comprehensive review of the functional elements of all communication satellites. Heavily drawn from the first edition, the expanded material gives additional background to allow readers to focus on the more critical aspects that affect users and operators alike. Newer approaches employing low Earth orbit satellite constellations and digital processing are introduced. To the author's knowledge, such a complete presentation on commercial satellites does not exist elsewhere in the literature or in electronic form. The objective here is to aid readers in understanding the key issues in satellites and is not appropriate for the detailed engineering design of satellite components and subsystems. As is customary, the physical piece of hardware is referred to as the spacecraft, which becomes an artificial Earth satellite (or just satellite) when in orbit. The major elements of the spacecraft are the repeater, the antennas, and the spacecraft bus (the supporting vehicle), which are covered in Chapters 6, 7, and 8, respectively. While those chapters are not essential to understanding how to use satellite communications, they will be of general benefit because the actual operation of a spacecraft affects the performance of the services rendered.

The complementary topics for the ground facilities used in conjunction with the satellite are reviewed in Chapter 9, "Earth Stations and Terrestrial Technology." Care has been taken to include only current classes of Earth stations, particularly those used for satellite control, television broadcasting, fixed-service very small aperture terminal applications, and mobile satellite services (GEO and non-GEO). Chapter 9 will be useful for those readers who plan to use satellite transmission, since ownership and control of ground facilities fundamentally are under the control of the user rather than the satellite operator.

Chapter 10, "Launch Vehicles and Services," covers topics that have become of great concern to operators and major users of satellites alike. The chapter is a complete review of the alternatives for placing satellites into Earth orbit and emphasizes that particular launch vehicle choices change over time. However, because reliability is based on a consistent experience record, much of the change is evolutionary rather than revolutionary. Chapter 10 also discusses the planning and operation of the mission, which is the sequence of events of launch and placement into operating orbit.

A new chapter in this second edition, Chapter 11, "Satellite Operations and Organization," is intended to help potential operators of satellites appreciate some of the special needs of this type of business. In some detail, it reviews the complete satellite control system, the communication network needed to support such a system, and the human resources that are appropriate to those functions. Chapter 12, "Economics of Satellite Systems," provides the underlying characteristics of satellites and Earth stations that are related to the cost of implementing and operating satellite networks. Our perspective is that of a commercial operator who is in business to make money (or reduce costs). The framework is useful for analyzing the economics of either a complete system or a portion of a system (e.g., one or a few Earth stations). Chapter 13, "Future Directions for Satellite Communications," makes a reasonable projection into the future from the year 2000 as a base. It is quite conservative, since developers have not yet thought of some of the more important uses to come. The comparable chapter in the first edition took the same approach and was not far off the mark.

The book can be read sequentially from cover to cover because the material follows a consistent thread. All the elements and uses of spacecraft and Earth stations are covered. Chapters also can be read out of sequence, if necessary, because each chapter explains the concepts relevant to it. References to other chapters are provided throughout. The material is completely current as of the time of publication (circa 1999), but care has been taken to emphasize concepts that are not likely to change quickly. That is the same approach taken in the first edition, a book that remained in print (and in demand) for 12 years. Once read, this book can be used as a reference because most of the terminology in current usage is defined and illustrated. This book is a mate to the author's previous work, *The Satellite Communication Applications Handbook;* together they provide a desk encyclopedia for the overall industry.

Acknowledgments

I wish to acknowledge my colleagues at Hughes Electronics with whom I have worked, studied, and traveled over the years. To them I add my friends and associates in the industry and around the world whom I have helped put satellite communication into practical use. My first introduction to satellites came at COMSAT Laboratories, where we examined many concepts and systems well before they were applied commercially. Since 1990, I have been affiliated with UCLA Extension, and I wish to express my special appreciation to Dr. Bill Goodin, manager of the Engineering Short Course Program, for his continued support. My technical editor, Ray Sperber, made countless contributions to the quality and accuracy of this material (something he also did so ably in the first edition). Finally, I wish to acknowledge my wife, Cathy, who provided the support and motivation necessary to start and complete the revision of this book, and my daughters, who every day make a dad proud.

1

Fundamentals of Satellite Systems

Satellite communications, no longer a marvel of human space activity, have evolved into an everyday, commonplace thing. By the year 2000, satellite systems will represent over $50 billion of investment and provide an essential ingredient to many businesses worldwide. Nearly all television coverage travels by satellite, today reaching directly to the home from space. Even in the age of wideband fiber optic cables and digital switching systems, satellites still serve the basic telecommunication needs of a majority of countries around the world. For example, domestic satellites have greatly improved the quality of service of the public telephone system and brought nations more tightly together. Satellites are adapting to developments in multimedia information and personal communication, with the cost of one-way and two-way Earth stations now within reach of many potential users. A unique benefit has appeared in the area of emergency preparedness and response. When a devastating earthquake hit Mexico City in September 1985, the newly launched Morelos 1 satellite maintained reliable television transmission around the nation even though all terrestrial long-distance lines out of the city fell silent.

This book is about the technologies that make up the field of commercial satellite communications; its intent is to provide a bridge between those who need a practical understanding and those whose business it is to design, implement, and operate such systems. The main objective of the first edition, published in 1987, was to give readers enough of that understanding to allow them to ask the right questions. The explanations and factual material provided in that work and in this second edition are directed toward that objective rather than offering a technical or historical reference. In addition to updating some of the more mature material, it is the purpose of this edition to reach further into the coming millennium and provide more options and some interesting new possibilities for this versatile technology.

In another more recent work [1], we addressed the principal applications that fuel this industry. Reviewed in that book are satellite television, including video distribution and direct-to-home (DTH) broadcasting, digital video and telephone compression systems, very small aperture terminal (VSAT) networks, fixed telephony networks, mobile satellite communication systems, and new broadband technologies. The book you are now reading provides the technical foundation for understanding how those applications are implemented through the design of the space segment and the ground segment.

All the relevant and active fields of commercial satellite communication are covered here.[1] Technical concepts are introduced as needed, without using excessively detailed formulas and design information. Such an approach has been used successfully in our university extension course for practicing engineers, for which this book is used as a text. The reader does not get lost in the forest of engineering equations and complicated diagrams found in other technically oriented textbooks and manuals.

To provide an overview of the field, the remainder of this chapter describes the features of satellite communication networks as well as the principal elements of an overall system. The brief history in Chapter 2 is intended to explain how the satellite industry developed and why things are the way they are. Chapter 2 can be read in sequence or deferred until other chapters have been read, to get a better feel for the technology and terms.

1.1 Basic Characteristics of Satellites

A communications satellite is a microwave repeater station that permits two or more users to deliver or exchange information in various forms. The initial class of Earth-orbiting satellites developed prior to about 1995 consists of those satellites that appear to be fixed above the Earth. As discussed later in this chapter and again in Chapter 6, a satellite in a geostationary Earth orbit (GEO) revolves around the Earth in the plane of the equator once in 24 hours, maintaining precise synchronization with the Earth's rotation. It is well known that a system of three satellites in GEO each separated by 120 degrees of longitude, as shown in Figure 1.1, can receive and send radio signals over almost all the inhabited portions of the globe. (The small regions around the North and

---

1. *Satellite communication* and *satellite communications* are used interchangeably throughout.

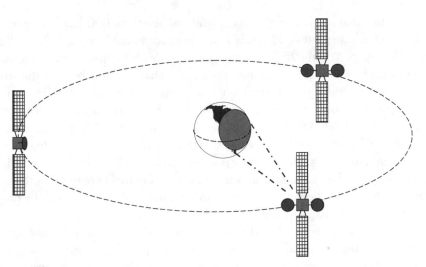

Figure 1.1 A system of three geostationary communication satellites provides nearly world-wide coverage.

South Poles that are not covered do not represent a significant loss of business.) A given GEO satellite has a coverage region, illustrated by the shaded oval, within which Earth stations can communicate with and be linked by the satellite. The range from user to satellite is a minimum of 36,000 km, which makes the design of the microwave link quite stringent in terms of providing adequate received signal power. Also, that distance introduces a propagation delay of about one-quarter of a second for a single hop between a pair of users.

The GEO is the ideal case of the entire class of geosynchronous (or synchronous) orbits, which all have a 24-hour period of revolution but are typically inclined with respect to the equator and/or elliptical in shape. As viewed from the Earth, a synchronous satellite in an inclined orbit appears to drift during a day above and below its normal position in the sky. While ideal, the circular GEO is not a stable arrangement, and inclination naturally increases in time. As discussed in Chapter 8, inclination is controlled by the use of an onboard propulsion system with enough fuel for corrections during the entire lifetime of the satellite. A synchronous satellite not intended for GEO operation can be launched with considerably less auxiliary fuel for that purpose. Orbit inclination of greater than 0.1 degree usually is not acceptable for commercial service unless the Earth station antennas can automatically repoint toward (track) the satellite as it appears to move. Mechanical tracking is the most practical (and cumbersome) approach, but electrical beam steering systems are expected in coming years.

Orbits that are below a mean altitude of about 36,000 km have periods of revolution shorter that 24 hours and hence are termed non-GEO. As illustrated in Figure 1.2, the Iridium system uses multiple satellites to provide continuous coverage of a given region of the Earth. That is simply because the satellites appear to move past a point on the Earth. The Iridium mobile satellite system employs low Earth orbit (LEO), in which satellites are at an altitude of approximately 1,000 km and each passes a given user in only a few minutes. In providing telephone services, users are relatively motionless compared to the satellite they are using. Hence, there is a need to hand off a telephone call while it is in progress. The advantage to using a non-GEO satellite network is that the range to the user is shorter; hence, less radiated power is required and the propagation delay is reduced as well.

The key dimension of a GEO satellite is its ability to provide coverage of an entire hemisphere at one time. As shown in Figure 1.3, a large contiguous land area (i.e., a country) as well as offshore locations can simultaneously access a single satellite. If the satellite has a specially designed communications beam focused on those areas, then any receiving antennas within the footprint of the beam (the area of coverage) receives precisely the same transmission. Locations well outside the footprint generally are not able to use the satellite effectively. The typical example in Asia is the Orion 3 satellite (Figure 1.4), which has 24 C-band and 24 Ku-band transponders. *Transponder* is the term used in the industry to identify one complete microwave channel of transmission from a satellite.

Many satellite networks provide two-way (full-duplex) communications via the same coverage footprint. Terrestrial communications systems, including copper and fiber optic cable and point-to-point microwave radio, offer that

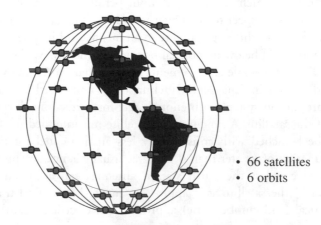

- 66 satellites
- 6 orbits

Figure 1.2   The non-GEO satellite constellation used by the Iridium system.

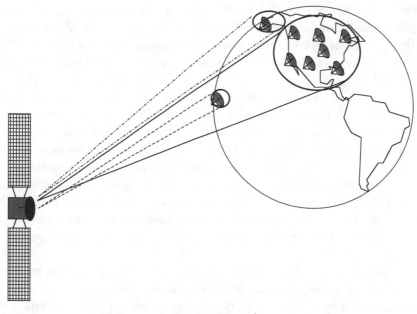

Figure 1.3 Typical footprint of a U.S. domestic communication satellite showing coverage of continental and offshore points.

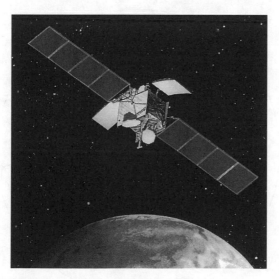

Figure 1.4 Artist's rendering shows the Orion 3, an HS-601 HP model spacecraft from Orion Asia Pacific Corporation manufactured by Hughes Space and Communications International, Inc. (*Source:* Rendering courtesy of Hughes Space and Communications International, Inc.)

capability between fixed points on the ground. Such systems will be around long into the future (as will satellites). Technology is always advancing, and satellite and terrestrial communications will improve in quality, capability, and economy. As shown in Figure 1.5, however, terrestrial systems must spread out over a land mass like a highway network to reach the points of access. They are most effective for connecting major locations, like cities, key suburbs, and industrial parks. The time, difficulty, and expense incurred are extensive, but once established, a terrestrial infrastructure delivers very low unit service costs (e.g., per phone call or data packet) and can last a lifetime. Satellites, on the other hand, are designed to last only about 15 years in orbit, because of the practical inability to service a satellite in GEO and replenish consumables (fuel, battery cells, and degraded or failed components). Non-GEO satellites at altitudes below about 1,500 km are subject to atmospheric drag and a harsh radiation environment and are likely to require replacement after five to eight years of operation.

Another way to view a satellite is as a bypass that can step over the existing terrestrial network and thus avoid the installation problems and service delays associated with initiating local telephone service. Figure 1.6 depicts the three means of long-haul communications used to connect two user locations.

Figure 1.5 An extensive network of cable and microwave links would be required to provide wide area coverage comparable to that of a satellite network.

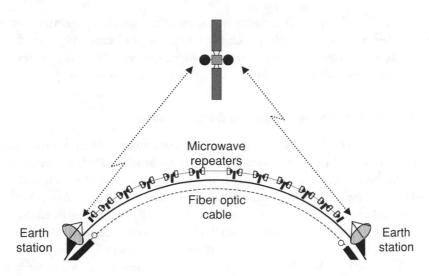

Figure 1.6 Terrestrial microwave and cable systems require multiple hops, while a satellite system provides the same capability with a single hop.

Using the satellite in a duplex mode (i.e., allowing simultaneous two-way inter-active communications), a user can employ Earth stations at each end, elimi-nating any connection with the terrestrial network. In a terrestrial microwave system, radio repeaters must be positioned at intermediate points along the route to maintain line-of-sight contact. It takes about 20 repeater sites to com-plete a microwave link of 1,000 km. That is because microwave energy, includ-ing that on terrestrial and satellite radio links, travels in a straight line with a minimum of bending over or around obstacles. In the case of a long-distance fiber optic cable system, a different form of repeater is needed to restore the dig-ital bits that convey the information. For that reason, cable systems are proba-bly the most costly to install and maintain. Only providers of local and long-distance telephone services and major users of communications services (e.g., government agencies, multinational corporations, railroads, and utilities) are able to justify the expense of operating their own terrestrial cable or microwave networks.

Satellite networks, on the other hand, are well within the reach of much smaller organizations because satellite capacity can be rented from a much larger company or agency. Examples of satellite operators include international consortia, like INTELSAT; public communications companies, like PT Telkom in Indonesia; and private satellite operating companies, like AsiaSat and PanAmSat. All three construct the systems and operate them as a business. The availability of small, low-cost Earth stations that take advantage of reasonably priced satellite transponder capacity has allowed the smallest potential users to apply satellite bypass networks to achieve economies and save time. The idea of a satellite dish on every rooftop is now possible. The rest of this section reviews important aspects of satellites as we enter the second millennium.

1.1.1   Eight Advantages of Satellite Communication

Satellites are used extensively for a variety of communication applications as a result of some well-recognized benefits. The eight benefits identified here derive from the basic physics of the system, the most important of which is that a satellite can "see" a substantial amount of geography at one time. Also, satellites employ microwave radio signals and thereby benefit from the freedom and mobility of wireless connections. Several advantages are interrelated, while others will become more important as the technology and applications evolve.

The point of these benefits is that satellite communication can represent a powerful medium when the developer of the system or service plays to its strengths. Businesses like cable TV networks (which deliver programming by satellite), national department store chains (which use VSAT networks to overcome the limitations of poor or fragmented terrestrial data communication networks), and ocean shipping lines (which demand reliable ship-to-shore communication) depend on satellites as their life's blood. If you can find that kind of connection through satellite technology, then you have a powerful bond on which to build and extend a business or strategic opportunity.

1.1.1.1   Mobile/Wireless Communication, Independent of Location

Any users with an appropriate Earth station can employ a satellite as long as they are within its footprint. A properly designed service establishes an effective radio link between the satellite and the user. If that is possible, the radio link is said to "close." The user can be stationary (*fixed* in the terminology) or in motion (i.e., *mobile*). As long as the link closes, the service will be satisfactory for the intended purpose. An example is COMSAT's PlaNet laptop satellite phone, designed to use the Inmarsat 3 satellites (Figure 1.7). The phone is oriented so that the back of the flip top is pointed toward the satellite position in the equatorial arc.

1.1.1.2   Wide Area Coverage: Country, Continent, or Globe

The coverage footprint of a satellite does not obey provincial, national, or continental boundaries. When designed properly, a satellite can serve any size region that can see it. As mentioned earlier, a GEO satellite can see about one-third of the Earth's surface. Satellites can be colocated at one orbit position to deliver greater quantities of communication channels, as illustrated for the Astra 1 series

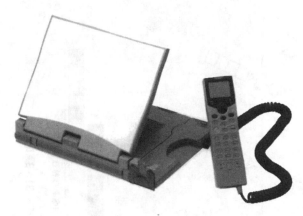

Figure 1.7 The PlaNet mobile satellite telephone offered by COMSAT for use with the Inmarsat 3 GEO satellites.

of satellites (Figure 1.8). A non-GEO constellation like Iridium extends coverage beyond what one satellite sees through a web of intersatellite links.

1.1.1.3 Wide Bandwidth Available Throughout

Frequency spectrum availability for satellites is quite good, and satellite users have enjoyed ample bandwidth, principally for fixed services. Bandwith is the measure of communication capicity in terms of hertz (Hz), either for the amount of radio spectrum used or for the input information that is delivered to the distant end of the link. For C and Ku bands, the available spectrum is of the order of 1 GHz each. That bandwidth can be made available to users across the footprint and can be multiplied through a variety of frequency-reuse techniques. (The bands are reviewed in Section 1.4.)

1.1.1.4 Independence From Terrestrial Infrastructure

By providing a repeater station in space, a satellite creates an independent microwave relay for ground-based radio stations. Installing Earth stations directly at the point of application allows users to communicate without external connections. That can be attractive in places where the terrestrial infrastructure is poor or expensive to install or employ. For example, small telephone Earth stations in China extend reliable voice and data services to remote western regions of the country.

1.1.1.5 Rapid Installation of Ground Networks

Once the satellite (or satellite constellation, in the case of a non-GEO system) is operational, individual Earth stations in the ground segment can be activated

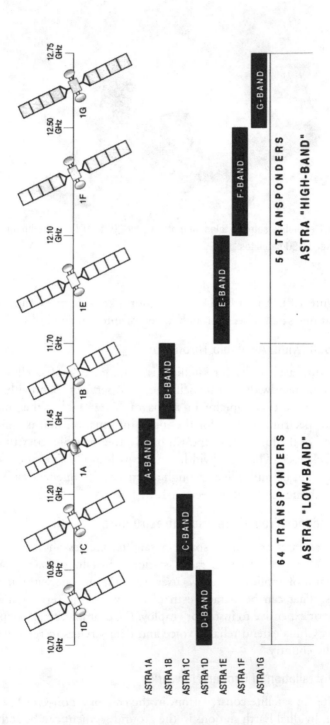

Figure 1.8 Television services are delivered throughout Europe on the Astra-1 GEO satellites, operated at 19.2° WL by Société Européenne des Satellites.

quickly in response to demand for services. Each Earth station can be installed and tested in a short time frame, depending on the degree of difficulty associated with the particular site. That is much simpler than for a terrestrial infrastructure, which requires an extensive ground construction program, including securing rights-of-way along cable routes or for intermediary towers in the case of microwave or other terrestrial wireless systems. Maintenance of a ground-based infrastructure also is substantially more expensive and complex, due to the greater quantity of working elements and the opportunities for failure (e.g., AT&T's long-distance service extended by GEO satellite to U.S. armed forces personnel in Saudi Arabia during the Gulf War).

1.1.1.6 Low Cost per Added Site

Along with rapid installation, the cost of constructing a single site can be quite modest. Simple DTH receivers are consumer items that even homeowners can install. Earth stations for two-way communication may require more care in preparation for secure mounting of the antenna and selection of a suitable location for indoor equipment. Alternatively, hand-held mobile satellite terminals can be purchased for as little as $1,000 each. In general, the cost per site is low enough to keep the particular application within the budget of the particular user. An example is the Iridium telephone made available in October, 1998 (Figure 1.9).

Figure 1.9 The Motorola satellite handphone used with the Iridium satellite system.

1.1.1.7 Uniform Service Characteristics

The footprint of the satellite defines the service area, and within that area services are delivered in precisely the same form. That overcomes some of the fragmentation of service that would result from connecting network segments from various terrestrial telecommunication operators. An example is WalMart's satellite network in the United States that interconnects its stores located in small towns and villages to the Arkansas headquarters.

1.1.1.8 Total Service From a Single Provider

A satellite network can be operated by a single company or by a government agency. The customer, which also could be a nationwide organization, can then deal with a single entity to arrange service. This is in contrast to telephone service in the United States, which is provided by regional companies so that no one company can deliver a consistent nationwide service. (The situation may change as AT&T, MCI, and Sprint seek to enter the local telephone market.) A satellite-based service can be provided nationally or regionally, without regard to fragmentation imposed by the local market situation. An example is the DirecPC system, which delivers Turbo Internet service from Hughes Network Systems to any location in North America (Figure 1.10).

Figure 1.10   DirecPC service using a 60-cm receive antenna allows a home PC user to connect to the Internet. (*Source:* Photograph courtesy of Hughes Network Systems.)

Figure 1.11 The new class of large spacecraft, exemplified by the Hughes HS-702.

HS-702, depicted in Figure 1.11. In contrast, the requirements of non-GEO systems have pushed spacecraft manufacturers to devise more cost-effective platforms. That is because of the vastly greater number of satellites required coupled with the need for quicker replacement. Figure 1.12 illustrates an Iridium satellite from Motorola and Lockheed-Martin, which is indicative of this class of spacecraft.

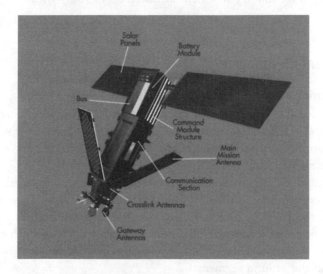

Figure 1.12 The Iridium LEO spacecraft, manufactured by Motorola and Lockheed-Martin.

Launch vehicles have had to keep pace with the demands of spacecraft builders and buyers. As discussed in Chapter 10, all the existing launch vehicle agencies have extended their systems to lift the largest class of GEO satellite into transfer orbit. Multiple launch systems also have been introduced so that non-GEO constellations can be created economically. Competition is the other factor working to the benefit of satellite users, as new rocket systems and launch services companies enter the world market.

1.1.5 Use of Non-GEO Satellites and Constellations

Early experiments with non-GEO orbits, like Telstar 1 and Project Relay, demonstrated that space-based repeaters were practical. Nevertheless, it was the GEO satellites of INTELSAT and Télésat Canada that proved that satellites could be used commercially. That GEO satellites changed the world is undeniable. As we proceed to the next millennium, it is possible that more non-GEO satellites than GEO satellites will be launched in years to come. One of the key reasons for that simply is that it takes many more non-GEO satellites to provide continuous service and reduced propagation delay to ground-based users. That places a much larger economic burden on developers of non-GEO systems, who need more satellites and must replace them more frequently. The other side of the coin is that the non-GEO satellites are individually smaller and possibly less complex and cost less than their GEO counterparts to put into their lower orbits. Service must be handed off from satellite to satellite in a seamless way, a technology first demonstrated in 1998 by Iridium.

1.2 System Elements

Implementation of a communication satellite system is a major undertaking. Fortunately for the great majority of current and future satellite users, several systems already exist and are open for business in every part of the world. Users can lease capacity on domestic satellites and from regional and international operators as well.

In some cases, the requirements may be so extensive that a dedicated satellite or constellation is justified. The purpose of this book is to define and describe all the parts of such a system because an understanding of satellite communication requires the development of a feel for the breadth of the technology. This section describes the system in terms of two major parts, the space segment and the ground segment.

1.2.1 Space Segment

Figure 1.13 is an overview of the main elements of the space segment. As can be seen from the figure, placing a satellite into orbit and operating it for six or more years involves a great deal. Placement in orbit is accomplished by contracting with a spacecraft manufacturer and a launch agency and allowing them about two years to design, construct, and launch the satellite. Chapters 6, 7, and 8 provide considerable detail on the design of a communication satellite, while Chapter 10 reviews the launch aspects.

After the spacecraft is placed in the proper orbit, it becomes the responsibility of a satellite operator to control the satellite for the duration of its mission (its lifetime in orbit). As discussed in Chapter 11, that is a fairly complex task and involves both sophisticated ground-based facilities as well as highly trained technical personnel. The complexity of the problem increases for a non-GEO satellite, even more for a constellation. The tracking, telemetry, and command (TT&C) station (or stations) establishes a control and monitoring link with the satellite. Precise tracking data are collected periodically via the ground

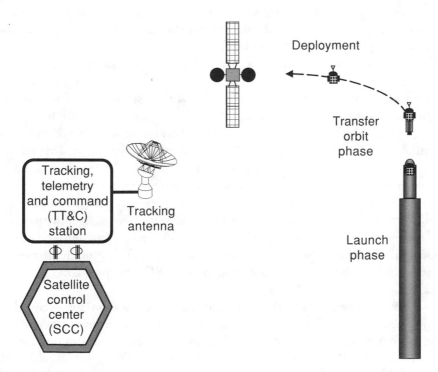

Figure 1.13 The elements needed to implement the space segment of a communication satellite system.

antenna to allow the pinpointing of the satellite's position and the planning of on-orbit position corrections. That is because any orbit tends to distort and shift with respect to a fixed point in space due to irregular gravitational forces from the nonspherical Earth and the pull of the sun and moon.

Eventually, the onboard propellant used for orbit correction is exhausted and the satellite must be removed from service. Retirement of a GEO satellite usually involves using a small reserve of propellant (say, 2 kg) to increase the velocity sufficiently to raise the orbit a few kilometers. The retired satellite will remain in orbit for eternity; with its repeater turned off, it cannot interfere with the operation of usable satellites in GEO. Satellites in LEO are subject to atmospheric drag, which, if not corrected, causes the orbit to decay and the satellite to reenter the atmosphere and burn up.

The second facility shown in Figure 1.13 is the satellite control center (SCC), which houses the operator consoles and data processing equipment for use by satellite control personnel and supporting engineering staff. Other common names for an SCC include operations control center (OCC) and satellite control facility (SCF). The SCC could be at the site of the TT&C station, but more commonly it is located some distance away, usually at the headquarters of the satellite operator. The actual satellite-related data can be passed between the sites over data and voice lines (either terrestrial or satellite). Adding a second TT&C site improves system reliability and would be required for non-GEO satellite operations. As mentioned previously, the level of training of the personnel at the SCC is quite high, involving such fields as orbital mechanics, aerospace and electronic engineering, and computer system operation.

In the launch phase, the satellite must be located and tracked from the ground as soon as it has been released from the last rocket stage. That service normally is obtained from the spacecraft manufacturer. The satellite operator could employ its own TT&C station to participate in transfer orbit maneuvers (actions prior to placement in GEO), and the station can be the point from which commands are transmitted to the satellite to deploy its antennas and solar panels. Some non-GEO systems can be launched directly into the operational orbit, avoiding the transfer orbit phase entirely. Touch-up maneuvers to fine-tune the orbit would be anticipated for GEO and non-GEO satellites alike.

Routine operations at the SCC and the TT&C station are intended to produce continuous and nearly uniform performance from each satellite. Actual communications services via the microwave repeater aboard the satellite do not need to pass through the satellite operator's ground facilities, although any suitably located TT&C station could serve as a communication hub for a large city. Often, the TT&C station includes separate communications equipment to access the satellite repeater for the purposes of testing and monitoring its performance.

One particularly nice feature of a GEO satellite is that the communications monitoring function can be performed from anywhere within the footprint (see Figure 1.3); for example, the SCC can have its own independent monitoring antenna not connected with the TT&C station. GEO and non-GEO satellites can be monitored by several ground antennas strategically positioned around the coverage region. Such antennas often are vital for measuring satellite repeater output and for troubleshooting complaints and problems (users may blame the satellite for poor or no reception when the difficulty lies with their Earth stations).

Another problem area for which monitoring is essential is dealing with harmful interference to communications services. Also called double illumination, it occurs when an errant Earth station operator activates a transmitter on the wrong frequency or even on the wrong satellite. In the congested orbital arcs serving North America, Europe, and eastern Asia, double illumination must be detected quickly and the source identified so disruption of valid communications services is minimized. While the vast majority of interference problems are unintentional, there have been a few instances of intentional jamming and possible piracy (e.g., unauthorized usage of satellite capacity). Satellite operators and users are motivated to act quickly on those rare instances (the author is aware of only three such acts: two in the United States and one in Asia). Such problems, while serious when they occur, are relatively infrequent; satellite transmission is perhaps the most reliable means of communications.

1.2.2 Ground Segment

The ground segment provides access to the satellite repeater from Earth stations to meet communications needs of users (telephone subscribers, television viewers, information network providers, Web surfers). A typical ground segment is illustrated in Figure 1.14; a single satellite is shown to indicate that the links are established through its repeater rather than directly from Earth station to Earth station. Incidentally, *Earth station* is an internationally accepted term that includes satellite communication stations located on the ground, in the air (on airplanes), or on the sea (on ships). Many commercial applications are through Earth stations at fixed locations on the ground; thus, the international designation for such an arrangement is fixed satellite service (FSS). Related to FSS is broadcasting satellite service (BSS), targeted to individual home reception of a variety of broadcast information (e.g., TV, radio, and data). Mobile satellite service (MSS) has been in operation for some time, offering interactive voice

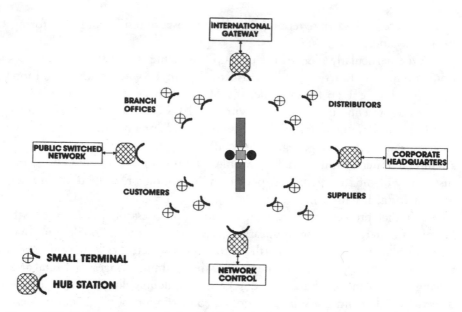

Figure 1.14 The ground segment of a satellite network providing two-way interactive services to a variety of locations.

and data services for ships, aircraft, and individuals. Fixed Earth stations have experienced a tremendous reduction in size at the same time that satellites have grown in size and power capability. Earth stations have evolved from the first international FSS behemoths with 30m antennas (Figure 1.15) to the inexpensive BSS systems, with their half-meter fixed dishes (Figure 1.16). For the new generation of MSS Earth stations, the requirements are being met with hand-held devices about the size of a second-generation hand-held cellular phone.

The majority of user fixed Earth stations (see Figure 1.14) are customer-premise equipment that the user can employ directly with a personal computer (PC), a telephone, or other terminal device. They are inexpensive enough to justify their placement at points of low traffic demand such as small towns, branch offices, or suppliers. Figure 1.17 shows an example of a VSAT, which is a compact and inexpensive Earth station intended for that purpose.

The ground segment, therefore, is not a single, homogeneous entity but rather a diverse collection of facilities, users, and applications. It is constantly changing and evolving, providing service when and where needed. Some networks, such as those used for BSS or MSS applications, may be homogeneous in the sense that a common air interface (CAI) is used for all links over the satellite. Satellites used in an FSS, on the other hand, can support a variety of applications within the common bandwidth of several MHz. The CAI concept

Figure 1.15 This INTELSAT Earth station antenna in Hong Kong was built at 30m to be able to operate with low-power C-band satellites such as INTELSAT III. (*Source:* Photograph courtesy of Hong Kong Telecom.)

still might apply to a subnetwork of VSATs used by a large bank in China or for telephone services to rural villages in Brazil.

One of the four large stations shown in Figure 1.14 is a control point for a portion of the network or perhaps even the entire network. Implemented as an automated network management center, the control Earth station is critical in networks in which traffic flows are dynamic (i.e., changing with time) and

Figure 1.16 The DSS equipment offered by Sony Electronics Corporation consists of a 45-cm satellite antenna, a decoder box, and a remote control. (*Source:* Photograph courtesy of DIRECTV, Inc.)

Figure 1.17   An example of a small customer-premise VSAT used in two-way data communication. (*Source:* Photograph courtesy of Hughes Network Systems.)

in which remote sites are unmanned. Using modern network management workstations and remote microprocessor technology, the network management center can measure and maintain service quality by remote control. Remote control is a particular feature of a satellite network because the satellite provides a common relay point for control messages and other management data. Realizing that type of management integration is a much more difficult problem in large terrestrial telecommunication networks.

1.3   Satellite Orbit Configurations

Having already introduced the concept of GEO and non-GEO satellites, we can now examine some of the technical particulars. Behavior of a satellite in Earth orbit follows Kepler's laws of planetary motion (Figure 1.18), which can be restated for artificial satellites as follows:

1. The orbit of each satellite is an ellipse with the Earth centered at one focus.

2. The line joining the satellite and the center of the Earth sweeps out equal areas in equal times. The area is bounded by the arc segment of the orbit and two lines that extend from the center of the Earth.

3. The square of the period of a satellite is proportional to the cube of its mean distance from the center of the Earth. That relationship is plotted in Figure 1.19 to give an idea of the relationship between altitude and orbital period.

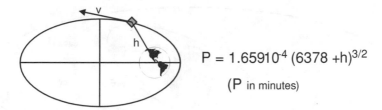

$$P = 1.65910^{-4} (6378 + h)^{3/2}$$

$$(P \text{ in minutes})$$

Figure 1.18 Illustration of Kepler's laws of planetary motion.

Figure 1.20 provides three classical Earth-orbit schemes and their basic properties. LEO systems employ satellites at altitudes ranging from 500 to 1,000 km. Over that range, the orbit period is between 1.6 and 1.8 hours, the higher orbit resulting in a slightly longer period of revolution. The reason for that small change in period is that it is the distance from the center of the Earth that determines the period, not the elevation above sea level. The altitude of a typical medium Earth orbit (MEO) system such as ICO is around 10,000 km (a period of about 6 hours). Between 2,000 and 8,000 km, there is an inhospitable environment for electronic components produced by the Van Allen radiation belt.

The principal advantage of LEO satellites is the shorter range that the radio signal has to traverse, requiring less power and minimizing propagation delay. Their short orbital period produces relatively brief durations when a given satellite can serve a particular user. For altitudes in the range of 8,000 to 10,000 km, a MEO satellite has a much longer period and thus tends to "hang" over a given region on the Earth for a few hours. Transmission distance

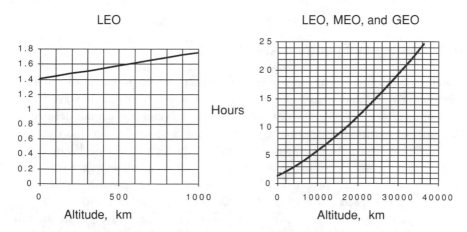

Figure 1.19 Orbit period (in hours) versus altitude, based on Kepler's third law.

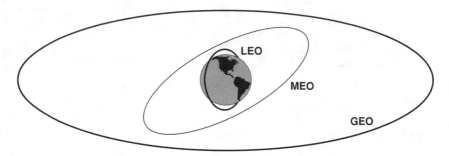

Figure 1.20 The three most popular orbits for communication satellites are LEO, MEO, and
GEO. The respective altitude ranges are 500 km to 1,000 km for LEO; 5,000 km to
12,000 km for MEO; and 36,000 for GEO. Only one orbit per altitude is illustrated,
even though there is a requirement for multiple orbits for LEO and MEO satellites to provide continuous service.

and propagation delay are greater than for LEO but still significantly less than
for GEO. In the case of the latter, there are really two classes of orbits that have
a 24-hour period. A geosynchronous orbit could be elliptical or inclined with
respect to the equator (or both). The special case of an equatorial 24-hour circular orbit, in which the satellite appears to remain over a point on the ground
(which is on the equator at the same longitude where the satellite is maintained), is called GEO. A 24-hour circular orbit that is inclined with respect
to the equator is not GEO becasue the satellite appears to move relative to the
fixed point on the Earth. A GEO satellite would not require ground antennas
that track the satellite, while an inclined geosynchronous orbit satellite might.

1.4 Frequency Spectrum Allocations

Satellite communications employ electromagnetic waves to carry information
between ground and space. The frequency of an electromagnetic wave is the
rate of reversal of its polarity (not polarization) in cycles per second, defined to
be units of hertz (Hz). Alternating current in a copper wire also has that frequency property; if the frequency is sufficiently high, the wire becomes an
antenna, radiating electromagnetic energy at the same frequency. Recall that
wavelength is inversely proportional to frequency, with the proportionality constant being the speed of light (i.e., 300,000,000 mps in a vacuum).

A particular range of frequencies is called a frequency band, while the full
extent of all frequencies for zero to infinity is called the spectrum. In particu-

lar, the RF part of the electromagnetic spectrum permits the efficient generation of signal power, its radiation into free space, and reception at a distant point. The most useful RF frequencies lie in the 300 MHz and 300,000 MHz range, although lower frequencies (longer wavelengths) are attractive for certain applications. While frequencies are all expressed either in hertz or meters of wavelength, they are not created equal. Some segments of the band can be transmitted with greater ease (e.g., can propagate through the atmosphere and space with less loss and temporary fading) than others. Also, differing levels of natural and man-made noise interfere with the transfer. Figure 1.21 indicates the relative merits of the various bands. We see that above about 30 GHz, propagation grows in difficulty due to higher levels of noise and atmospheric attenuation effects. Not shown in the figure is that manmade noise becomes a significant detriment below about 1 GHz.

The optimum piece of the spectrum for space-to-Earth applications lies between about 1 and 4 GHz, which we call the *noise window* for microwave transmission. Above about 12 GHz, there is a significant amount of signal absorption by the atmosphere. There are points of near total absorption at 22 GHz and 60 GHz, due to resonance by water vapor and oxygen, respectively. Those frequencies cannot be used for Earth-to-space paths but can be applied for intersatellite links.

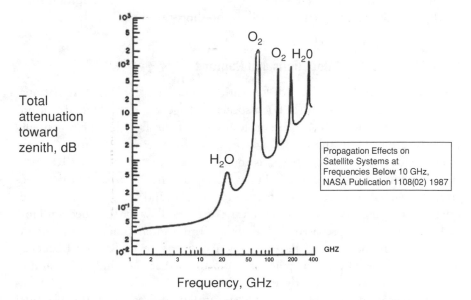

Figure 1.21 Atmospheric attenuation effects for space-to-Earth paths as a function of frequency (clear air conditions).

The most detrimental atmospheric effect above 4 GHz is rain attenuation, which is a decrease of signal level due to absorption of microwave energy by water droplets in a rainstorm. Due to the relationship between the size of droplets relative to the wavelength of the radio signal, microwave energy at higher frequencies is absorbed more heavily than that at lower frequencies. (Sections 1.4.4 and 1.4.5 provide more background.) Rain attenuation is a serious problem in tropical regions of the world with heavy thunderstorm activity, as those storms contain intense rain cells.

An RF signal on one frequency is called a carrier, and the actual information that it carries (voice, video, or data) is called modulation. The types of modulation available are either analog or digital in nature. Once the carrier is modulated, it develops sidebands that contain a representation of the original information. The modulated carrier therefore occupies an amount of RF bandwidth within the frequency band of interest. If two carriers are either on the same frequency or have overlapping bandwidths, then radio frequency interference (RFI) may occur. To the user, RFI can look or sound like background noise (which is neither intelligible nor particularly distressful), or it can produce an annoying effect like herringbone patterns on a TV monitor or a buzzing sound or tone in the audio. In digital modulation, interference and noise introduce bit errors that either degrade quality or produce information dropouts. When the interfering carrier is comparable in power level to the desired (wanted) carrier, the interference effect is classed as harmful, a condition similar to the radio jamming encountered in the shortwave broadcast band.

1.4.1 ITU Spectrum Allocations and Regions

Frequency bands are allocated for various purposes by the International Telecommunication Union (ITU), a specialized agency of the United Nations that has its headquarters in Geneva. Members of the ITU include essentially every government on the planet, which in turn are responsible for assigning radio frequencies in allocated bands to domestic users. Because the RF spectrum is a limited resource, the ITU has reassigned the same parts of the spectrum to many countries and for many purposes around the globe. The consequence is that users of radio communications must allow for specified amounts of RFI and must be prepared to deal with harmful interference if and when it occurs. When there are disputes between countries over RFI or frequency assignments, the ITU often plays the role of mediator or judge. It cannot, under present treaties, directly enforce its own rules. The rules and the frequency allocations are subject to review and modification at the biannual World Radiocommunication Conference (WRC), conducted by the ITU and all its members. The spelling of the word *radiocommunication* was contrived by the ITU itself to reflect the vital connection between the *radio* spectrum (a limited resource that the members must learn to share) and *communication* (perhaps the must valuable application of that spectrum).

The spectrum of RF frequencies is depicted in Figure 1.22, which indicates on a logarithmic scale the abbreviations that are in common usage. The bottom end of the spectrum from 0.1 to 100 MHz has been applied to the various radio broadcasting services and is not used for space communication. The frequency bands of interest for satellites lie above 100 MHz, where we find the VHF, UHF, and super high frequency (SHF) bands. The SHF range has been broken down further by common usage into subbands with letter designations, the familiar L, S, C, X, Ku, and Ka bands being included. Generally, Ku band and those below it are the most popular because of the relative low cost of available equipment and the more favorable propagation characteristics. The Ka, V, and Q bands employ millimeter wavelengths and are potentially useful for very high bandwidth transmissions into small receiving antennas. There are difficulties in applying the millimeter-wave bands because of increased rain attenuation and hardware expense. The letter designations, in general, first were used during World War II as classified code words for the microwave bands applied for radar and other military or government purposes. Today, they are simply short-hand names for the more popular satellite bands, all lying in the range of 1 GHz (1,000 MHz) to 60 GHz.

A typical satellite band is divided into separate halves, one for ground-to-space links (the uplink) and one for space-to-ground links (the downlink). That separation is reflected in the design of the satellite microwave repeater to minimize the chance of downlink signals being re-received and thereby jamming

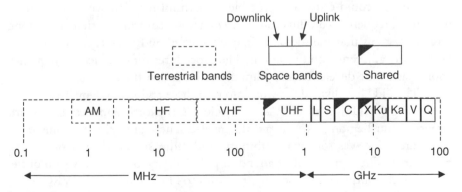

Figure 1.22 The radio frequency spectrum identifying commonly used frequency bands and their designations.

the operation of the satellite. Some satellite systems (notably Iridium, discussed in [1]) transmit and receive on the same frequency but not at the same time.

Uplink frequency bands allocated by the ITU are typically slightly above the corresponding downlink frequency band, to take advantage of the fact that it is easier to generate RF power within an Earth station than it is onboard a satellite. It is a natural characteristic of the types of RF power amplifiers used in both locations that the efficiency of conversion from alternating current (ac) power into RF power tends to decrease as frequency increases. Along with that, the output from the Earth station power amplifier usually is greater than that of the satellite by a factor of between 10 and 100. More recently, satellite systems that make extensive use of VSATs and mobile terminals allow less uplink power, so the cost of the Earth station can be minimized.

An important consideration in the use of microwave frequencies for satellite communication is the matter of coexistence. Figure 1.22 indicates that most of the satellite bands are shared, which means that the same frequencies are used by terrestrial microwave links. Parts of the Ku and Ka bands, on the other hand, are not shared with terrestrial links, so only satellite links are permitted. When sharing is invoked by the ITU Radio Regulations (the so-called Red Books), the users of different systems must take standard precautions to reduce the possibility for RFI. That involves keeping the radiated power below a prescribed threshold and locating transmitting and receiving antennas so as to minimize the amount of interference caused. A band that is not shared, therefore, is particularly valuable to satellite communications, since terrestrial microwave systems can be totally ignored.

Even if the band is not shared, we still have to allow for multiple satellite and Earth station operators, who could interfere with one another. We employ the process of frequency coordination in cases where a new satellite or Earth station operator could cause unacceptable or harmful interference to an existing operator. Frequency coordination is necessary to control interference among satellite systems that use the same frequency band and operate in adjacent (or even nearly adjacent) orbit positions. The procedures for conducting frequency coordination are developed and administered through the ITU.

The ITU has divided the world up into three regions: Region 1 is Europe and Africa (including all of Russia); Region 2 is North America and South America; and Region 3 is Asia and the Pacific. The regionalization, illustrated in Figure 1.23, was somewhat arbitrary, and radio signals do not obey such boundaries. The regions reflect a kind of political/economic subdivision of the world for the purposes of allocating spectrum to ITU members. At WRC meetings, all regions are combined into one, while there can still be Regional Radiocommunication Conferences (RRCs) that are attended by members of a single region. Such RRCs are rare, however.

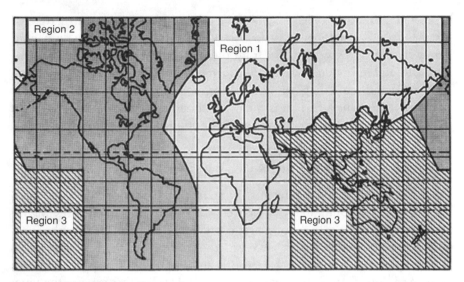

Figure 1.23 The ITU Regions.

The rest of this section reviews the major segments of the satellite allocations and identifies the particular ranges and some of the key features and applications.

1.4.2 VHF and UHF Frequency Ranges

The VHF and UHF bands are defined to be in the ranges 30 to 300 MHz and 300 to 3,000 MHz, respectively. For the purposes of satellite communication, we are concerned with application of the band between about 400 MHz and 1,000 MHz. That range already is highly contested for terrestrial wireless applications, notably cellular; mobile data; and various radio systems used for government and emergency services. In the past, this range was used by the early scientific satellites and for telemetry and command purposes. More recently, some of the small non-GEO systems (referred to as "little LEOs") adopted this range because of the ease of generating power and the convenience of using simple antennas on the satellites and user terminals. The amount of bandwidth available is limited simply because of the low range of frequencies. That, as well as the competition from terrestrial applications, limits the types of services to those that can be delivered by low data rates.

While the satellite hardware and user terminal design are relatively straightforward, the propagation at these frequencies certainly is not. There are advantages to using longer wavelengths to bend around obstacles and to penetrate nonmetallic structures. That is why some of the little LEO systems have chosen these bands for a variety of low-speed data services, particularly paging. However, there is a severe impact from the ionosphere, which can twist, bend, attenuate, and reflect these wavelengths. Any system that uses VHF/UHF must allow for significant fading and other disruption of the transmission, often on a random basis. The situation is more pronounced in the tropics around the geomagnetic equator, particularly during the spring and fall equinox seasons.

1.4.3 Microwave Bands: L and S

The frequency of 1 GHz represents the starting point for microwave applications that involve communication between Earth stations and satellites. Signals travel nearly by line-of-sight propagation and are less hampered by the ionosphere. There are other factors to consider, of course, but the basic fact is that these frequencies have greater bandwidth and more stable propagation under most conditions than do frequencies below 1 GHz. We first explore the range up to about 3 GHz, which includes L and S bands.

With available spectrum much greater than VHF/UHF but less than C or Ku, the L and S bands are particularly effective for providing rapid communications by way of mobile and transportable Earth stations. Two-way voice and data service can be provided through a ground antenna as simple as a small Yagi (TV-type antenna), a wire helix, or even a quarter-wavelength rod (which would be about 4 cm long). The use of relatively high satellite transmit power for each individual channel of communication (voice or data) also helps to reduce the size and cost of the receiving terminal. The trade-off is in the number of satellites per hemisphere and the quantity of voice channels per satellite: instead of being measured in the hundreds of thousands for one C- and Ku-band satellite, capacity of such lower frequency satellites ranges from a thousand to tens of thousands of channels. That can be adequate for high-value mobile telephone communications or such innovative services as two-way paging and personal Web browsing on the move.

The use of such simple antennas on the ground, taking advantage of high power per channel in the satellite, restricts the number of satellites that can operate at the same time within the same footprint of coverage. That applies to GEO and non-GEO satellite systems alike. Figure 1.24 illustrates how an Earth station's antenna beam influences satellite orbit spacing. Satellite A is at the center (peak) of the oval-shaped antenna beam emanating from an Earth station. Operation at a relatively low frequency causes that beam to be broad enough to interfere with the operation of a closely spaced satellite at position B. The chords drawn within the beam show that the strength of the antenna beam (its gain) in the direction of A is greater than in the direction of B, indicating that

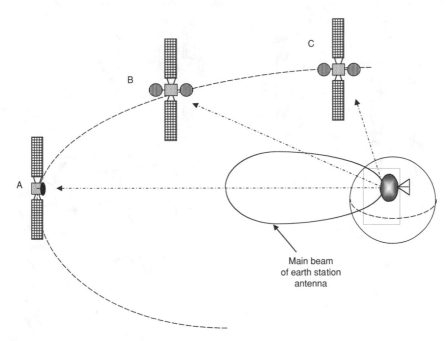

Figure 1.24 Effect of wide-beam Earth station antennas on GEO arc utilization.

RFI probably would be harmful (although not the maximum possible). A simple way to control RFI under that condition would be to segment the frequency band by using nonoverlapping bandwidths for the RF carriers of the respective satellite systems.

For the satellite at position C, the main beam of the ground antenna provides no measurable performance, and proper isolation is achieved. That means satellite C can operate on the same frequencies as satellite A without causing or receiving harmful RFI. Compare that to the situation illustrated in Figure 1.25 for a higher frequency like C or Ku band. Automatically, the ground antenna beamwidth is narrowed to the point that three satellites can be moved closer together while maintaining the desired isolation. The same result is obtained at L or S band if the diameter of the Earth station antenna is increased. Whenever we use a narrower beam to allow more satellites to operate, we introduce a need to point the ground antenna precisely. Typically, we would use a reflector-type antenna, familiar to home DTH installations. The advantage to the more broad beams for simple L- and S-band antennas is that they do not have to be pointed toward the satellite, thus facilitating mobile and personal communication applications.

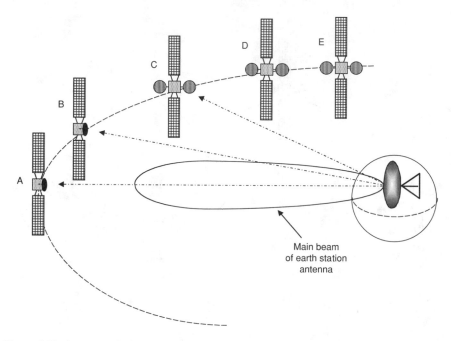

Main beam
of earth station
antenna

Figure 1.25 Improved GEO arc utilization from narrow Earth station antenna beams.

Larger parabolic reflector antennas are practical for L-band ship-to-shore satellite communications. These satellites are operated by Inmarsat, the global provider of international maritime communications. Until 1990, ships at sea depended primarily on HF radio communication using Morse code; that was replaced by Inmarsat terminal usage as the service achieved 100% penetration on large commercial vessels. An Inmarsat terminal can pass one or more digital telephone channels plus two-way data. The ship's directional antenna is protected from the elements by an umbrella-shaped "radome." To compensate for the rolling and pitching of the ship, the antenna is attached to a controlled mount that centers the beam on the satellite. The satellite repeater translates the link to C band for transmission to a fixed Earth station. Telephone calls and data messages then can be routed to distant points over the public telephone and data networks. The maritime mobile satellite system is being expanded, but the number of satellites that can operate simultaneously around the world is limited to fewer than 25. Such mobile services provide reliable communications with ships, airplanes, vehicles, and individuals as they move at will.

S band, nominally centered at 2 GHz (just below C band), was used for the downlink on the first experimental geosynchronous satellite, SYNCOM. It is even closer than C band to the optimum frequency for space communications and is one of the bands preferred by the U.S. National Aeronautics and Space Administration (NASA) for communication with scientific deep-space probes. However, the amount of bandwidth is much less than that afforded by C and Ku bands.

Non-GEO mobile satellite networks like ICO and Globalstar employ S band for the downlink to mobile users. That choice was driven by the need for adequate bandwidth at a frequency that can adapt to the wide variety of situations that users will experience in their day-to-day lives. It was foreseen that L band would not support the simultaneous operation of existing and planned GEO satellites along with the coming generation of non-GEO constellations. S band, like L band, can bend around some obstacles and penetrate nonmetallic buildings. There also is somewhat more bandwidth available for such global applications.

Sharing S band between satellite networks and terrestrial radio services such as industrial and educational television and studio-to-television transmitter links has been extremely difficult to accomplish. NASA, in fact, had to place their deep-space tracking stations in remote parts of the country and the world. Because of the attractiveness of S band for mobile satellite services, terrestrial users gradually are being relocated to other bands. Radio astronomy also coexists with these S-band services.

1.4.4 Microwave Bands: C, X, and Ku

The most popular microwave bands exploited for satellite communication are the C, X, and Ku bands. Without a doubt, C band is the most popular of those bands and has become the mainstay of such industries as cable TV in North America and satellite DTH in Asia. On a worldwide basis, C band is very popular for domestic and international telephone services for thin routes to small cities (and countries) and to reach rural areas of such large countries as China, Russia, Brazil, and Mexico. X band, on the other hand, is the established medium for fixed services for government and military users. The ground terminals that have been developed for these applications are designed to operate under all types of weather by soldiers and security forces in a variety of countries. Ku band was first exploited for high-capacity trunking links between North America and Europe but was later adopted in western Europe for domestic services, particularly TV distribution. It continues as the band of choice for DTH and VSAT networks in developed countries.

C band was the first part of the microwave spectrum to be used extensively for commercial satellite communication. Another common designation we use is 6/4 GHz, which identifies the nominal center of the uplink frequency band (5.925 to 6.425 GHz) followed after the slash (/) by the nominal center of the downlink frequency band (3.700 to 4.200 GHz). (The uplink/downlink convention is not an industry standard, and others have reversed the order to express the numbers in ascending order.) Additional C-band spectrum was allocated by the ITU in the 1990s. Those bands, termed the expansion bands, add about another 500 MHz in each direction. To this day, C band remains the dominant frequency band for commercial satellite communication, even as the higher bands (i.e., Ka) come into greater use. What were the reasons for the selection of C band in the first place, and why does it remain so popular?

The principal benefit of C band is the low level of natural and man-made noise coupled with a very small degree of attenuation in heavy rain. The simplest means of overcoming noise is to increase the level of received signal power from the radio link, either by transmitting more power toward the receiving antenna or by increasing the size of the receiving antenna. Considerably more detail on the subject of radio link engineering is given in Chapter 4. All other things being equal, C band requires less signal level to provide good-quality communications. With nearly 1 GHz of bandwidth available in the uplink and the downlink, C band continues to dominate the widest range of commercial satellite applications. We could even see C band begin to expand its role in DTH and mobile applications, which up to now have concentrated on Ku and L bands, respectively.

Equipment technology and availability were factors in the favor of C band's initial selection for satellite services. In the early years (1965 to 1970), C-band microwave hardware was obtainable from other applications such as terrestrial microwave, tropospheric scatter communications systems (which use high-power microwave beams to achieve over-the-horizon links), and radar. No breakthrough in the state of the art was necessary to take advantage of the technical features of C-band technology. Today, the equipment has been made very inexpensive (relatively speaking) because of competition and high-volume production for the global market.

With all its benefits, C band still is constrained in some ways because of the international requirement that it be shared with terrestrial radio services. Traditionally, C-band Earth stations were located in remote places, where terrestrial microwave signals on the same frequencies would be weak. The potential problem runs in both directions: the terrestrial microwave transmitter can interfere with satellite reception at the Earth station (i.e., the downlink), and RF energy from an Earth station uplink can leak through antenna side lobes toward a terrestrial microwave receiver and disturb its operation. That incompatibility is addressed through the ITU's processes of (1) requiring both types of stations to control emissions through technical sharing criteria and (2) conducting terrestrial frequency coordination when signals could traverse national borders.

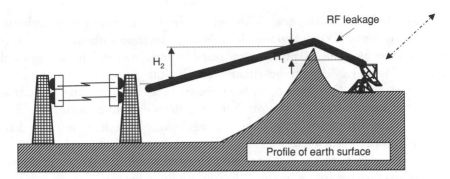

Figure 1.26 The use of terrain shielding to block RFI between a terrestrial microwave link and an Earth station.

Shielding is the technique by which sharing can be made to work (Figure 1.26). Let's say that an Earth station must be located about 30 km from a terrestrial microwave station that either transmits or receives on the same C-band frequency. The signals could be blocked or at least greatly attenuated by a natural or man-made obstacle located very near the Earth station antenna but between it and the terrestrial microwave station. As stated at the beginning of this chapter, microwave signals travel in a straight line, and one would expect that an obstacle would block them entirely. However, microwave energy can bend over the top of such an obstacle through electromagnetic diffraction. Visible light, being electromagnetic in nature, also will bend over a knife-edge obstacle. What diffraction does is cause the wave to propagate over the top of the obstacle and thereby potentially interfere with reception on the other side. The amount of bending can be predicted and is a function of the distances between the source, the obstacle, and the receiver, as well as of the height differences (indicated in Figure 1.26 by H_1 and H_2). If the height differences are large, causing the antenna or antennas to lie well below the top of the obstacle, little signal will reach the receiver and good shielding is therefore achieved. Note that shielding is equal for both directions of propagation (i.e., from Earth station to terrestrial microwave tower and vice versa).

A distance of greater than 50 km usually provides adequate natural shielding from the curvature of the Earth augmented by foliage and man-made structures. Obviously, if the microwave station is on top of a high building or a mountain, the Earth station siting engineer will have to look long and hard for adequate natural shielding. Man-made shielding in the form of a 10–20m metal or concrete wall has proved effective in such difficult situations.

C band must live with yet another aspect of sharing: the ITU regulation that attempts to protect terrestrial radio receivers from direct satellite radiated signals. The level of such signals is low relative to that emanating from an Earth station; however, if satellite power is allowed to increase without bound, it is theoretically possible to produce measurable interference into a terrestrial receiver. Figure 1.27 depicts the situation that has the greatest potential for such interference: when the antenna of the terrestrial station is pointing at another terrestrial station along the horizon but on a direct path to a satellite in orbit. To deal with that potential problem, the regulations limit the power flux density (power per square meter on the surface of the Earth produced by the satellite) to an amount that would not cause significant interference to any terrestrial microwave receiver no matter where it might be located.

Government and military satellite communications systems in the United States and some other countries employ X band and, on a limited basis, Ka band. With an uplink range of 7.90 to 8.40 GHz and a downlink range of 7.25 to 7.75 GHz, X band is used extensively for military long-haul communications links, much as C band is used on a commercial basis. In highly specialized cases, Ka band is being applied because very narrow spot beams can be transmitted to and from the satellite.

The frequency band that has done more to interest new users of satellite communications is Ku band, a part of the spectrum lying just above 10 GHz. Portions of Ku band are not shared with terrestrial radio, which has some

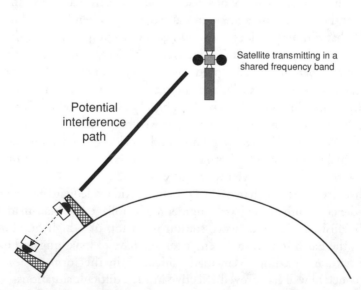

Satellite transmitting in a
shared frequency band

Potential
interference
path

Figure 1.27 The potential for direct RFI from a satellite into a terrestrial microwave receiver where the frequency band is shared.

advantages over C band particularly for thin-route and receive-only services using inexpensive terminals with small-diameter antennas. The precise uplink and downlink frequency ranges allocated by the ITU vary to some degree from region to region.

Effectively, three sections of Ku band have been allocated to different services on an international or a domestic basis. The most prevalent is FSS, which is the service intended for one- or two-way communications between fixed points on the ground. All of C band and much of Ku band are allocated to FSS for wide application in international and domestic communications. Part of Ku band is subject to the same coordination and siting difficulties as C band. This particular part of Ku band is referred to as 14/11 GHz, where the uplink range is 14.00 to 14.500 GHz, and the downlink range is 10.95 to 11.7 GHz (minus a gap of 0.25 GHz in the center). Only the downlink part of the allocation is actually subject to sharing.

A portion of the Ku allocation for FSS that is not shared with terrestrial services is referred to as 14/12 GHz (the uplink range again is 14.00 to 14.50 GHz, and the downlink range is 11.70 to 12.20 GHz). The availability of 14/12 GHz is limited to Region 2, which comprises North America and South America, and can be used only for domestic communications services. North America, in particular, has seen wide use of 14/12 GHz since about 1982; services in this band were first introduced in South America in 1992. Power levels from these satellites are not subject to the same restrictions as at C band, although there is an upper limit, to minimize interference between satellites. Ku-band satellite operations in the rest of the world (i.e., Regions 1 and 3) are restricted to the 14/11 GHz shared allocation. In some instances, a Region 1 (Europe and Africa) or Region 3 (Asia) country can make 14/11 GHz appear like 14/12 GHz simply by precluding domestic terrestrial services from that band. However, terrestrial radio services in adjacent countries are not under their control, so international coordination still must be dealt with in border areas.

A third segment of Ku band, referred to as 18/12 GHz, is allocated strictly to BSS. As with the 14/12 FSS band in Region 2, the BSS band is not shared with terrestrial services. Its intended purpose is to allow television broadcasting and other DTH transmissions from the satellite. There are two regulatory features of this band that make direct broadcasting to small antennas feasible. Because the band is not shared with terrestrial radio, the satellite power level can be set at the highest possible level. Adjacent satellite interference could be a problem but is precluded by the second feature: BSS satellites are to be spaced a comfortable 6 or 9 degrees apart, depending on the region. In comparison, while there is no mandated separation between FSS satellites, a 2-degree spacing has become the standard in crowded orbital arc segments such as North America, Europe, and East Asia.

The operational advantages of 14/12 and 18/12 GHz lie with the simplicity of locating Earth station sites (without regard to terrestrial radio stations) and the higher satellite downlink power levels permitted. The latter results in smaller ground antenna diameters than at C band, all other things being equal. However, as discussed earlier in this section, Ku band is subject to higher rain attenuation, which can increase the incidents and duration of loss of an acceptable signal. Figure 1.28 indicates the relative amounts of extra downlink power, measured in decibels of margin, needed to reduce the outage time to four hours a month (on average). Ka band (30/20 GHz) is included for completeness and is discussed in Section 1.4.5. The decibel (dB) is a logarithmic scale (like the Richter scale used to measure earthquake intensity) and is discussed in Chapter 4.

As shown in Figure 1.28, the amount of margin to overcome a fade is also a strong function of the elevation angle from the Earth station to the satellite in orbit. Rain cells, which are large atmospheric volumes (wider than they are high), produce the greatest attenuation. Because of the flattened shape, radio waves that take off at a low elevation angle relative to the ground must pass through a greater thickness of rainfall. Consequently, elevation angles of 20 degrees or greater are preferred for Ku-band frequencies and higher. Another important variable is the local climate: deserts are less affected than tropical regions. In general, the need for greater power margin at Ku band tends to reduce some of the benefits obtainable by virtue of the higher satellite powers

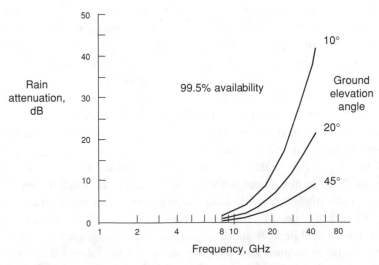

Figure 1.28 Rain attenuation as a function of frequency for different ground elevation angles.

permitted by international regulations. A proper system design at Ku band still can deliver a quality TV, voice, and data signal that is maintained for a satisfactory period of time. Only for those regions with the highest rainfall, such as a tropical rain forest, and where there is a high incidence of thunderstorms, would Ku band tend to be less practical than C band.

1.4.5 Millimeter Wave and Higher: Ka, Q, V, and Optical Bands

Beginning at around 20 GHz and extending to suboptical frequencies, we find the millimeter-wave spectrum range. These frequencies were viewed as the domain of the researcher until around 1970, when satellite planners began to look ahead to the time when the C and Ku bands would be heavily used. Also, they saw some unique applications that might employ the greater bandwidth and smaller wavelengths that came with these allocations. First to be considered was Ka band, with its 30/20 GHz allocations and about 2.5 GHz of available spectrum. The frequencies above about 40 GHz, designated as Q and V band, are now being evaluated for space-to-ground applications because of the wide bandwidths and ability to transmit through very narrow spot beams. During heavy rainstorms, the atmosphere will appear to be nearly opaque at these frequencies, so applications must allow for some periods of total outage.

There is a potential for the use of Ka band (30/20 GHz) for commercial applications in cases where C and Ku bands are fully occupied. During the 1970s, the Japanese space agency, NASDA, in conjunction with the government-owned telecommunications company, NTT, launched some experimental communications satellites that operate in the Ka-band frequency range. The experimental links generally showed that the ample bandwidth that was allocated could in fact be used under certain circumstances. Severe rain attenuation at these high frequencies would require that a given Earth station include two receiving sites separated sufficiently so that only one is heavily affected at a time. This principle of diversity reception allows the downlink to be switched to the site with the acceptable signal. That works because the most intense part of the storm is smaller than the spacing between receiving sites. In the United States, NASA has been pursuing a precursor 30/20 GHz program called the Advanced Communications Technology Satellite (ACTS) to advance the technology base. The ACTS demonstrated a series of interesting and possibly useful applications in the area of high-speed data transfer and mobile communications, all at Ka band. A kind of frequency "land rush" developed in the mid-1990s in the United States, when several companies filled applications for spectrum and GEO orbit positions with the Federal Communications Commission (FCC). None of these systems was actively in construction at the time of this writing. However, experiments with Ka band have continued unabated as government and private organizations work to identify the parameters for a successful Ka-band commercial application. An exception is the Iridium system, which currently uses Ka band for intersatellite and gateway Earth station links. Prospects for advanced Ka-band systems are reviewed in Chapters 2 and 13.

Reference

[1] Elbert, B. R., *The Satellite Communication Applications Handbook*, Norwood, MA: Artech House, 1997.

2

Evolution of Satellite Communication

The satellite communication industry has grown tremendously since the first edition of this text, and with it the number of professionals and the range of activities have grown as well. With many of the originators of the basic technical and operating principles having reached retirement, it is likely that a significant fraction of industry participants lack a historical perspective. The purpose of this chapter is to repeat the history that appeared in the first edition and to add much that occurred over the next decade. What follows is an overview of key developments and events; it is not meant to be exhaustive. It can only be said that satellite communication is an extremely rich field of endeavor, encompassing space technology, telecommunications systems, information technology, and multimedia entertainment. Add to that the introduction of new megamarkets in satellite-based wireless personal communication, and you begin to comprehend the possibilities for new historical chapters after 2000.

Figure 2.1 is a basic timeline of many of the critical events. A familiarity with the past satellite programs and applications will aid the reader in understanding why today's satellite systems and services are the way they are. This chapter can be read in sequence or deferred until the basic concepts covered in Chapters 1 and 3 through 8 have been absorbed.

2.1 Source of the Original Idea

While various individuals in the early part of this century recognized the existence of the 24-hour geostationary orbit, it is the eminent science fiction author

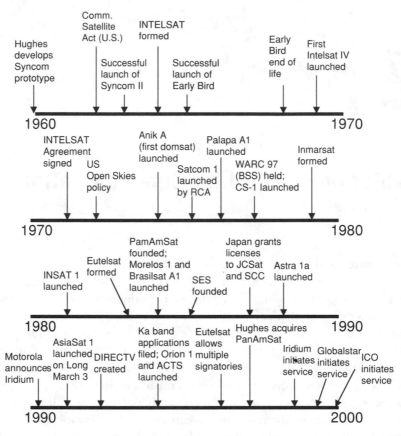

Figure 2.1 Timeline of critical events in the satellite communications industry.

Arthur C. Clarke who is credited with conceiving of the application to communication satellites. His article on extraterrestrial relays in *Wireless World* magazine back in 1945 specifically describes a three-satellite worldwide network, including a drawing similar to Figure 1.1. In recent years, some leaders in the satellite communication field have proposed that the GEO be renamed the Clarke Orbit, in recognition of Mr. Clarke's forethought. He knew at that time that the requisite electronic and spacecraft technology would take a few years to develop; in fact, it was not until the early 1960s that the first workable spacecraft was built and launched into geosynchronous orbit.

Following the launch of the Soviet Sputnik satellite in 1957, the United States accelerated its space program and many proposals were developed for space relay satellites. The geosynchronous satellite was thought by many to be too speculative, and efforts were centered instead on LEO passive reflector and active repeater technology. Programs like Echo, Advent, Telstar, Courier, and Relay were pursued by NASA (and AT&T in the case of Telstar) in hopes of coming up with a viable system that would be competitive with terrestrial radio, buried cable, and transoceanic cable. The feeling in that community was that LEO was the appropriate orbit because of the reduced time delay and the difficulty of constructing and operating a workable GEO spacecraft. It is interesting that non-GEO systems have come back into favor after lying dormant for many years.

2.1.1 SYNCOM

In 1959, Dr. Harold A. Rosen and two of his colleagues at Hughes Aircraft Company, Tom Hudspeth and the late Don Williams, identified the key technologies for a simple and lightweight active repeater communication satellite for launch into GEO. Dr. Rosen was the product champion and systems engineer; Tom Hudspeth provided the microwave know-how that produced a compact but effective repeater for space application; and Don Williams, a brilliant Harvard-trained mathematician, figured out how the satellite could be placed into orbit and oriented for satisfactory operation.

The basic concept used a drum-shaped spinning body for stability with tiny gas jets to alter the attitude (orientation) in space. In 1960 at Hughes Laboratories in Los Angeles, the team built a working prototype that demonstrated the feasibility of spin stabilization and microwave communication. Supported by Dr. Allen Puckett, then vice president and now retired chairman of the board of Hughes Aircraft, they convinced NASA and the Department of Defense to go ahead with the launch of SYNCOM, an acronym for "synchronous orbit communications satellite." On the second attempted launch of the Thor Delta rocket in July 1963, SYNCOM II became the first operational geosynchronous satellite providing intercontinental communications. Little did the team know that SYNCOM, shown in Figure 2.2, would be the precursor for a powerful telecommunication medium and the foundation of whole new industries.

2.1.2 COMSAT

The idea of developing a commercial communications business based on the use of satellites can be attributed to the administration of President John F. Kennedy, which issued ground rules in 1961 for the U.S. operation of an international communications satellite system. President Kennedy favored private ownership and operation of the U.S. portion of an international system that would benefit all the nations of the world, large and small. The

Figure 2.2 Historic photo taken from high in the Eiffel Tower shows the world's original synchronous communications satellite, which Hughes Aircraft Company demon-
strated at the Paris Air Show in 1961 in an effort to find a sponsor to launch it.
"This is as high as it will ever get," observed one skeptic. Scientists pooh-
poohed the satellite as unworkable until NASA backed Hughes in the SYNCOM
program and paved the way for the current age of space communications. The
original satellite embodied design features now used in every commercial satellite in service. Thomas Hudspeth (left) designed the spacecraft's electronics,
and Dr. Harold A. Rosen (right) was the leader of the group that conceived and
hand-built the satellite. (*Source:* Hughes Space and Communications Company.)

Communications Satellite Act of 1962 resulted from that initiative, establishing the charter for the Communications Satellite Corporation (COMSAT).
Under the leadership of its first chief executive, Dr. Joseph Charyk, the company developed a workable satellite system and helped to encourage worldwide
expansion. Over the years, COMSAT gradually moved away from that core and
into a variety of related and unrelated businesses, including ownership of the

Denver Nuggets basketball team. Nevertheless, many of the innovations in satellite communication networks and operation were COMSAT's creations.

COMSAT raised money through the sale of common stock both to the public and to the U.S. common carriers, particularly AT&T. At the time, it was assumed that subsynchronous satellites would be used basically for two reasons. First, the U.S. communications companies felt that the time delay associated with the propagation of radio waves, which travel at the speed of light (approximately one-quarter of a second for the GEO uplink and downlink combined), would prove to be unacceptable to a significant percentage of telephone subscribers. Second, there existed a lingering doubt about the viability of the technology necessary to deliver and operate the spacecraft. Those fears were substantially dissolved when SYNCOM demonstrated the feasibility and quality of synchronous orbit satellite communications. That success allowed COMSAT to create INTELSAT and, at the same time, to save more than half its original $200 million of capital (a lot of money at the time) that otherwise would have been needed to launch the required LEO constellation.

2.2 Evolving Satellite Coverage

Once Hughes and COMSAT proved that GEO satellites were the wave of the future, it remained a difficult task to bring the first systems into practical existence. That was the role of the global satellite operators, along with regional and domestic systems. All three demonstrated that GEO satellites could fill a wide variety of telecommunications needs, in line with the eight advantages of satellite communications outlined in Chapter 1.

2.2.1 Global Service: INTELSAT, PanAmSat, and Orion

Clearly, the first and most demonstrable need for the commercial satellite was to provide international communications links. At the beginning of the international GEO satellite era, most telecommunications companies around the world were government owned or controlled (with the exception of North America). In time, a trend toward privatization and even competition appeared, which eventually affected the structure of international satellite operation. The following paragraphs outline the evolution of the now well-established international satellite system.

2.2.1.1 Early Bird

In March of 1964, COMSAT contracted with Hughes Aircraft Company for the construction of two spin-stabilized satellites using C band (SYNCOM used

S band, which was authorized only for NASA experimentation). Early Bird, the well-recognized name for this first commercial satellite, was launched on a McDonnell Douglas Delta rocket in the spring of 1965. The satellite worked perfectly for six years, well past its design life of two years. The second spacecraft was never launched and now resides in the Air and Space Museum of the Smithsonian Institution in Washington, D.C. Early Bird was positioned over the Atlantic Ocean and was first used to link stations in Andover, Maine, and Goonhilly Downs, England, providing voice, telex, and television service. Having gained such renown, *bird* continues to be used as a nickname for all GEO satellites.

2.2.1.2 INTELSAT

The intended international nature of the system led the U.S. government and COMSAT to establish a new organization in 1964 called the International Telecommunications Satellite Consortium (INTELSAT). Later, *Consortium* was changed to *Organization*, and INTELSAT grew to become the preeminent satellite operator in the world. The INTELSAT Agreement, a treaty among member countries, was signed in 1971. The members of INTELSAT are the domestic telecommunications operators or selected satellite operator, depending on the business and regulatory approach taken in a given country. The organization was established through a treaty that gives INTELSAT a unique status in global telecommunications politics (at the time of this writing, INTELSAT was going through a restructuring called privatization). COMSAT continued as the U.S. signatory, while the Government Post Office (now British Telecom) fulfilled that role for the United Kingdom. In time, INTELSAT evolved into more of a business, purchasing spacecraft based on the global demand for telephone and TV services and other noninternational applications.

INTELSAT operates as a wholesaler, providing space segment services to end users who operate their own Earth stations. Users have had to arrange with their respective signatories for satellite capacity and pay a charge for all space segment services. The charges vary depending on the type, amount, and duration of the service. Any nation, whether or not it is a member, can use the INTELSAT system. Some INTELSAT member nations have chosen to authorize several organizations to provide INTELSAT services within their own countries. Currently, INTELSAT has more than 300 authorized customers. Most of the decisions that INTELSAT's member nations must make regarding the INTELSAT system are accomplished by consensus, a noteworthy achievement for an international organization with such a large and diverse membership.

This basic approach served INTELSAT and the world community well for three decades. The signatories and their respective users employ the capacity to implement international telephone and TV links to augment terrestrial transmission means like fiber optic cable and microwave links. One important application is the use of INTELSAT links for fiber optic cable restoration. Also, INTELSAT offers excess transponders for use by signatories for domestic communication purposes. That can allow a given country to forestall the acquisition of its own domestic satellite. In 1984, INTELSAT introduced the INTELSAT Business Service (IBS), an innovative service for business communication private lines. In some countries, corporate users can install and operate their own Earth stations for direct access to the INTELSAT space segment. The service became very popular, because it was based on digital transmission and the pricing was attractive relative to public leased lines. INTELSAT then proceeded to digitize all its international telephone links, resulting in the extensive use of intermediate digital rate (IDR) carriers. The widespread use of IDR carriers at capacities of 2 Mbps and higher has improved the efficiency and quality of telephone services over INTELSAT satellites.

INTELSAT goes through a deliberate process to establish the requirements for its satellites and to procure them from manufacturers. Each generation of spacecraft, indicated in Figure 2.3, is marked by a significant upgrade in capability. In the following discussion of generations of INTELSAT's satellites, only the first letter is capitalized (e.g., Intelsat I) to distinguish the name of a satellite series from the name of the organization itself.

NASA provided the motivation for the construction by Hughes Aircraft of the Intelsat II series to support the worldwide communications requirements of the Apollo program, which placed men on the moon. The design of the spacecraft was similar to Early Bird with the exception that transmitter power was increased (supported by more solar cells and batteries; those aspects of the design are covered in detail in Chapter 8). Intelsat I and II employed a simple antenna that radiated the signal 360 degrees around the satellite as it spun, obviating the need to point or control the physical antenna structure in space. The first Intelsat II did not reach geosynchronous orbit because of failure of an internal rocket called the apogee kick motor; however, the remaining three were successful, and the satellites provided worldwide service for five years.

An innovation in the Intelsat III spacecraft, built by TRW of Redondo Beach, California, was the despinning of a directional antenna aboard the satellite so as to maintain a more intense beam on the Earth. This type of global coverage beam (see Figure 1.1) allows access to the satellite repeater by any Earth station that lies within the hemisphere facing the satellite. Eight of those satellites were built and launched, although the first launch failed and the apogee motor on the last malfunctioned. A common characteristic of Intelsats I, II, and III was their limited capacity in terms of the number of individual transmitters because electrical power generated by their small solar panels could support only one or two power amplifiers.

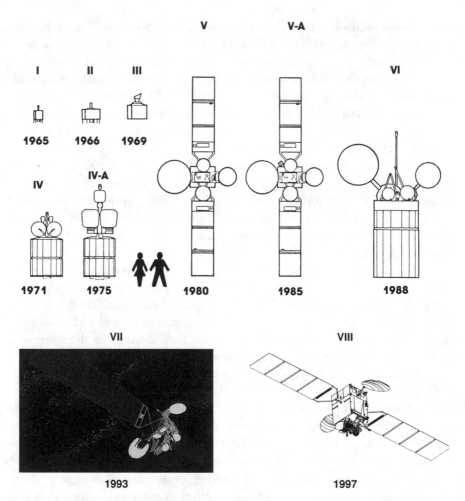

Figure 2.3 Eight generations of INTELSAT satellites.

The big step up in satellite capability happened with the introduction of the Hughes Aircraft–built Intelsat IV (Figure 2.4), a spacecraft many times larger than its commercial forerunners. Intelsat IV employed the same basic spacecraft configuration as a military satellite called Tacsat, also built by Hughes Aircraft. A major innovation came in the way the repeater and the antenna were attached. Instead of spinning the repeater with the spacecraft and despinning the antenna, the repeater with the antenna were despun as a package. The configuration resulted from the work of Dr. Anthony Iorillo and is called the GYROSTAT (described in Chapter 8). With considerably more power and communications equipment, the C-band repeater on Intelsat IV contained 12

Figure 2.4 The Intelsat IV spacecraft. (*Source:* Photograph courtesy of Hughes Aircraft
Company.)

individual channels of approximately 36 MHz each. The selection of the now
standard 36 MHz RF bandwidth resulted from the technical trade-off involving the bandwidth required for a single TV channel (using FM) and the maximum number of amplifiers that could be carried and powered. The term
transponder has come into common usage to refer to the individual RF channel, typically of 36 MHz bandwidth. The first Intelsat IV was launched in
1971, and a total of seven have each provided 10 or more years of service. Only
one Intelsat IV was lost, and that was due to the failure in 1975 of the Atlas
Centaur rocket to reach orbit. The Intelsat IV series became the backbone of
international communications, and the last of the series was retired at the end
of 1985. The transponder concept has been used in every subsequent commercial satellite operating at C and Ku band.

INTELSAT saw the need to go beyond the design of Intelsat IV to increase the capacity in the Atlantic Ocean region (AOR). The ideal solution in almost all cases is to have one large satellite at which all countries point their antennas. However, the 1970 traffic requirements among 50 Earth stations exceeded the capability of one Intelsat IV, and INTELSAT needed a solution. Dr. Rosen of Hughes Aircraft developed a practical approach to increase satellite capacity while retaining the basic configuration of the Intelsat IV spacecraft. The concept, called frequency reuse, doubles the 500 MHz of allocated bandwidth at the satellite by directing two independent beams toward the visible face of the Earth (the east and west hemi beams in Figure 2.5). Shaping of each beam causes the footprint to cover the land mass of the eastern or western hemisphere, leaving the open sea and some land extremities outside the coverage area (that is necessary to prevent the signals in one beam from interfering with signals on the same frequency in the other beam).

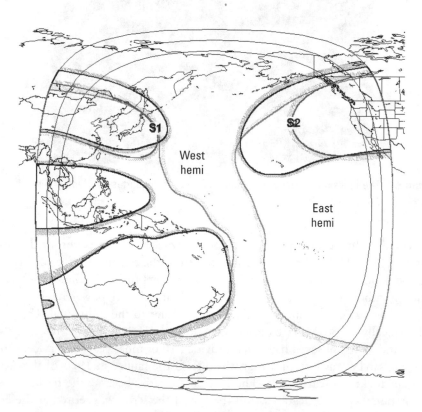

Figure 2.5 Intelsat VII Pacific Ocean region frequency reuse coverage; hemi and zone beam coverage at C band and spot beam (S1 and S2) coverage at Ku band.

The redesignated Intelsat IV-A spacecraft contained nearly 24 transponders of equivalent capability, with half connected to each of two hemispherical beams, one directed at North America and South America and the other at Europe and Africa. Additional improvements in solar cell and amplifier efficiency helped yield a spacecraft of the same size and design as Intelsat IV that was able to do the job of two satellites. That was the forerunner of satellite designs with several such beams (i.e., multiple beams), each narrowed sufficiently to cover a geographic region of high communications demand. A global coverage beam with a few transponders was retained to allow access by stations not within the "hemi" beams. (Bandwidth used in the global beam is not actually available for frequency reuse.)

Because of continued expansion of the INTELSAT system, in terms of both the number of member/users and the traffic demands, INTELSAT found that the problems of traffic overload in the AOR and also the Pacific Ocean region (POR) required another doubling of capacity per satellite. The 14/11 GHz portion of Ku band was selected as the means of carrying heavy links in the two regions, freeing up C band for general connectivity among countries. A satellite with both C and Ku bands generally is referred to as a hybrid satellite. The Ku-band transponders are connected to spot beams that point toward heavy traffic regions, such as Western Europe and the northeastern United States (ES and WS in Figure 2.6). Having conceived the hybrid approach, COMSAT proceeded to conduct an international tender for the supply of Intelsat V spacecraft.

The Intelsat V program was begun, with Ford Aerospace and Communications Corporation, in Palo Alto, California, winning the development and production contract. (The company was later purchased by Loral and continues as Space Systems/Loral.) Ford chose the three-axis approach (also called body-stabilized), whereby an internal spinning wheel and small jets stabilize the satellite to keep the footprint over the proper portion of the Earth (more on this in Chapter 8). With Intelsat IV and IV-A providing excellent service throughout the 1970s, the first Intelsat V was launched in 1980.

The C-band repeater on Intelsat V is very similar to that of Intelsat IV-A, with the exception that frequency reuse was taken to another dimension with the use of polarization discrimination. In that approach, two signals on the same frequency and in the same footprint are isolated from each other by cross-polarizing their electromagnetic waves. A more complete explanation can be found in Chapter 4. This allows a doubling of bandwidth in the Intelsat V global coverage beam. Having a Ku-band repeater onboard provides the extra capacity desired and even allows cross-connection between stations using C band and those using Ku band. A modification to Intelsat V, called the V-A, added an L-band repeater for maritime communications similar to that pro-

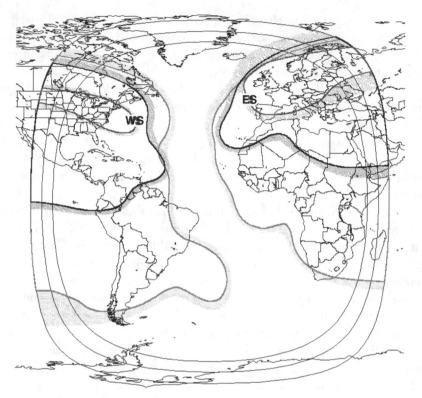

Figure 2.6   Intelsat VI AOR frequency reuse coverage; hemi and zone beam coverage at C band and spot beam (west and east) at Ku band.

vided by Marisat (discussed later in this chapter). Altogether, 12 Intelsat V/V-A satellites were constructed and eventually took over the bulk of international satellite communications during the 1980s.

Conceived in the late 1970s to provide for the widest possible expansion of international satellite services, the massive Intelsat VI spacecraft was defined by INTELSAT to continue the decline in cost per circuit. Hughes Aircraft won the contract in 1982 for the design and production of this very capable satellite. Again, it was a hybrid design (C and Ku bands), but frequency reuse was carried to an even greater stage to triple the capacity of the satellite compared to Intelsat V/V-A. A new feature on Intelsat VI is satellite-switched time division multiple access (SS/TDMA), which increases the efficiency of spot beams for high-capacity links and allows their reconfiguration with minimal impact on the Earth stations.

The 1990s saw INTELSAT expand its global system through the construction and launch of the Intelsat VII and VIII series of hybrid spacecraft. A total of nine Intelsat VII satellites were constructed by Loral, and eight reached operational status, beginning with the 701 satellite in October 1993. Each satellite has a capacity of 26 C-band transponders and either 10 or 14 Ku band, depending on whether it is a VII or VII-A model (the VII-A design was added for the last group of spacecraft constructed). These satellites provide both international services through fixed area coverage beams and domestic services through movable beams. The next production contract was with GE Astro (now Lockheed Martin) for Intelsat VIII, which concentrates on delivering greater C-band capacity for overall system expansion. Each of these spacecraft contains 38 C-band transponders and 6 Ku-band transponders. The next spacecraft to enter the production cycle was referred to as FOS-II at the time of this writing. Constructed by Loral and to be launched in 2000, the FOS-II series will replace the Intelsat VI satellites that will be nearing their end of life in the first decade of the twenty-first century.

In addition to funding technology development and investment in new satellites, INTELSAT created many innovative ground-based systems of communication. The initial system used analog transmission of telephone and video signals. However, it was clear to the engineers at COMSAT who provided most of the technical foundation for the international system that digital technologies would provide substantial benefits. Through the efforts of COMSAT Laboratories, the internal research organization, key transmission technologies like time division multiple access (TDMA) and demand assignment multiple access (DAMA) were invented and perfected. The 200,000 VSATs that are in operation at the time of this writing all employ those technologies, which grew up along with INTELSAT during its first 30 years.

INTELSAT is changing its approach to the market and structure, in response to global trends toward privatization and competition. As a first step, they contracted for a truly commercial satellite, K-TV, for the purpose of entering the DTH market in Asia. This spacecraft provides high-power Ku-band capacity for specific country markets, notably China and India. Under construction at Matra Marconi Space, Intelsat K-TV will help INTELSAT realign itself with commercial markets where they must compete with private global operators like PanAmSat and strong regional players like AsiaSat.

With its changing structure and mission, INTELSAT will have to meet many challenges in coming years. The organization appears to have the technical, financial, and political strength to weather the storm of competition. Many developing countries, which depend on INTELSAT for their primary international communications links, fear they will be left behind, allowing the more powerful signatories to manage the organization for maximum profits. Time will tell how well INTELSAT performs as a truly commercial organization.

2.2.1.3 PanAmSat

Founded in 1984 by legendary satellite entrepreneur Rene Anselmo, PanAmSat is a powerhouse in the global marketplace for space segment services. Anselmo started the company by using his own personal funds to purchase a spacecraft from GE Astro (now part of Lockheed Martin), coupling it with an inexpensive launch on the first Ariane IV rocket. The real challenge that the company overcame was the prohibition on any U.S. company competing with COMSAT and INTELSAT in the international satellite communication market. PAS-1 was successfully placed in orbit in 1988 to provide services in the AOR. PanAmSat's first customer was CNN, which selected PAS-1 to transmit programming to Latin America.

Anselmo broke through the political barrier and built his company from the ground up. That involved considerable work in Washington to obtain support from the U.S. government to force COMSAT and INTELSAT to allow a competitive international satellite operator to enter the business. Prior to that time, INTELSAT used the powerful provisions of the INTELSAT Agreement to preclude signatories for encroaching on its territory. After PanAmSat reached profitable operation, Anselmo became famous for his steadfast courage against difficult odds. PanAmSat's advertisements in various satellite industry publications featured a cartoon dog named Spot, which indicated Anselmo's disregard for the entrenched monopolies then prevalent. His acclaim continues because SPOT is the ticker symbol for the publicly traded stock of his highly successful company.

After Anselmo's death in 1995, the company entered into a merger with Hughes Communications, to form a new PanAmSat Corporation, which is the largest satellite operator in the world. On May 16, 1997, after the merger, PanAmSat owned and operated a fleet of 14 satellites: four PanAmSats, designated PAS 1, 2, 3, and 4; seven Galaxy satellites, designated as Galaxy 1R, 3R, 4, 5, 6, 7, and 9; and three Satellite Business Systems (SBS) satellites, designated SBS 4, 5, and 6. The SBS satellites were assets acquired by Hughes Communications from MCI Corporation in 1991 and will be out of fuel before the year 2000. The Galaxy satellites generally serve the U.S. market (see Figure 2.7) and were contributed by Hughes to the merged company. Galaxy 1 was launched in 1983 for use by the leading cable networks in the United States, including HBO, CNN, the Disney Channel, and ESPN. Subsequently, Hughes Communications built the Galaxy System into the most financially successful satellite operation in history.

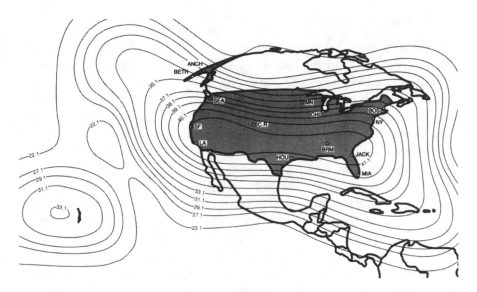

Figure 2.7   U.S. domestic satellite footprint coverage in C band (99° WL). (*Source:* Hughes Communications, Inc.)

PAS 1, the first satellite launched by Mr. Anselmo's startup company, contained 12 C-band transponders, split between 8.5W and 16.2W amplifiers, and 6 Ku-band transponders, each with 16.2W amplifiers. The satellite served the AOR, in direct competition with INTELSAT's primary region of business. The initial business was to serve Latin American TV customers, but PanAmSat was also successful at penetrating the U.S. to U.K. market. That new market entry required that PanAmSat obtain landing rights, which is the permission of the local government to deliver or originate transmissions within its border. Landing rights acquired cheaply (the way PanAmSat did it) eventually can be worth a lot of money. A second and a third satellite were added to the lineup in the Atlantic. Pacific Ocean service was initiated in July 1994 with the launch of the Hughes-built PAS 2 satellite. That bird carries 16 C-band and 12 Ku-band transponders and has become very popular for trans-Pacific TV transmission, Asia program distribution, and VSAT networks. Service was extended to the Indian Ocean region in January 1996 with the launch of PAS 3.

PanAmSat is aggressively expanding its system by having acquired additional satellites for international services. Those satellites are built by Loral and Hughes and will offer higher power and more focused service to specific markets. Also, the company is building on the base offered by the Galaxy System to interconnect domestic U.S. customers more effectively with the rest of the world.

2.2.1.4 Orion

The official name of the second private international satellite startup company is Orion Network Systems. Formed in October 1982 and initially the brainchild of John Puente, another famous satellite entrepreneur, Orion had a slow beginning. Like PanAmSat, Orion selected the AOR market and the strategy of gaining market access in the United States and the United Kingdom. Orion chose to build the company as a joint venture with British Aerospace (BAe) and thereby obtained access to spacecraft technology and funding. Orion contracted with BAe for two satellites but completed only the purchase of Orion 1, launched in November 1994. Subsequently, Orion selected Matra to complete Orion 2, and it is scheduled for launch in 1998.

Mr. Puente, formerly president of Digital Communications Corporation, the company that eventually became Hughes Network Systems (HNS), had the vision of Orion providing international business networks using VSATs. That remains the principal vision, but the Orion satellites also serve video customers. At the time of this writing, Orion 1 was in operation at 37.5 degrees west longitude with a projected life to 2005. Orion 2 would be placed at 12 degrees west longitude. The first Pacific region bird, Orion 3 (Figure 1.4), was contracted with Hughes in 1997 for launch in 1999 and placement at 139 degrees east longitude. A unique business feature of Orion 3 is that a Korean telecommunications company, Dacom, acquired a block of 12 of its transponders at the time of launch. Orion, which had plans at the time of this writing for up to six additional satellites, was acquired in 1998 by Loral Space and Communication, the parent company of Space Systems/Loral.

2.2.2 Regional Coverage: EUTELSAT

The international systems, notably INTELSAT, were created to meet long-haul requirements for telecommunications transmission. Because the satellites must provide broad coverage, they typically are not optimized to the needs of particular countries or affiliated groups of countries. The first step in meeting those more focused needs was the creation of the regional system. The first to do this on a significant scale were the Western European countries, in the formation of EUTELSAT. The parallels with INTELSAT are clear, including the basic strategy of a space segment cooperative whose members provide their own Earth stations.

The EUTELSAT project appeared in 1977 as a regional provider along European lines, including the focus for development of the indigenous manufacturing base. The first EUTELSAT satellite was launched in 1983 by the

European Space Agency (ESA). The organization was formally established as an operating company in 1985 with an initial membership of 17 Western European countries. Its first director general was Andrea Caruso, a well-recognized figure in INTELSAT circles. With the opening of Eastern Europe, EUTELSAT has grown remarkably and the membership now stands at 44. Table 2.1 is a complete listing of signatories, indicating the breadth and extent of this important satellite communication cooperative. Bulgaria, Estonia, Kazakhstan, and Slovenia were completing membership procedures at the time of this writing. In time, EUTELSAT has become the leading provider of satellite capacity over the now-extended Europe. This role is extremely vital to a wide range of public and private users.

Like INTELSAT, EUTELSAT is a treaty organization that operates like an intergovernmental body. The official government members, called Parties, meet once a year in an Assembly of Parties to make decisions on long-term objectives. Each party designates a Signatory to the Operating Agreement (unless it assumes that role itself) to be responsible for the investment in EUTELSAT in proportion to their utilization. The Signatories generally are national private or public telecommunications operators. The Board of Signatories is the main decision-making body and meets at least four times a year to discuss commercial and financial strategies, design and deployment of satellites, and planning of new services. Signatories' shareholding is determined annually in proportion to their use of capacity.

In January 1996, EUTELSAT's Assembly of Parties approved an amendment to the convention, permitting the designation of more than one Signatory per EUTELSAT member country. That first step to allow limited forms of direct access to the space segment could be critical to long-term viability. This is a difficult issue for the members of EUTELSAT, who enjoy a local monopoly in terms of applying the space segment.

2.2.3 Domestic Systems: Telesat, Westar, and Palapa

The technology and economics of commercial satellite communication were established during the expansion of INTELSAT. However, the real opportunities for business development appeared as domestic satellite systems came into being during the 1970s and multiplied in the 1980s. New operators that based their businesses on commercial services and customers grew rapidly, extending from developed regions to the rapidly developing parts of the world. The history and worldwide activity of domestic satellite systems are reviewed in the following paragraphs.

Table 2.1
The EUTELSAT Signatories

Albania	Albanian Telecommunications Enterprise
Andorra	Servei de Télécommunications d'Andorra
Armenia	Ministry of Telecommunications
Austria	Austrian Administration of Posts and Telegraphs
Azerbaijan	Ministry of Communications
Belarus	Ministry of Posts, Telecommunications, and Informatics
Belgium	Belgacom
Bosnia-Herzegovina	Public Enterprise of PTT Transport
Bulgaria	Bulgarian Telecommunications Company Ltd.
Croatia	Ministry of Maritime Affairs, Transport, and Communications
Cyprus	Cyprus Telecommunications Authority
Czech Republic	Ceske Radiokomunikace a.s
Denmark	Tele Danmark Ltd
Finland	Telecom Finland Ltd
France	France Telecom
Georgia	Georgian Ministry of Communications
Germany	Deutsche Bundespost Telekom
Greece	OTE
Hungary	Hunsat
Iceland	Post and Telecommunications Administration of Iceland
Ireland	Telecom Eireann
Italy	Telecom Italia
Latvia	Lattelekom
Liechtenstein	Government of the Principality of Liechtenstein
Lithuania	Lithuanian Ministry of Communications and Informatics
Luxembourg	Administration Luxembourgeoise des Postes et Télécommunications
Moldova	Ministry of Communications and Informatics
Monaco	Gouvernement de la Principalité de Monaco
Netherlands	Royal PTT Nederland NV (KPN)
Norway	Telenor Satellite Services AS
Poland	Polish Telecommunications SA
Portugal	Companhia Portuguesa Radio Marconi (CPRM)
Romania	Romanian Ministry of Communications
Russia	Ministry of Posts and Telecommunications
San Marino	Government of the Republic of San Marino

Table 2.1 (continued)

Slovakia	Slovak Telecommunications
Spain	Telefonica
Sweden	Telia AB
Switzerland	Direction Générale de l'Entreprise des Postes, Télégraphes et Téléphones Suisses
Turkey	Türk Telekomunikasyon AS UKBT
Ukraine	KRRT
Vatican	Government of the Holy See
Yugoslavia	Community of Yugoslav Posts, Telegraphs and Telephones

2.2.3.1 Canada

The first country to proceed with its own domestic satellite system was Canada, which contracted with Hughes Aircraft for a 12-transponder HS-333 satellite similar in capability to Intelsat IV but weighing half as much. The spacecraft, launched in 1974 and shown in Figure 2.8, had a beam that produced a single

Figure 2.8 The Hughes HS-333 spacecraft. (*Source:* Courtesy of Hughes Aircraft Company.)

Canadian footprint. A company called Télésat Canada was established in much the same way as COMSAT; however, its purpose was to offer satellite communication services within Canadian boundaries. The system has proven useful in extending reliable communications into the remote and northern reaches of Canada as well as demonstrating the attractiveness of using a domestic satellite to distribute television programming. The name *Anik,* meaning *brother* in the native Inuit language, was adopted for each of the satellites. Télésat went on to introduce Ku band into domestic service with Anik B, built by RCA Astro Electronics, and Anik C, built by Hughes Aircraft. The original Anik A satellites were replaced in 1983 with C-band 24 transponder satellites named Anik D, built by the team of Spar Aerospace Limited of Canada and Hughes Aircraft. The two Anik E hybrid C- and Ku-band spacecraft, under contract to Spar and RCA Astro, replaced the satellites that used separate C and Ku bands. Construction of the next-generation Anik F spacecraft was begun at Hughes in 1997. Télésat is highly recognized both as a competent satellite operator and a consultant to other companies that require specialized expertise in spacecraft acquisition.

2.2.3.2 U.S. Domestic System

The policy issues in the United States took longer to settle, and it was the administration of President Richard M. Nixon that allowed the industry to start through the "Open Skies" policy of 1972. That meant simply that any U.S. company with the financial resources and demonstrable need for telecommunications could be granted the requisite authorization to construct a system and operate its satellites in orbital slots assigned by the FCC. The coverage footprint of most U.S. domestic satellites is similar to that shown in Figure 2.7.

Six companies applied to the FCC under the Open Skies policy and one of them, the Western Union Telegraph Company, was the first to proceed. Their first two Westar satellites, of the same Hughes Aircraft design as Anik A, went into operation in 1974, augmenting Western Union's existing nationwide terrestrial microwave system. Only slightly delayed were COMSAT and AT&T with their Comstar satellites, which provided 24 transponders using polarization frequency reuse and a spacecraft based on the Hughes Aircraft Intelsat IV-A. This system, also used by GTE, provided long-distance telephone circuits to expand the nationwide network and offered both diversity and flexibility in meeting a variety of future requirements. Replacement of three Comstars was accomplished in the early 1980s with satellites of the Telstar 3 series, built by Hughes Aircraft but owned and operated by AT&T alone. GTE went ahead with its own successor system.

The polarization frequency reuse technology of Intelsat V and Comstar was adopted by RCA American Communications (Americom), now a subsidiary of General Electric Company, for their own domestic satellite system. RCA had pioneered three-axis stabilization for LEO weather satellites and adapted the technology to a GEO communications spacecraft. Designed and built by RCA Astro Electronics, of Princeton, New Jersey, the spacecraft was particularly compact and efficient; it achieved the capability of Comstar with significantly less weight and volume. That allowed the spacecraft to fit on the Delta instead of the much larger and more costly Atlas Centaur. The Satcom series of satellites, first launched in 1976, proved to be particularly popular for video programming distribution to cable television systems, a market pioneered by HBO, a subsidiary of Time Warner. Americom continued to launch satellites, including in 1985 two of their Satcom K and in 1996 their GE series. Now extending its business to the global market, Americom is one of the most successful satellite operating companies in the world.

The U.S. market underwent further expansion in terms of the number of operators and satellites. With Hughes, AT&T, and GTE establishing themselves as powerful forces, the market became nearly glutted with C- and Ku-band transponders by 1987. That proved to be too much for Western Union, which decided to exit the business by selling its in-orbit satellites to Hughes. Another satellite venture, SBS, the Ku-band pioneer in the United States, tried to gain a foothold in the rapidly growing field of private corporate telecommunications networks. However, being ahead of its time did not prove adequate for financial success, and SBS, too, folded its satellites into the Hughes Communications constellation. The consolidations produced just two strong operators in the United States: Hughes (to become PanAmSat) and GE Americom (part of GE Capital Corporation). The AT&T Telstar system was acquired by Loral and became the core of its entry into the business of operating satellites as well as constructing them.

2.2.3.3 Indonesia

The year 1976 also witnessed the introduction of a major domestic satellite system; however, this time it was halfway around the world in the Republic of Indonesia. Implemented in only 18 months under the direction of the late Mr. Soehardjono, director general of Post and Telecommunications, it used two Hughes Aircraft satellites, which were of the same design as Westar and Anik A. A ground segment of 40 Earth stations was installed and turned up in parallel, allowing President Suharto to demonstrate its operation to his people on the thirty-first anniversary of the independence of the Republic. The Palapa system was named for a mythical fruit that the ancient king, Gadjamada, pledged not to eat until all of Indonesia was united.

The project was completed on time and within budget by a coordinated team consisting of Perumtel of Indonesia (the domestic telephone operator),

Hughes (the spacecraft supplier, system architect, and integrator of 10 of the initial 40 Earth stations), ITT (the telephone switch supplier and integrator of 15 of the initial Earth stations), Ford Aerospace (the integrator of the remaining 15 stations), and TRT (supplier of all terrestrial transmission links). This highly integrated system, illustrated in Figure 2.9, provides telephone and television service throughout this archipelago nation spanning approximately 5,000 km. The cost of implementing and operating the system proved to be a small fraction of what it would have taken to do an even lesser job with terrestrial approaches.

The Palapa system has been expanded greatly to more than a thousand two-way Earth stations and also provides space segment services to neighboring Southeast Asian nations. The system moved into its second generation in 1984 with larger Palapa B satellites, each with twice the capacity of the Palapa A. The second Palapa B failed to reach proper orbit, was recovered by the space shuttle, and subsequently was replaced in 1987 by a similar spacecraft. The operator of the Palapa B satellites, PT Telkom (the new name for Perumtel), is now privatized and still provides most of the domestic telephone service in Indonesia. Another satellite operator, Satelindo, continued with the Palapa C series, which is composed of hybrid C- and Ku-band satellites.

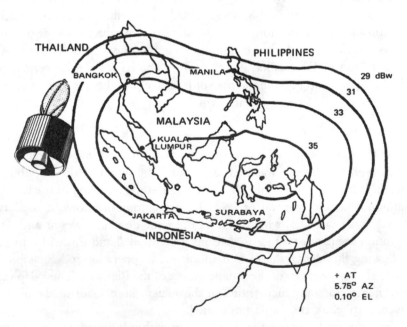

Figure 2.9 The Palapa A system.

2.2.3.4 Applications Technology Satellite

As the U.S. government agency responsible for space system research, NASA has been involved in several geostationary communication satellite projects since SYNCOM. The Advanced Technology Satellite (ATS) series consisted of six satellites that were used for communications demonstrations throughout the United States and the world. Hughes Aircraft built ATS-1 through 5 to demonstrate a variety of technologies and applications. ATS-1 had a VHF repeater that was used for basic communications on a push-to-talk basis (like a taxi radio). However, the satellite provided some essential services in the Pacific for medical consultation, experimental courses, and emergency communications. Services were maintained through 1983 with ATS-3.

ATS-6, the most advanced and the largest of the six, was built by Fairchild Industries of Germantown, Maryland, and was launched in May of 1974 directly into synchronous orbit by a Titan 3C rocket. With its 9m unfurlable mesh antenna, ATS-6 was used in the first direct broadcasting experiments, with the emphasis placed on educational programming. After providing service in the United States to isolated regions, the satellite was moved into position to be used in India. These experiments were very successful in that the benefit of using satellites to reach remote areas with educational and public interest programming was clearly shown.

Another NASA experimental satellite program called Communications Technology Satellite (CTS), also named Hermes, was jointly undertaken with the Canadian government. Designed and built in Canada but launched in 1976 from the United States on a Delta, CTS operated at Ku band (14/12 GHz) and demonstrated the capability of high-power satellite broadcasting in this new frequency band. One-way and two-way services were provided with small-aperture ground antennas. Pioneering work was done in the area of new applications, one being the first experience by nontechnical users with interactive video teleconferencing and the use of such facilities for delivering medical treatment and education. Of particular note is that CTS was used in the restoration of communications during the Johnstown, Ohio, flood disaster in 1977, proving the effectiveness of compact, transportable ground stations operating at Ku band.

NASA's more recent experimental satellite was ACTS, which was built by a team of GE Astro, Motorola, and COMSAT. As with previous NASA communication satellite experiments, ACTS set out to explore new applications in an underdeveloped part of the radio spectrum. In this instance, it was Ka band. The unique feature of ACTS is that its communication payload incorporates onboard switching and steerable, spot-beam antennas that allow routing of signals within the spacecraft repeater. The Ka-band frequencies used on ACTS are new capabilities for U.S. communications satellites (the Japanese, however, were first to launch a 30/20 GHz satellite). The ACTS system was made available for experiments by industry, universities, and other government agencies as well as for demonstrations that evaluate new service applications. Around 100 experiments were executed. (The discussion of advanced Ka-band satellites and systems continues at the end of this chapter.)

2.2.3.5 Europe

Regional satellite systems have always been of interest to provide services within and between countries that could not individually justify a dedicated satellite system. The countries of Western Europe wished to pursue a regional system, both for the services that could be rendered and to provide markets for their own aerospace and communications industries. Programs such as the orbital test satellite (OTS) and European communications satellite (ECS) were initiated in the late 1970s, leading to the creation of EUTELSAT.

The European space industry includes a variety of different types of high-technology companies: those that build entire spacecraft, like Aerospatiale of France, Matra Marconi Space of France and the United Kingdom, and Daimler Aerospace (DASA) of Germany; those that provide launch vehicles and associated services, notably Arianespace; and those that design and build major components, like Alcatel Althsom of France and Saab Ericsson of Sweden. In 1998, the French government reorganized its satellite and aerospace businesses (resulting in Alratel and Aerospatiale as the only corporate entities).These satellites have Ku-band (14/11 GHz) communications systems, and the users tend to be the domestic post, telephone, and telegraph (PTT) government agencies. More recently, the U.S.-based manufacturers have made inroads into the previously insular European market. RCA and Hughes have sold spacecraft to SES of Luxembourg, and Hughes has sold spacecraft to British Satellite Broadcasting (BSB), Telenor of Norway, and the Swedish Satellite Company, NSAB.

As a country, France has done more in an integrated way than anyone else outside the United States. French companies operate satellites (EUTELSAT and France Telecom), construct spacecraft (Aerospatiale and Alcatel), build and operate launchers (Arianespace), construct communication payloads (Alcatel and Thomson), and design and construct satellite communication networks (Alcatel). Those activities are backed up by a network of specialized government-sponsored institutes, universities, financial institutions, and standards bodies. Taken together, the French infrastructure can take a new satellite operator through every phase of building a system and a business.

Perhaps the most exciting satellite development in Europe has been the introduction of a competitive regional system within its own borders. Luxembourg, a central figure in the European Union and long a promoter of TV and radio broadcasting, entered into its own satellite system in spite of the fact that the entire country can be covered from a suitably tall tower. What the government and private interests foresaw was an opportunity to lead the expansion of satellite-based TV delivery throughout Western Europe. After a few fits and starts, a joint venture company called Société Européenne des Satéllites (SES) was created and with it the basis of launching the now-famous Astra satellites. Founded in March 1985, SES is the first privately owned satellite operator in Europe. Focusing on the needs of European TV viewers and satellite broadcasters, SES has introduced a new approach to satellite television. As a service- and market-oriented company, its aim is to offer entertaining and high-quality packages of television channels and radio stations for the different European language markets, available on attractively priced and easy-to-install reception equipment. SES has been leading the transition to digital satellite broadcasting in Europe by operating satellites dedicated to digital technology since 1995. Astra is Europe's most popular satellite system, reaching 94% of all satellite and cable households in the region.

Initialized with its first satellite launch in 1988, the Astra fleet has grown to eight satellites, with the first one nearing the end of its life at the time of this writing. SES pioneered DTH services in the United Kingdom and elsewhere on the continent, effectively leapfrogging several domestic programs in countries like the United Kingdom, Ireland, Germany, and France. The first Astra satellite allowed Rupert Murdoch's Sky services, discussed later in this chapter, to enter the U.K. market ahead of BSB. SES operates six satellites at the orbital position of 19.2 degrees east (Figure 1.8). Colocation ensures that the several hundred TV and radio programs carried on Astra both in digital and analog can be received on small, single, fixed dishes. By the end of 1998, SES plans to have 10 satellites in orbit, including two satellites at a second orbital position.

2.2.3.6 Japan

As the leading industrialized country in Asia, Japan has aggressively sought to be a power in the field of satellites and satellite communication systems. The Japanese government funded several satellite projects for experimental and industrial development purposes. The Communications Satellite (CS) series, launched in the late 1970s, emphasized experimentation with Ka band; however, that was at a time when C- and Ku-band spectrum and orbit resources were relatively plentiful, raising doubts about the economic viability of that particular strategy.

The first true Ku-band direct broadcast satellite (DBS) satellite was developed by GE for the national broadcast network, NHK. There were technical problems with high-power traveling wave tube amplifiers (TWTA) and their associated power supplies, so the spacecraft failed to meet expectations. A follow-on DBS program in Japan included satellites designed and built by RCA

Astro Electronics and Hughes. Interestingly, the Japanese government became one of the most successful DBS operators in the world (in terms of the number of viewers) by offering a package of U.S. programming material. Regular cable channels like ESPN and Financial News Network (now merged into CNBC) and TV networks like ABC and CBS were combined into a continuous feed of non-Japanese material that was available only on DBS. Another unique feature of the system is that it is used to transmit high-definition TV (HDTV) on a trial basis. The NHK channels were joined by private service providers who used the same space segment to deliver subscription services like WOWOW.

Satellite communications began to be privatized in 1987, when licenses were granted to Japan Communications Satellite Corporation (JCSAT) and Space Communications Corporation (SCC). JCSAT was an extremely interesting company because of its ownership: 40% Itochu, 30% Mitsui and Company (both highly respected Japanese trading companies), and 30% Hughes. The combined company successfully launched two Ku-band satellites and largely cornered the domestic market. SCC, heavily backed by Mitsubishi, the largest trading group in Japan, was plagued by problems with its first satellites supplied by Ford Aerospace (later Space Systems/Loral) but eventually went into operation using new satellites built by Loral and Hughes. A third satellite operator, Satellite Japan, never went into business, but the partners ended up buying out Hughes' interest in JCSAT (renamed JSat).

2.2.3.7 India

India has conducted an extensive scientific and technological research program in space, under the stewardship of the Indian Space Research Organization (ISRO). Headquartered in Bangalore, India's fifth-largest city and the capital of the southern state of Karnataka, ISRO has created a domestic industrial base capable of design and building spacecraft and placing them in orbit with an indigenous rocket. The first domestic satellite for India, called INSAT 1, was launched in 1981 and performed the dual function of providing communication services at C and S band (for TV broadcast) and meteorological imaging. The first-launched of these complicated three-axis spacecraft, designed and built by Ford Aerospace, reached orbit but experienced technical problems with deployment of antennas and operation of propulsion. A modified second spacecraft was launched in 1983 and met its service objectives.

ISRO successfully transferred technology to develop INSAT 2, the second-generation spacecraft that continued to offer domestic transponders and weather imaging. A trend toward limited privatization of telecommunications started in 1995, and there is the prospect of other satellites being introduced into the domestic market. In April 1997, ISRO announced the INSAT 3 series, to consist of five satellites. At least three of those satellites are to be constructed in India and launched by their indigenous launch vehicle.

2.2.3.8 Middle East

After many years of planning and discussion, the Arab League countries embarked in 1976 on a regional system called the Arab Satellite Communications Organization. It was structured along the lines of EUTEL-SAT, with membership including Algeria, Bahrain, Egypt, Iraq, Djibouti, Jordan, Kuwait, Lebanon, Libya, Mauritania, Morocco, Oman, Palestine, Qatar, Saudi Arabia (the largest and most powerful member), Somalia, Sudan, Syria, Tunisia, the United Arab Emirates (UAE), and Yemen. The satellite and system were to become known as Arabsat.

Design and construction of the first series of satellites were contracted to Aerospatiale, with a significant portion of the three-axis technology provided by Ford Aerospace. The satellites, first launched in 1985, have communication capability similarly to Insat in that C- and S-band services are possible. What ensued was a number of years of technical and political difficulties, which neg-atively affected the initial business prospects. Eventually, Arabsat developed a more functional and businesslike structure and became profitable. Arabsat 2C, for example, provides TV and telecommunication services to the region. All services are delivered at Ku band, including DTH television. The Arabsat 3 series of spacecraft were contracted again to Aeorspatiale in 1996.

A number of other companies in the Middle East have embarked on their own domestic and regional satellite programs. Nilesat, under development by Matra Marconi Space for the Egyptian Satellite Company, will be established by an Arabsat member. Turksat, from Turk Telekomunikasyon AS of Turkey, was launched in August 1994, providing Ku-band capacity to the region. Another program of interest is Thuraya, a GEO mobile system to be supplied by Hughes. Thuraya is being developed by Etisalat, the public telecommunica-tions operator in the UAE, and a consortium of telecommunication orperators in the Middle East. Capacity is also provided to the Middle East by such estab-lished operators as INTELSAT, EUTELSAT, AsiaSat, and PanAmSat. A coop-erative spacecraft program called Amos 1 involved the Israel Aircraft Industries along with Daimler Aerospace and Alcatel. This Ku-band spacecraft was suc-cessfully launched on May 15, 1996, on an Ariane 4 rocket. In addition to a Middle East beam, Amos 1 has a Central Europe beam that delivers HBO pro-gramming to Hungary, Poland, and the Czech Republic.

2.2.3.9 Latin America

Latin America embraced satellite communication almost immediately after the launch of Early Bird. With vast expanses of underdeveloped territory and difficult natural obstacles between its countries, Latin America proved to be a loyal supporter of the INTELSAT system. Mexico (technically in North America but close to South America by virtue of a common heritage) and Brazil (the largest and most industrialized economy in the region) were among the first in the world to construct Standard A Earth stations near their capital cities. Later on, Mexico and Brazil initiated their satellite systems in 1982 using Hughes Aircraft designs. In the case of Brazil, the first two satellites were assembled in Canada by Spar and provided 24 C-band transponders each. The Mexican Morelos 1 and 2 satellites, named for a martyred priest who led a rebellion against the Spanish colonial government, are shown in Figures 2.10 and 2.11. Each satellite incorporated a hybrid C- and Ku-band payload, intended to provide a variety of communications services to large- and small-diameter Earth stations. One important Ku-band application was the extension of tele-education to remote villages in Mexico, something that was previously started with terrestrial microwave.

Figure 2.10   The Morelos satellite fully deployed in orbit. (*Source:* Rendering courtesy of Hughes Space and Communications International, Inc.)

Figure 2.11 The completed Morelos 1 spacecraft prior to start of thermal-vacuum testing at the manufacturer. (*Source:* Photogragh courtesy of Hughes Space and Communication Company.)

Both Mexico and Brazil were successful in implementing their space segments in 1985. Other Latin American nations have taken action to develop their own orbit resources. Argentina initiated its own satellite program called Nahuel in the 1990s, contracting with DASA for the system. The unique aspect of Nahuel is that the contract grants DASA the right to operate the satellite to provide services in Argentina and around the region. Colombia has applied for orbit positions beginning in 1980 but had not initiated construction of its own satellite at the time of this writing. Another regional project named Simon Bolivar (later changed to AndeanSat) has been under development by the nations of the Andean Pact, which includes Colombia, Venezuela, Ecuador, Peru, and Bolivia.

In the 1990s, Mexico and Brazil were in a position to maintain and expand their respective satellite systems. Telecomunicaciones de Mexico (Telecomm), the government agency responsible for Morelos, contracted with Hughes Space and Communications International for two medium-power hybrid spacecraft, Solidaridad 1 and 2. An Ariane 4 rocket launched Solidaridad 1 on November 20, 1993, resulting in the spacecraft being placed at 109.2 degrees west longitude. The satellite was the first to carry L band in addition to C and Ku bands. Telecomm offers vehicular mobile services to Mexico using the same network platform and terminal types as American Mobile Satellite Corporation (AMSC) (discussed in Section 2.3.2). Further, Telecomm contracted with Hughes in 1996 for a replacement of Morelos 1. The Mexican government spun off the satellite business as SatMex, which was acquired by Telefonica Autry and Loral Space and Communications. Embratel, the Brazilian long distance telephone company and operator of the Brasilsat system, was sold to MCI of the United States in 1998.

2.2.3.10 Africa

The African continent, with the exception of the Middle East, still remains the area with the lowest level of economic and telecommunications development in the world. However, it cannot be ignored when considering markets for satellite capacity and, ultimately, domestic spacecraft. According to Euroconsult [1], "The region has recently begun to wake up, however; encouraging signs include the ongoing or imminent privatization of several African telephone carriers, the development of several pay-television ventures in South Africa and the Middle East, and the fairly high level of interest shown by the region's few potential investors in mobile satellite projects." Telephone density in the region is the lowest in the world at around 5 per 100, although an acceleration in the rate of installation of new lines is emerging.

African nations have long been members and users of the INTELSAT system. That contributed to the telecommunications development of many countries, allowing the more politically stable to attract foreign investment in natural resource and transportation industries. The pooling of demand for those transponders led to the creation of Regional African Satellite Communications (RASCOM), a 42-member cooperative of public operators. RASCOM also has been studying the possibility of acquiring and operating its own satellite as a way to reduce costs and to better manage resources for the benefit of its members.

2.3   Specialized Systems: DTH and Mobile

Satellites that employ the FSS spectrum allocations are intended for a variety of telecommunication services that can satisfy basic infrastructure needs of a country or a region. Such systems are employed on a global basis and form the core of the industry. In time, existing and new satellite operators saw an opportunity to create satellite-based business in the areas of DTH television and mobile communications. As discussed in detail in our other work [2], these systems reached operational status in the 1990s and are today the fastest growing segments. We review the systems that took the risk of being first to market.

2.3.1 DTH Development

The concept behind DTH is to deliver TV and other broadcast services directly to the individual user's home. Doing that required a step improvement in satellite performance to reduce the size and the cost of the home dish installation to something that users could afford. In addition, the service must be controlled through scrambling or encryption so that users are encouraged to pay for services. The other key ingredient is the programming, which must be attractive enough (and affordable) so that people acquire the equipment and subsequently pay for the service. The ITU initiated the flurry of interest by conducting a landmark World Administrative Radio Conference (WARC), as the WRCs were known prior to 1995, for broadcasting satellites. The conference was held in Geneva in 1977, at a time when only NHK of Japan had initiated a DTH program. WARC-77 made assignments of Ku-band channels and GEO positions to ITU members in Regions 1 and 3 (Region 2 completed action on its assignments at the 1983 Regional Administrative Radio Conference).

Several U.S. companies approached the DTH market in the early 1980s, but unfortunately none survived long enough to establish a permanent service. COMSAT was the most ambitious, forming Satellite Television Corporation (STC) to utilize the new assignments from WARC-77. They contracted with RCA Astro in 1983 for two small DBS satellites (small by today's standards but fairly sizable in their time). However, they could not put a viable business together, lacking programming resources from a TV network partner. Another startup, United States Satellite Communications Inc. (USSCI), obtained FCC approval in 1983 to use the Canadian Anik C satellite to deliver a limited DTH service to a segment of the U.S. market. USSCI ceased operation in 1985, when it ran out of money before achieving adequate revenues. By 1986, the real development of DTH in the United States came in the form of the C-band backyard dish business as an adjunct to normal cable TV program distribution. The large networks, like HBO, CNN, and Disney, sought to increase their revenues overall by gaining backyard dish subscribers. That involved the introduction of scrambling and conditional access control, two key technologies that any DTH system demands. A firm foothold in the U.S. DTH market was eventually established by DIRECTV, Inc., a wholly owned subsidiary of

Hughes Electronics Corporation. They started service in 1994, using an array of 150 digitally compressed TV channels.

The introduction of a truly commercial DTH business on a standalone basis had to wait until 1987, when Sky was launched in the United Kingdom by Rupert Murdoch's News Corporation. Using analog technology and delivering its signals from the newly launched Astra 1a satellite, Sky started along a path that proved both successful and unstoppable. The officially mandated U.K. DTH operator, BSB did not get off the ground until 1989. The prestigious backers of BSB, included Pearson, owners of the *Financial Times*, and Reed and Granada Television, famous in the United States for the excellent dramatic series shown on the Public Broadcasting System. The U.K. government awarded BSB the WARC-77 assignments, requiring also that they adopt the D-multiplexed analog components (D-MAC) transmission standard that had been mandated by the European Commission. The ensuing delay in getting its DTH platform, generally regarded as technically superior to that of Sky, off the ground proved to be fatal to the business prospects of BSB. On November 2, 1990, BSB and Sky merged to form British Sky Broadcasting (B-Sky-B). The combined service was directed by News Corporation, which held a 40% interest in the company. Subsequently, B-Sky-B reached a high degree of acceptance in the United Kingdom and became a very successful business in financial terms.

DTH systems in Asia were centered in two markets: Japan and East Asia. The Japanese approach to DTH is best illustrated by the efforts of NHK to establish early preeminence in the field. Two other Japanese DTH operators, PerfecTV and DIRECTV Japan, entered the market using FSS Ku-band transponders supplied by JSat and SCC, respectively. The rest of Asia was to be the broadcasting domain of Richard Lee and his company, Star TV. That venture rode along with AsiaSat 1 when it was launched in 1991. Star TV assembled a package of Hong Kong–developed programming channels in English and Mandarin Chinese, along with imported programming from the BBC and U.S. movie studios. The target was primarily the C-band DTH market, which could be developed within the AsiaSat 1 footprint. Richard Lee was definitely the starter-upper, but what really made him famous was the sale of the entire operation and licenses to Rupert Murdoch for over $600 million. The finances of the service were based on advertisers. StarTV has shown that there is indeed a place for pan-regional programming and for packages targeted to the attractive country markets. Financial success was yet to be achieved at the time of this writing.

2.3.2 MSS Development

Development of MSS, introduced in Chapter 1, started with Marisat and evolved through the various satellites used for maritime applications. Inmarsat, a truly international organization, was established in 1979 to provide worldwide mobile satellite communications for this community. At the time of this writing, it had 79 member countries and was the only GEO operator to provide commercial, emergency, and safety applications on land, at sea, and in the air. Inmarsat services are extended to land-based users with the Inmarsat M and Mini-M standards. The Inmarsat 2 and 3 satellites have sufficient power to support portable user terminals, which in the case of Mini-M are the size of a laptop computer (Figure 1.7).

Domestic MSS satellite services began to appear in the mid-1990s, when Optus Communications decided to include an L-band payload on its second-generation Ku-band satellites. The Optus MobileSat team designed and implemented the ground segment to make mobile telephone calls possible from conventional vehicles. In the United States, AMSC was formed from eight applicants for domestic licenses. The consortium floated its own series of stock on the public stock market to raise sufficient capital to fund a satellite and the associated gateway Earth station. Even with a dedicated high-power satellite, the AMSC network could serve only vehicular and portable terminals (similar to the mini-M terminal of Inmarsat). It would take satellites either much higher in power or, alternatively, closer to the Earth to close the link to a hand-held type of MSS telephone instrument.

The real excitement in mobile satellite services was created by Motorola when it announced the Iridium project in 1990. As discussed in Chapter 1, Iridium employs a LEO constellation of 66 satellites (the original proposal was for 77 satellites, hence the selected name Iridium, which is the element in the periodic table with 77 orbiting electrons). Loral followed suit with Globalstar, which no doubt will give Motorola a run for its money. Not to be outdone, Inmarsat created and then spun off the ICO Global system, which uses 12 MEO satellites to provide global services. All these systems are based on providing direct access for hand-held mobile phones.

The investment in the global MEO and LEO systems could equal or surpass that of the GEO satellites worldwide, but the following technical and financial questions remain to be answered at the time of this writing:

1. How big is the market for hand-held MSS services? Will it support the investment of more than one of these systems?

2. How will LEO operators maintain service in the face of the large number of satellites required and their relatively short individual lifespan? Will reentry into the Earth's atmosphere prove to be a serious problem?

3. Can MEO systems demonstrate superior economic performance compared to that of the LEOs? Will they have adequate link margin to overcome terrain blockage and other obstacles?

4. Is there a place for GEO systems, which require fewer satellites but are subject to greater propagation delay?

5. Will the rapid development of terrestrial cellular and other wireless technology provide too much competition to allow any MSS system to reach profitability and be sustained in the long term?

These questions and others can be answered only in time (some thoughts on this are expressed in Chapter 13). If MSS proves to be the next big market for satellite services, we could see even greater development of space-based applications in the future.

2.3.3 Digital Information Broadcasting

There remains one other application segment for satellite communication, that of providing direct to user transmission of information. Digital audio broadcasting (DAB) represents an interesting possibility for a significant domestic or global market. Experiments were conducted by NASA using the S-band payload of the tracking and data relay satellite (TDRS). The experiments proved that digitized radio signals can be transmitted to vehicular terminals with a special antenna and receiver. The difficulty was that terrain blockage caused a break in reception, something that probably would not be acceptable to the typical audience. That can be overcome either by transmitting the same signal from multiple diversely placed satellites or through terrestrial broadcast towers that fill in gaps. Considerable effort has been expended in Europe on a DAB standard that would be resilient to brief link outages. The FCC authorized DAB service in the United States by granting a license in 1997 for two satellite systems both at GEO.

On the nonaudio side of the picture, several U.S. companies have provided data broadcasting since the late 1980s. Examples include continuous feeds of financial information, like stock market and commodity prices. The typical receiver in that case would be a fixed dish of anywhere from 50 cm to 1.5m and a simple receiver and data demultiplexer. Early participants in the field include Equatorial Corporation, which was first in 1983 to employ spread spectrum for a commercial application, and United Video, which in 1985 used data broadcasting to deliver an on-line program guide for cable TV before that became a standard facility of DTH services. The most dramatic development in the field was the DirecPC service from HNS, which allowed a standard PC to receive a 12-Mbps data broadcast through a 60-cm dish.

2.4 Expansion at Higher Frequency Bands: Ka Band

Ka-band frequencies long have promised much greater bandwidths than are available at C, X, and Ku bands. Many years of experimentation in Japan, Italy, and the United States have produced a wealth of useful information. NTT of Japan, in particular, was first to employ a commercial Ka-band payload using the CS-1 satellite, also known as Sakura (Cherry Blossom). CS-1, launched in December 1977, was contracted with Ford by the National Space Development Agency (NASDA) of Japan and flew Japanese components made by NEC and Mitsubishi Electric Company A second generation, comprising CS-2a and CS-2b, was launched in 1983 to maintain a Ka-band capability. Those satellites provided the important feature of connecting Japan's remote islands into an integrated telecommunications network. All three satellites were spin-stabilized with mechanically despun horn antennas. NTT conducted numerous experiments using fixed and transportable Earth stations in Japan. Later, NTT acquired transponders from SCC and launched its own N-Star satellites to maintain Ka-band capacity to continue its development of applications.

As discussed in Chapter 1, the technical issues at Ka band are reasonably well understood; however, there remains the challenge of finding applications that can justify making the requisite investments. Almost without regard to the expected challenges, many established and startup satellite operators have made application to the FCC and the ITU for Ka-band satellite systems. Table 2.2 is a summary of the U.S.-based applications for GEO satellites.

Table 2.2
Ka-Band Applications Submitted to the FCC in 1995

Company	System name	Number of satellites	Repeater type
AT&T	VoiceSpan	12	Demod/remod
EchoStar	EchoStar Ka	2	Demod/remod
GE Americom	GE*Star	9	Bent pipe
Hughes Communications	Spaceway Direct	20	Demod/remod
KaStar Satellite Communications	KaStar	2	Demod/remod
Lockheed-Martin	AstroLink	5	Demod/remod
Loral Aerospace	CyberStar	3	Demod/remod
Morning Start Satellite Co.	Morning Star	5	Bent pipe
Motorola	Millennium Celestri	4	Demod/remod
Netsat 28 (DBXSAT)	Netsat 28	1	Bent pipe
Orion	Orion F2, F7, F8, F9	5	Bent pipe
PanAmSat	PAS 10/11	2	Bent pipe
Teledesic	Teledesic	LEO (840)	Demod/remod
Vision Star	Vision Star	1	Bent pipe

Applications from literally every country with an existing system or plans to implement a system have been filed with the ITU. They include the United States, Canada, Luxembourg, Hong Kong, France, Norway, Korea, Japan, Russia, Sweden, and India. That has resulted in a glut of applications and many volumes of work to be done to resolve how the band can be used on a global basis.

Two major Ka-band systems, at the time of this writing, are good prospects. The Spaceway project by Hughes was, on first sight, the most ambitious GEO satellite system ever imagined. With 15 operating Ka-band satellites and an advanced digital processing payload, Spaceway promises to be nearly all things to all people. Hughes was the first to propose a satellite that could dynamically process and route high-speed digital links between multiple Ka-band spot beams. This type of capability supports two-way broadband communications for applications as varied as basic telephony on one end of the spectrum all the way to video teleconferencing and advanced Web surfing on the other. Taking even a greater leap forward would be Teledesic, a LEO project supported by Bill Gates, Craig McCaw, and Boeing. As planned, Teledesic would employ 288 satellites to form a global mesh capable of broadband services comparable to Spaceway. In 1998, Teledesic selected Motorola as their prime contractor, which also resulted in the incorporation of Motorola's Celectri system. Another such GEO system was proposed by AT&T (see Table 2.2), but AT&T withdrew its application to the FCC in 1997. Ka band, therefore, still represents uncharted waters for satellite operators, although the prospects are exciting.

References

[1] Euroconsult, *World Satellite Communications and Broadcasting Markets Survey—Prospects to 2006*, Paris, August 1996.

[2] Elbert, B. R., *The Satellite Communication Applications Handbook, Norwood*, MA: Artech House, 1997.

3

Satellite Network Architectures

Any communications satellite performs the function of a microwave relay in a telecommunications network. The most basic type of relay is the bent-pipe style of satellite, which does not alter the nature of the transmissions between Earth stations. The Earth stations in this case must organize their transmissions to efficiently use the satellite's repeater resources and to allow the various points on the ground to transfer or exchange the required information. That then determines the network architecture for the particular application. In more advanced repeaters with onboard processing, the satellite performs other network functions such as switching the connections between stations or even modifying the format of transmission to further improve the efficiency of transmission. All that is carried out whether there is a single GEO satellite or a constellation of non-GEO satellites serving the network. From the users' perspective, the specific nature of the space segment should be irrelevant. However, the design, operation, and, more important, the required investment, all depend heavily on the way the architecture is put together.

The first part of this chapter reviews the features and generic arrangements of networks independent of specific uses. This provides a cross-reference with regard to the applications that are reviewed at the end of the chapter. More detail on these applications can be found in our other work [1].

The purpose of this chapter is to introduce readers to satellite communications, so we have taken a nontechnical approach in reviewing the range of network architectures and services. Chapter 4, the first technically oriented chapter, explores the fundamentals of satellite microwave links, including such important principles as the decibel (dB), radiowave propagation, and the link budget.

3.1 General Features of Satellite Networks

The manner in which a satellite network provides links between users is called *connectivity*. The three generic forms of connectivity are point-to-point, point-to-multipoint, and multipoint interactive, shown in Figures 3.1, 3.2, and 3.3, respectively. The first uses of satellites were for point-to-point links between fixed pairs of Earth stations. As shown in Figure 3.1, communication from one station to the other is on a dedicated path over the same satellite. Therefore, two links are needed to allow simultaneous communication in both directions. That is how the typical telephone conversation or interactive data link functions. The link can remain in place for an extended period, called preassigned service, or be put into place for the duration of a conversation, called demand-assigned service.

The point-to-point link allowed satellites to create a worldwide network of telephone circuits, providing a solid foundation for international telecommunications development. That was before the days of high-capacity fiber optic cable, a technology that is pushing satellites out of this role. In the 1980s and early 1990s, satellites were used more and more for point-to-multipoint connectivity, also called broadcast. As shown in Figure 3.2, a single uplink station can transmit a continuous stream of information to all receiving points within the coverage area. Satellites like GE 1 and Galaxy 1 in the United States, Astra 1A and Hot Bird 1 in Europe, and JCSAT and Superbird in Japan all use that broadcast mode to transmit television programming to large user

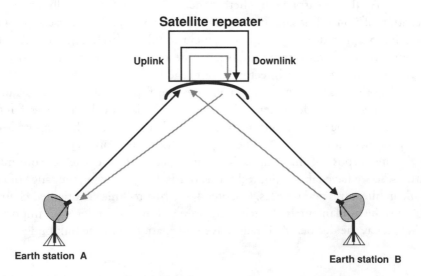

Figure 3.1 The most basic two-way satellite link provides point-to-point connectivity.

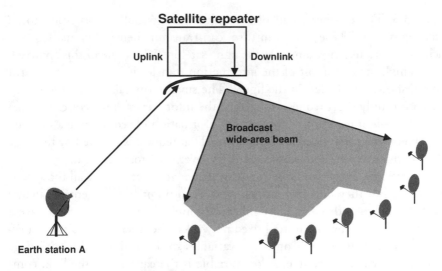

Figure 3.2 Point-to-multipoint (broadcast) connectivity delivers the same information over the satellite coverage footprint.

populations, thus using the wide-area coverage of a GEO satellite to its greatest extent.

Adding a transmit capability to each receive terminal is a technique to convert the broadcast system into a multipoint interactive network, shown in

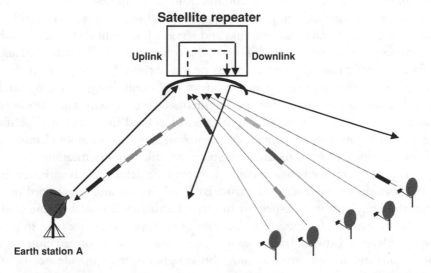

Figure 3.3 Multipoint-to-point (interactive) connectivity supports a star network of low-cost VSATS.

Figure 3.3. The broadcast half of the link transmits bulk information to all receiving points. Those points, in turn, can transmit their individual requests back to the originating point over the same satellite. The value of this approach is at its highest when most of the information transfer is from the large Earth station, shown on the left in the figure. The small terminals on the right need to transmit only short requests to change the information flow. For example, a subscriber might request to download an information file containing a software program or short movie, all in digital form. That request is received by the large Earth station, which then transmits the file over the broadcast link.

Terrestrial copper and fiber optic cable networks can create all those connectivities, but they do so by stringing point-to-point links together. Satellite networks, on the other hand, are inherently multipoint in nature and have a decided advantage over terrestrial fixed networks whenever multipoint connectivity is needed. It depends on many variables to determine if this advantage exists in a particular example. A key variable is the capacity of the link, commonly called the *bandwidth*. In digital communication, bandwidth is measured in bits per second (bps). What controls those variables are the particular aspects of the telecommunications application that the satellite network delivers. The rest of this section reviews many of those variables, relating them to how they are addressed in a particular satellite network design.

3.1.1 Dedicated Bandwidth Services

The commercial history of satellite communication is founded on the use of the GEO, in which satellites and Earth stations are fixed with respect to each other. The principal approach to satellite links and services is to dedicate the bandwidth to each transmission service. In this mode, the communication is established and maintained continuously for an extended period of time. The user must make a request to the network operator well in advance of need; that gives the operator enough time to plan ahead for the required satellite capacity and associated ground resources. Bandwidth usually is assigned as fixed time slots of a specific duration. They may repeat periodically to provide a near-continuous channel of communications, which is required for voice and video communication.

Figure 3.4 shows a basic arrangement for a dedicated bandwidth service. Each transmission comes from a particular Earth station and is provided on a noninterfering basis with respect to all other Earth stations on the same satellite. As discussed in Chapter 5, techniques of multiple access are used to prevent interference between transmissions. The transmission is formed either as a continuous stream or a noncontinuous burst signal. It may provide a constant bit rate (CBR) type of service, suitable for literally any communication application. Voice and conventional TV give optimal performance under CBR

Transmission from:

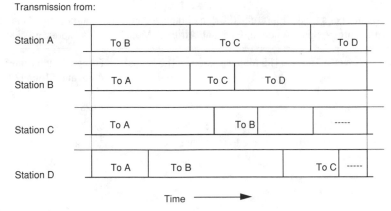

Station A | To B | To C | To D

Station B | To A | To C | To D

Station C | To A | To B | -----

Station D | To A | To B | To C | -----

Time ⟶

Figure 3.4 Transmission timelines for multidestination TDM carriers that connect four stations, using point-to-point links.

transmission; for intermittent services like electronic mail (e-mail), CBR is much less beneficial. Once established, the CBR link is maintained for an extended duration, sometimes measured in years for cable TV channels or hours for temporary wideband data channels.

Dedicated bandwidth links are either analog or digital in nature. Originally, all satellite transmissions were analog and the bandwidth was fixed at time of installation. The introduction of digital transmission permitted the bandwidth to be rearranged more conveniently, which was one of the big reasons for making the conversion. A given link could support a single digital data stream of 64 kbps, which is sufficient for a standard telephone channel or a combination of low-speed data channels. More likely, many 64-kbps channels are assembled into a higher speed stream through the technique of time division multiplexing (TDM). TDM is different from TDMA, in which the transmission is made as a burst to allow several stations to share the same RF channel. All these aspects are covered in Chapter 5.

TDM channels are specified in terms of a digital hierarchy that follows standard bandwidth group sizes. Table 3.1 lists the familiar North American T and European E hierarchies, respectively. One T1 provides 1.544 Mbps of data transmission, 24 individual 64-kbps user channels, or an appropriate mix of both. Likewise, the European E1 channel is 2.048 Mbps in capacity, suitable to transmit 30 user channels at 64 kbps each. While the T and E hierarchies are somewhat different, they both employ the 64-kbps basic channel of communication. It is up to the operator or ultimate user to determine how the channels are applied. For example, if the user is a telephone company, the T1 link can carry 24 simultaneous voice conversations. In a data communication application, the

Table 3.1
The Plesiochronous Digital Hierarchy (PDH) as Employed in North America and Japan (1.5 Mbps) and Europe, South America, Asia, and Africa (2 Mbps) (Entries Expressed in kbps)

Level	North America (1.5 Mbps)	Japan (1.5 Mbps)	Europe, South America, Asia, and Africa (2 Mbps)
1	1,544	1,544	2,028
2	6,312	6,312	8,448
3	44,736	32,064	34,368
4	97,728	139,264	

E1 channel in useful in connecting many remote clients to a mainframe computer used as a large server.

Greater bandwidth and service flexibility is offered by the synchronous digital hierarchy (SDH) approach, reviewed in Table 3.2. Although the North American scheme, called synchronous optical network (SONET) is structured slightly differently, the combined system of global digital tranport provides a seamless way to connect fiberoptic and advanced broadband satellite networks at the Gbps level.

There are many other ways that a dedicated bandwidth channel is employed. In DTH systems, the capacity of a single link might be 28 Mbps, which is broken down through multiplexing into 5 to 10 digital TV channels with sound and control information. A given DTH satellite with 16 high-power transponders will carry 16 of those data streams, which is equivalent to

Table 3.2
SONET (North America) and SDH (Worldwide) (Entries Are Expressed in Mbps)

SONET level	SONET designation	Bit rate (Mbps)	SDH level	SDH designation
STS-1	OC-1	51.840		
STS-3	OC-3	155.52	1	STM-1
STS-9	OC-9	466.560		
STS-12	OC-12	622.080	4	STM-4
STS-18	OC-18	933.120		
STS-24	OC-24	1,244.160		
STS-36	OC-36	1,866.240		
STS-48	OC-48	2,488.320	16	STM-16

between 80 and 160 TV channels. With analog TV, the satellite would support only 16 TV channels (e.g., one per transponder). The dedicated digital bandwidth approach, therefore, is popular for providing many TV channels and the associated sound programming.

One potential difficulty with dedicated bandwidth is that it may require a great deal of planning effort to ensure that the capacity is allocated as efficiently as possible. The network operator must assess (1) the quantity of individual channels of communication (64 kbps or digital video) and (2) where they are to be provided. The second point means that the Earth stations must be equipped with the proper equipment to land their respective carriers and demultiplex the individual information streams. A way around part of that problem is to use dynamic bandwidth allocation, also referred to as statistical multiplexing, to adjust individual usage of channels to near-instantaneous need. As illustrated in Figure 3.5, statistical multiplex reduces the total peak capacity by using bandwidth that otherwise would have been wasted.

There are two ways to apply dynamic bandwidth allocation: time slot reassignment (TSR) and packet switching. In TSR, the fixed total channel capacity of a given link is divided into fixed time slots each corresponding, for example, to an 8-bit byte of capacity. A standard 64-kbps telephone channel requires 1 byte to be transmitted in each 125-ms time frame. Other types of applications require a different quantity of bytes to be transmitted within that interval. The key behind TSR is that the time slots are assigned to individual users they need them and not on a dedicated basis. One might say that is not dedicated bandwidth because of the dynamic nature of allocation. However, the reason it remains in that category is that the satellite link has a fixed total bandwidth and connectivity.

Packet switching, covered in more detail in Section 3.1.3, also can be used to add dynamic bandwidth allocation to an otherwise dedicated bandwidth link. Instead of fixed bytes of capacity, the packet approach transfers the user

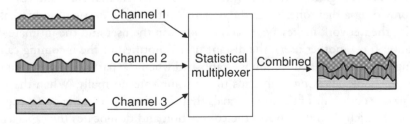

Figure 3.5 An illustration of the principle of statistical multiplexing, demonstrating how total channel bandwidth is reduced.

data to the link as it arrives at the point of access. Buffering is required to adjust for differing arrival rates of different user information. Variable-length packets are used by standard protocols like TCP/IP and X.25. A fixed-length packet, called a cell, is recognized as being appropriate for using satellite links, whether they be dedicated bandwidth or of a type to be discussed in Section 3.1.2. Asynchronous transfer mode (ATM), an international standard for cell switching, is being adopted for many applications where dynamic bandwidth allocation provides significant improvement over static approaches.

Non-GEO systems are not amenable to dedicated bandwidth services because the link is interrupted each time a new satellite must be acquired by the Earth station. Communication that lasts longer than one satellite pass must be handed over to another satellite without loss of connection. That forces the non-GEO operator to employ one of the other transmission modes to be discussed in the following paragraphs.

3.1.2 Circuit-Switched Services

Many communication applications are active only when a user has actual need for service. The best example of that is the telephone network, which gives a CBR type of service but only for the limited duration of a call. Other CBR applications, like video teleconferencing and high-quality fax, can employ circuit-switched service. Users like this type of service because they pay for only what they need. While network capacity is only provided on demand, each end of the circuit needs a dedicated termination facility so calls can be made and accepted on demand. The common telephone and the local loop to the local telephone exchange are examples of such a termination facility. For video teleconferencing, the termination facility might consist of television cameras, video compression equipment, and a special type of Earth station that can access the satellite directly.

Figure 3.6 illustrates the basic approach used to establish and release a circuit. The user makes a request for service to the network by lifting the telephone receiver (going "off hook"); the network detects that state and responds by providing a dial tone. After the user has entered the destination by dialing digits, the network makes a connection between the user and the ultimate destination (i.e., another user); the distant user is notified of the incoming call by a ringing signal; the connection is established when the party on the distant end goes off hook, allowing both ends to communicate normally. When the need for service ends, one of the users signals that situation by hanging up (going "on hook"), which then deactivates the connection and deallocates the capacity. In a typical commercial telephone network, many other details must be addressed properly. For that to happen, the network requires a distributed computer intel-

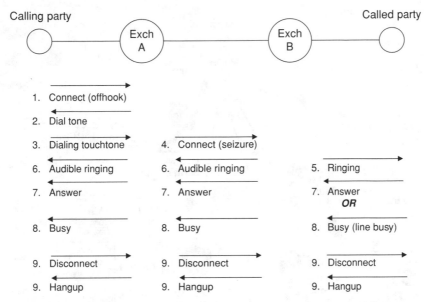

Figure 3.6 The basic steps used to establish an ordinary telephone cell.

ligence to be able to recognize the request, to track and control available capacity, to signal the initiating and accepting ends of the connection, and to provide service management facilities like billing and fault detection and correction. Few, if any, of those capabilities are needed in the case of dedicated bandwidth communications.

In satellite communication, the Earth station network responds to call requests that are given to it either by directly connected telephone users or the public telephone network itself. This type of application is referred to as *demand assignment* because capacity for the circuit is assigned to the particular connection on demand. The capacity that this comes from is a pool of narrowband frequency channels or of time slots within a common timeframe. Control of such a network can be a complex problem. For a traditional GEO satellite with a single footprint, a single control station can manage an entire demand assignment network. That is possible because the control station can receive requests directly from all user stations and transmit control signals to them.

Figure 3.7 illustrates the principle for a telephony network using a variety of single-user and shared multiple-user Earth stations. The Earth station at the upper left responds to a locally connected telephone (a) that goes off hook by sending a service request to the master station. If the master is able to provide satellite capacity, it signals the requesting station; that results in a dial tone

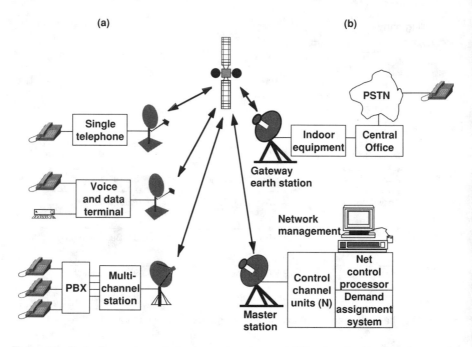

Figure 3.7 Typical arrangement of a demand assignment FTS network with central network control and management. (*Source:* Courtesy of Hughes Network Systems.)

being transmitted back to the originating telephone instrument. The calling party can then enter a dialing sequence that informs the network of the destination city, local exchange, and called subscriber. Depending on the particular implementation of the control network, the dialing digits are transferred to the control station, which determines the destination Earth station (indicated at the upper right) and signals it that a connection is to be established. Satellite capacity is allocated to the connection, and the two Earth stations perform a quick check that there is a workable signal path. The called station then signals the local telephone network to complete the connection to the final called party (b), with the final step being the ringing of that telephone instrument. Once answered, conversation can commence. Taking down of the connection works in the opposite direction.

The process becomes more complicated for a multiple spot beam satellite, either with or without an onboard processor. The control station may not be able to monitor all satellite transmissions, so additional steps are added. Also, a processor-type repeater may need to be configured for the connection. In a non-GEO system, the call may involve connection through multiple satellites and cross-links. Also, handover from satellite to satellite could add complexity.

Whether GEO or non-GEO, call setup time can be an issue if it takes more than about 10 sec. Apart from the possible sources of complexity, a satellite circuit switched network can provide exceptional service in terms of flexible connection for users who could be located anywhere in the coverage area.

3.1.3 Packet-Switched Services

Data communication networks largely have adopted packet switching as the predominant mode of transferring information, either on a point-to-point or point-to-multipoint basis. In the 1980s and 1990s, X.25 and TCP/IP reached critical mass in private and public network applications. In addition, the ATM protocol is a form of packet switching. All these technologies are based on the concept of the data protocol, in which a stream of digital information is broken down (or assembled, as the case may be) into packets that satisfy certain constraints as to length. To each packet is added a prescribed number of bits called the *header*, used in the process of reliably transferring the packet from source to destination. Once the packet has reached the destination, the header is removed and the original information stream reformed. The protocol often is used to transmit acknowledgment back to the source or, if the data transfer was corrupted, to transmit a request for retransmission.

There are two fundamental schemes for packet switching: the virtual circuit (also called connection-oriented) and datagram (also called connectionless). Details on the operation for many of the common standards can be found in [2]. The virtual circuit behaves much like circuit switching, except that there is neither a physical connection nor CBR service. There is a setup sequence, but that is only to establish a virtual circuit number (also called a virtual circuit identifier) and to inform the switching nodes that packets with this number in their headers should end up at the same destination. The connection is virtual in that there is only a logical record of its existence. Any particular point-to-point path within the network likely will have packets from many different virtual circuits passing over it.

Datagram packets are addressed for a particular destination as opposed to a particular circuit number. Therefore, no preparation is necessary to set up a virtual circuit, and packets are handled independently. The network routes each packet as if it were a dedicated message from a particular source to a particular recipient. The most classic datagram protocol is the IP network layer of TCP/IP. Interestingly, TCP uses the IP datagram service to perform the information transfer function. It goes through a circuit connection routine before actually offering packetized data from a particular stream of data onto the link.

The connection-oriented and connectionless paradigms represent two distinct ways of implementing a packet switched network. As with TCP/IP, they can be combined to further optimize the performance of the network. What we are looking for is (1) reliable information transfer, (2) no loss of any portion of the input, and (3) acceptable delay (also called latency). This all must occur as the links and network are loaded by user information, a process that often is dynamic. That is because it usually is not possible to predict the precise demand for data transfer, in terms of the number of simultaneous users and their individual data communication needs.

3.1.4 Flexibility Features

Satellite communication networks offer a variety of benefits to the system engineer and the user alike. The following paragraphs summarize several of those benefits and put them in context of the overall advantages of the medium introduced in Chapter 1.

3.1.4.1 Implementation of Satellite Networks

Installation of a fiber optic cable system requires first that the right-of-way be secured from organizations such as governments, utility companies, and railroads. Hundreds or even thousands of sites must be provided with shelter and power (and even access roads in the case of terrestrial microwave). After the entire system has been installed and tested, all the equipment must be maintained to ensure continuous service. Still, one outage along the route probably will put the entire chain out of service until a crew and equipment can arrive on the scene to effect repair. When the primary high-capacity fiber optic cable is cut, that can be devastating to the communications of a major city or industrial site. The way around that is to add redundant links, which can be costly in terms of capital and maintenance expenses.

In contrast, the time to install an Earth station network is relatively short, particularly if the sites are close to where service is provided. In the past, implementation times for Earth stations were lengthened not because of site construction but rather because electronic equipment had to be special-ordered and then manufactured. The low production volumes (because satellite communications require less equipment in general than terrestrial) discouraged manufacturers from mass producing standardized equipment and holding inventory for future sales. In today's larger and more competitive Earth station equipment market, higher manufacturing volumes along with the greater use of standardized digital systems have allowed equipment suppliers to reduce cost and maintain on-the-shelf inventory. The time to implement satellite networks and add stations has been reduced from 1–2 years down to 1–2 months.

3.1.4.2 Expansion of the Network

With a proper network architecture, new Earth stations can be added without affecting the existing stations. That reduces the expansion timeframe to a few months or weeks, since all that needs to be done is to purchase the equipment, prepare the site, and install the station(s). Increasing the number of receive-only Earth stations (e.g., DTH receivers) is particularly easy and economical because operation of existing stations is not affected.

Dealing with two-way communication Earth stations is a bit more complicated because some setup and coordination are needed to expand an operating network. Satellite networks of the 1970s that provide point-to-point links cannot be modified easily because of the old, inflexible analog technology employed. To add an Earth station to an old analog point-to-point network would require dismantling the equipment at each old station to be linked with the new station. That major drawback of the older systems has been eliminated with programmable digital technology. These more flexible digital approaches, described later in this chapter, can be assumed in virtually every future application involving two-way communications. VSATs and the newer multimedia terminals are designed for easy introduction into a working network, which is vital when one considers how large such networks have become.

3.1.4.3 Simplification of Network Routing

Many corporations and government agencies operate private backbone networks for use in enterprisewide information processing and dissemination. Client/server networks and corporate intranets are popular, representing a large commitment in financial and human terms. Once a major public network has been modernized, it can be expanded or modified by running new local cable loops and by making appropriate wiring changes in the telephone offices. However, time delays of many weeks to months are still involved, beginning from the moment when a user submits a significant service order. The public telecommunication service provider (common carrier) often must perform some network engineering, install equipment if necessary, make the required wiring changes, and then test the resulting circuits for proper operation. If the circuit or circuits cross the boundaries between the terrestrial networks of different common carriers, the process must be run simultaneously between the various organizations and coordination between them handled in some manner. (The reliability aspects of this particular problem are discussed in Section 3.1.5.)

After AT&T divested its local telephone company operations in 1984, it began to concentrate on long-distance service. A nationwide network of a corporation like General Motors traversed the networks of all of the now-independent regional Bell holding companies, along with the long-distance networks of AT&T, MCI, and US Sprint. That made life both complicated and difficult, as there no longer was one service provider to deal with. A similar situation exists in developed regions of Europe and East Asia.

In a modern satellite network, only the end connections are involved because the satellite itself provides all the intermediate routing. Unlike terrestrial networks that must deliver multipoint connectivity by extending terrestrial links to each and every point to be served, the satellite network reaches all points within a common footprint. There are terrestrial radio techniques that mimic a point-to-multipoint satellite network by placing broadcast transmitters on towers or mountaintops. AM and FM radio and TV work on a point-to-multipoint basis, and cellular mobile telephone is an excellent multipoint-to-point system. However, all such terrestrial techniques are severely restricted as to range because of line-of-sight radio propagation. To extend well beyond that geographical limitation, point-to-point links must be established between the radio towers to chain the broadcast or cellular stations together. Until that is accomplished on a uniform basis (something inherent in the GSM terrestrial cellular standard), the satellite network remains the most practical means of having a unified approach.

3.1.4.4 Introduction of New Services

Expansion of a satellite network can add new services ahead of terrestrial means. Perhaps the clearest example is the long-distance transmission of full motion color television, which could not be carried over the first generation of analog transoceanic telephone cables. It was not until the advent of terrestrial microwave radio in North America that coast-to-coast TV transmission was possible. Satellite repeaters in the FSS have sufficient bandwidth to carry several TV channels, along with an array of voice and data traffic. The telephone local loops, which bring voice and data services into the office and home, are still limited in their capacity. A family of analog multiplexing technologies called digital subscriber line (DSL) is under development in the United States to address this shortcoming. Home cable television can be enhanced with a separate coaxial cable that promises two-way interactive services comparable to what satellites currently deliver. Any and all of those services can be included in or added to networks of small Earth stations, particularly the VSAT operating at Ku and Ka bands.

A given satellite network can achieve all three connectivities individually or simultaneously, while a terrestrial network usually is restricted to a point-to-point capability. It is not uncommon for a user to implement a point-to-point satellite network (see Figure 3.1) involving 10 to 50 Earth stations and then add a broadcast capability to extend the network to hundreds or even thousands of receiving points. Any one of the point-to-point stations could then be used as an uplink site to broadcast digital information or video programming (see Figure 3.2) on a full-time or occasional basis. The multipoint-to-point capability (see Figure 3.3) can be installed in the future by adding a transmit "retrofit" package to many of the smaller receive-only stations.

3.1.5 Reliability of Satellites and Links

The remaining features are more difficult to explain and quantify; however, they ultimately can be the factors that decide in favor of satellite transmission over terrestrial transmission. The mere fact that the typical satellite link requires only one repeater hop tends to make the satellite connection extremely reliable. Complexity is increased for dual-hop intercontinental links over GEO satellites, as well as for multiple hops and intersatellite links for non-GEO systems. Nevertheless, the number of such transfers is substantially less than for terrestrial alternatives that must extend the same distance.

The reliability of a communication link typically is specified by the percentage of time it is capable of providing service, called the *availability*. Figure 3.8 indicates the number of outage hours per year that results from a given availability value. The engineering of the link, described in Chapter 4, must properly take into account the frequency bands, fade margin requirements, and orbit geometry. When that is done and an established satellite system is employed, the link will be up and usable well in excess of 99% of the time. In fact, satellite engineers normally talk of link reliabilities of 99.99% for FSS applications, which equates to an outage or downtime of 9 hours a year.

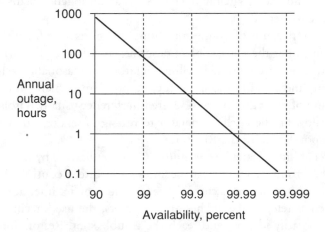

Figure 3.8 Service availability in percentage versus equivalent outages per year in hours.

Normally, outage is segmented into durations of a few minutes distributed mainly through the rainiest months. In the case of MSS systems, availability is reduced by the impact of mobile fading due to obstructions and, in the case of non-GEO constellations, satellite motion and handover.

A single buried cable or microwave system is relatively unreliable due to the inevitable breakage or failure. Long-distance terrestrial systems normally provide reliabilities in the range of 98 to 99.5%; outage can be produced by fades on any of its radio paths (in the case of terrestrial microwave) and by equipment outage at any of the hundreds of repeater sites along the route. Cable systems are susceptible to accidental breakage or deterioration of the cable itself, and outages of several hours or even days at a time do occur. Therefore, proper terrestrial system design requires that diverse routes be provided. Buried cable with diverse routing and automatic switching can satisfy the 99.99%-level requirement, although the cost of implementation would be within the reach only of relatively wealthy organizations (national telephone operators, governments, and major industrial corporations).

Equipment failures on satellite links do occur, and for that reason backup systems are appropriate. A communications satellite contains essentially 100% backup for all its critical subsystems to prevent a catastrophic failure. Double failures, for example of primary and backup equipment, have occured, so it is vital that procedures be in place to routinely check operation. The individual transponders or transmitters within the repeater section usually will not be backed up 100%, so a fractional loss of capacity is possible at some time in the useful orbital life. Experience with modern commercial satellites has been excellent, and users have come to expect near perfection in the reliability of the spacecraft. However, because total satellite failures (including launch and on orbit) can and do occur, satellite operators need contingency plans of the type discussed in Chapter 10.

A common cause of communications outage is not failure of satellite hardware but rather the RFI problem described in Chapter 1. Harmful RFI is a fairly routine occurrence, and satellite operators are reasonably well equipped to respond to and identify the source of the problem (which almost always is accidental and of short duration). A severe interference usually is isolated to one frequency, allowing the satellite operator to reassign users to different frequencies or transponders until the RFI is removed.

The reliability of satellite communication is enhanced by the fact that virtually all the ground facilities can be under the direct control of one using organization. If a problem occurs with equipment or its interface with other facilities such as telephone switches or computers, the user's technical support personnel can easily identify and reach the trouble spot. Restoration of service thus can be accomplished conveniently and quickly. In contrast, terrestrial linkups can involve many organizations that provide segments of the overall service, complicating the necessary troubleshooting and follow-up.

3.1.6 Quality Features and Issues

The quality of satellite communications or any communications service must be defined in user terms. The following paragraphs identify different approaches to measuring quality of transmission as perceived by humans. If the information is analog in nature, quality is measured objectively by using a signal-to-noise ratio or subjectively by surveying people (subjects) who have been exposed to the service. Quality in data communications boils down to the quantity of valid data that reaches the distant end (the bit or packet error rate), along with the time it takes to get there (the latency).

3.1.6.1 Signal Reproduction

For a single point of transmission, a satellite is nearly ideal for delivering a signal of the highest quality. Modern GEO satellites radiate sufficient power into the geographical footprint to be received by ground antennas of diameters in the range of 45 cm to 10m (18 inches to 32 feet). The GEO and non-GEO MSS systems being introduced in the late 1990s can reach to antennas the size of your thumb. Because satellites use line-of-sight transmission in directions perpendicular to the atmosphere, the frequency and duration of link fades are reduced compared to terrestrial microwave.

The predominant mode of transmission over satellites is digital, in which information is transformed from its analog form into a data stream. Such a process alters the ideal quality of the signal, introducing a certain level of distortion due to quantization (assigning the analog signal to discrete voltage levels that can be represented by numerical values) and compression (altering the actual information to reduce the required number of bits). As long as the bits are transferred through the link with little or no error, the quality remains exactly as it was after the initial conversion process. Whether a satellite or a terrestrial link is used, signal quality is maintained at all times.

Many terrestrial networks suffer from man-made noise and various kinds of short interruptions ("glitches") while satellite links experience primarily receiver noise (explained in Chapter 1), which is constant and easily compensated for with power. Satellite-based mobile communication is affected by various forms of blockage, which makes these services somewhat more glitch-prone than terrestrial cellular networks. That can cause data dropouts that will affect the consistency and continuity of the delivered information.

When these factors are properly taken into account, satellite communication can be engineered for high circuit quality and reliability. Designers can then select appropriate Earth station equipment that will provide that quality with confidence. The communication application in which these aspects of quality play the greatest role is in point-to-point and point-to-multipoint video. The quality of the delivered video signal is for all practical purposes identical to that of the signal created at the studio or played from the originating video tape machine.

3.1.6.2 Voice Quality and Echo

The issue of quality of voice transmission over satellites has received a lot of attention because of heavy competition among satellite and terrestrial communication companies. As discussed in Chapter 2, the use of the GEO for communications relays was a controversial topic prior to the first use of SYNCOM in 1962 because of the quarter-second delay introduced by the long transmission path. The impact of that delay on voice communications continues to be debated even today, particularly as non-GEO satellite systems regain a position of acceptance. Voice communications over a GEO satellite can be made acceptable to over 90% of telephone subscribers, as has been proved by numerous quality surveys. Terrestrial and LEO systems that span continents and oceans do not suffer from noticeable delay and hence are potentially more desirable from that standpoint alone.

The delay itself is not as detrimental as the presence of a returned echo to the speaking party. Both terrestrial and satellite links are capable of producing echo, as illustrated in Figure 3.9. At the left of the figure, speech from a female talker is converted into electrical energy in the voice frequency range (300 to 3400 Hz) by the handset and passes over the single pair of wires to the telephone equipment used to connect to a long-distance circuit. The same pair of wires allows the speech from the distant end (where the male talker is listening) to reach the female talker.

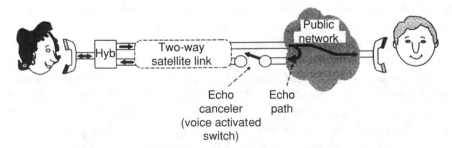

Figure 3.9 Telephone echo over a satellite is caused by electrical reflection at the distant end, where it can be eliminated by an echo canceler.

The connection point onto the long-distance circuit breaks the two directions in half, segregating the sending and receiving wire pairs. The device that routes the energy properly between the two-wire local loop and the four-wire trunk line is called a *hybrid*. The typical configuration has a hybrid on each end; however, Figure 3.9 shows that the male talker is connected through an undefined terrestrial network within which several hybrids could exit. The echo path is produced within one or more of the unseen hybrids, allowing some of the female talker's speech energy to make a U-turn and head back over the link toward the female talker. Because the echo is the result of uncontrollable factors in the terrestrial network, it must be blocked or removed in some manner. Obviously, a satellite circuit with its quarter-second delay is subject to an echo that is heard approximately one-half second after the talker utters his or her first word. Undersea cable circuits might introduce an echo return delay of only 50 ms or less, but that still would be unacceptable for commercial telephone service.

The simplest and most effective way to block the echo is to use a voice-activated switch at the opposite end of the circuit, as shown in Figure 3.9. Whenever the female talker is speaking, the control circuitry detects the presence of the incoming speech on the upper wire pair, and the switch on the lower wire pair is opened. When she stops talking, however, the switch closes automatically and the male talker is free to speak and be heard by the female talker. A similar switch would have to be placed on the female talker's end to protect the male talker from his own echo. Other features are necessary to make the switch respond to characteristics of human conversation, such as when one party needs to interrupt the other (double talking). One of the biggest problems with satellite voice circuits of past years has been the difficulty of getting old-fashioned echo suppressors to work correctly. Fortunately, echo suppressors were retired from service as the networks of the world were converted to digital.

All long-distance telephone networks now achieve excellent echo control with the digital *echo canceler*, which is included with a digital telephone exchange or satellite network. An echo canceler works with the digitized version of the speech and mathematically eliminates the echo from it. It does that initially by measuring and characterizing the echo path through the hybrid and terrestrial network. From that information, the canceler determines how to subtract a sample of the incoming speech from the return path to the distant talker. After an initial training interval, the echo is suppressed by a factor of about 10,000 (e.g., 40 dB; to be discussed in Chapter 4). Additional echo reduction is accomplished with a precise echo suppressor that can totally block the return path when only one person is speaking. To allow double talking, only the echo cancelation function is operating to permit both sides of the circuit to work simultaneously. While the details of how this technology works are beyond the scope of this book, the important point is that there is strong evidence that modern echo cancelation produces satisfactory link quality for even single-hop GEO satellite networks with a total delay of 400 ms or less. For extended double-hop circuits greater than 500 ms, even the echo canceler cannot overcome the effect of latency on the interactive quality of a telephone circuit.

3.1.6.3 Data Communications and Protocols

The key to successful computer-to-computer communication over satellite is to employ the right protocols and coding schemes. Computers that communicate over terrestrial links transmit data in blocks of data, or *words*, as shown in Figure 3.10. (The specific structure of protocols varies widely, as in some cases the word is the same as a packet.) A protocol provides a set of rules that the two ends follow in assembling the words for transmission and determining if the received words are valid. Provision is also made to automatically request retransmission if an error is detected at the distant end. The "conversation" illustrated in Figure 3.10 shows what happens when the receiving end detects an error and requests retransmission from the sending end (on the left). The sequence starts at the upper left and moves left to right and right to left, following the direction of the arrows. As long as errors are infrequent and the resulting delay is short, the protocol does its job effectively. However, satellite

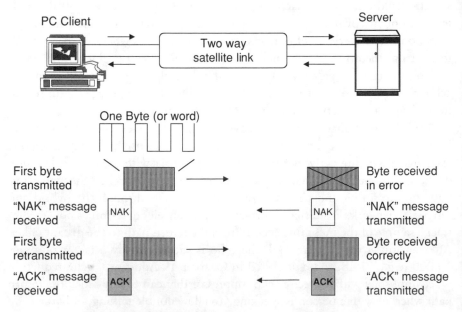

Figure 3.10 Inefficient "word-oriented" data protocols are subject to greater delay on a satellite link when errors in transmission are experienced.

delay could force the sending end to halt the transmission of new data for as much as 2 seconds as the necessary requests and retransmissions are executed. A terrestrial connection would not exhibit that difficulty, particularly for high-quality fiber optics.

This problem can be corrected in satellite data links by giving the sending end the acknowledgment that it expects even before the distant end has received the word or group of words. The piece of equipment that accomplishes the task of pacifying the sending end is called a delay compensation unit (DCU). Another term that is used to describe this is protocol *spoofing*. Errors are corrected by the DCU at the other end using forward error correction (FEC) coding by the transmitting DCU. Another feature that adds to the reliability of end-to-end communication is the cyclic redundancy check (CRC). To apply the technique, the sending DCU or communication device performs a calculation on the data word that results in a CRC bit sequence. The receiving end makes a similar calculation on the incoming word and compares this version of the CRC sequence against the one attached by the sending end. If they do not agree, then the receiving end requests retransmission.

Recognizing that the DCU approach was only a provisional fix, computer manufacturers and software developers have adopted standard computer protocols that do not use repeated acknowledgment messages. For example, the IBM system network architecture/synchronous data link control (SNA/SDLC) protocol transmits a long series of blocks of data words before expecting an acknowledgment. If an error word is detected on the receive end (using CRC or another technique), the request for retransmission is made only for a specific word. Such modern protocols can make satellite links nearly as efficient as the best-quality terrestrial links for the transmission of high-speed data. The performance is further enhanced by increasing the bandwidth on the satellite link and hence, the throughput of data as provided on an end-to-end basis.

The packet approach, discussed earlier in this chapter, provides another effective means to control errors on terrestrial and satellite networks. The first packet protocol specifically designed for satellite links is called ALOHA (discussed in Chapter 5), developed by Dr. Norman Abramson at the University of Hawaii. In ALOHA, a given station transmits its packet of data immediately and in the blind. That ensures that there is the least possible delay through the satellite link, provided there is no overlap between two or more transmissions that happen to arrive at the satellite at the same time. Such an overlap is termed a collision, causing those particular packets to become corrupted. When the receiving station gets a good packet, it sends an acknowledgment message. The protocol of ALOHA provides for automatic retransmission if no acknowledgment is received, with the transmitting stations each waiting a randomly selected time period to prevent a second occurrence. Most VSAT data networks employ versions of ALOHA and packet transmission to greatly improve throughput and responsiveness of satellite data transmission.

The issues of reliability and quality can, in balance, weigh in favor of satellite communications when the engineering, installation, and operation are done properly. The difficulty often is that the terrestrial network exists first, and transferring service over to a satellite network becomes a complex and difficult task. In cases where terrestrial service did not exist first or was totally inadequate, the new satellite network was very well accepted (INTELSAT and the Indonesian Palapa A system are good examples). However, terrestrial telephone communications always have been very good, making it more difficult for satellites to gain wide acceptance. The growth of MSS to hand-held terminals adds a new dimension for satellites, and so the balance could shift in their favor. Non-GEO MSS systems offer a unified worldwide mobile telephone service and could command more than their share of the global mobile phone and data service markets. In addition, GEO and non-GEO broadband fixed point-to-point systems at Ka band could become a significant contributor to the global information infrastructure.

3.2 Point-to-Multipoint (Broadcast) Networks

Broadcast links take full advantage of the wide area coverage of the satellite's footprint. Figure 3.11 provides a typical architecture for how satellite broadcasting is accomplished, using one transmitting Earth station (called the *uplink*) and many receive-only (RO) Earth stations. The GEO satellite repeater retransmits one or more RF carriers containing the information to be distributed. This particular network is used to distribute TV programming to local rebroadcasting stations that serve viewers in urban areas.

As the number of ROs increases into the millions common in DTH applications, the optimum transmitter power to use in space becomes much larger than that permitted at C band by the ITU. The BSS segment of Ku band is available for such high-power broadcast applications, as discussed in Chapter 1. The cost of the more expensive BSS satellite is shared among more and more users, who then save substantial amounts on the cost of their ground equipment. That is an economic tradeoff between the cost of the satellite and that of the ground segment, as discussed in Chapter 12.

Achieving point-to-multipoint connectivity with a terrestrial network is extremely expensive, since the cost of adding cable or microwave facilities to reach service points is roughly proportional to the number of points. The only economy of scale in terrestrial broadcasting comes if many services can

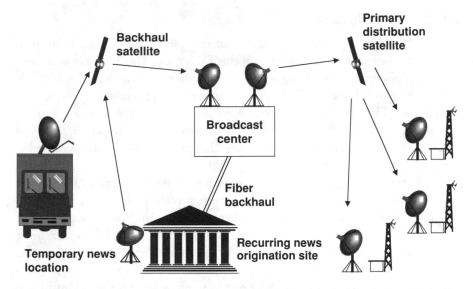

Figure 3.11 The use of satellite transmission in the commercial television industry for backhaul of event coverage and program distribution to affiliates.

share a common cable distribution network. Chaining terrestrial links together tends to be less reliable on an overall basis because a chain is no stronger than its weakest link. Cable TV systems, on the other hand, are very effective in urban and suburban environments because the cable infrastructure has the potential to serve a large and economically attractive community. In data communications, a terrestrial chain of that type using telephone circuits is called a *multidrop line*.

This section reviews the typical satellite applications for the point-to-multipoint broadcast connectivity.

3.2.1 Video Distribution

Television or video service (which are one and the same) is perhaps the most popular source of entertainment and information for the public. The broadcasting industry has embraced satellite communication as the primary means of carrying programming from the program originator (TV networks, cable TV programmers, content producers, and program syndicators) to the final point of distribution (broadcast TV stations, cable TV system operators, home dishes, and PCs). In this section, we review the way in which programmers and distributors use satellites and Earth stations in their business.

3.2.1.1 TV Broadcasting

As an example of the importance of the current role of satellites, we begin with a review of the general characteristics of the TV broadcasting industry as it exists in North America. Broadcasting is the commonplace medium whereby local TV stations employ VHF or UHF frequencies to transmit programming to the local community. The range of reception usually is limited by line-of-sight propagation to approximately 50 to 150 km. The Nielsen Media Research organization specifies the local coverage regions in the United States as the defined market areas (DMAs). There are approximately 200 DMAs in the continental United States, corresponding more or less to coverage of local TV stations and their respective advertising. To conserve frequency spectrum, the same channel is assigned by the government to another station some safe distance away. Individuals use directional antennas (yagis and reflector dipoles) to maximize signal strength and to suppress reception of unwanted distant stations operating on the same or adjacent channels. A given station transmits only a single channel and hence, is constrained to offer only one program at a time. This limitation can change when digital television (DTV) is introduced.

Television is analog in nature at the time of this writing, providing transmission of one of the three global standards: National Television Standards Committee (NTSC) in North America and South America, Japan, the Philippines, and Korea; Phase-Alteration-Line (PAL) in much of Europe and Asia; and *Sequential Couleur a Memoire* (SECAM) in France, several other European countries, Russia, and many countries in Africa and Central Europe. DTV broadcasting is on the horizon, and it is likely that a conversion will be underway by the year 2000. It promises to add more channels and features, such as data distribution and interactivity over the air. The FCC decided that U.S. TV stations must adopt a digital HDTV standard known as the Advanced Television Committee (ATC) or face the loss of their broadcast rights.

Networks, Affiliates, and Independent Stations

National television networks (ABC, CBS, NBC, and Fox in the United States) provide programs to affiliated TV stations for broadcast over their assigned frequency channel either in real time or by replay from video tape. In the majority of other countries, all local broadcasting is via network programming. Stations in the United States also can obtain programming from the outside from syndication companies, which sell programs either individually or as packages. Networks around the world often purchase programs from syndicators, which themselves are networks. While most syndicated programs are in fact old network programs (reruns), syndicators often deliver new programs and movies. For example, "Wheel of Fortune" and "Entertainment Tonight" are two popular syndicated programs not offered by the networks.

Stations operate their own studios so that they too can originate programs, particularly local news, special events, and, most important to the success of the station, advertising. The revenues of the stations and the networks are derived primarily from the sale of advertising because individual viewers do not pay for the right to watch over-the-air television (except of course when they buy the advertised product or service). Subscription television (STV) employs scrambling to control viewing of the broadcasts and to ensure that monthly fees are paid to the service provider and ultimately the studio or other copyright holder. In the United States, STV was successful for only a short while between 1980 and 1982 until competition from videocassette recorders (VCRs) and cable television undermined their profitability. A new version of STV has appeared in the form of an S-band terrestrial microwave broadcasting system called multichannel microwave distribution service (MMDS), along with the local multichannel distribution service (LMDS), which operates at even higher frequencies in Ka band.

Networks offer advertisers a nationwide audience, which is important to products like GM automobiles and Sony televisions. The revenues of the stations and networks are tied to the relative size of their respective audiences, which is evaluated by respected polling organizations such as Nielsen. Therefore, the programmers need to deliver programming of sufficiently high quality to attract the largest possible audience. Their profitability is constrained, however, by the cost of producing this programming and of delivering it to the affiliated stations. More recently, viewing of major networks has declined due to competition from cable TV networks, DTH, pay-per-view (PPV) services, and other video media such as prerecorded videocassettes and disks, video games, and PC-related network services (such as the Internet).

Satellite Content Distribution

That brings us to the importance of satellites in providing the needed low cost and highly reliable means of delivering the content. A single satellite can employ point-to-multipoint connectivity to perform that function on a routine basis. To receive programming, every TV station in the United States owns and operates at least one receiving Earth station, and many own Earth stations usable as uplinks. To achieve very high reliability during an extremely high value (in terms of advertising dollars) event such as the Olympics or the Super Bowl, a network will "double feed" the program on a second satellite or a fiber optic link.

The Public Broadcasting Service (PBS) is the U.S.-based nonprofit television network that distributes programming to local public television stations. Most of the funds for PBS and the stations are raised through individual and corporate contributions rather than advertising. Stations also pay PBS and each other for program production and rights for broadcasting. That has allowed the development of a narrower slice of programming (i.e., not of mass appeal) that caters to an audience more interested in education, public affairs, and cultural events. Because of budget constraints, PBS was the first to adopt satellite delivery in 1976, using the Westar 1 satellite. Having demonstrated the benefits of satellite delivery, the commercial networks then began to move quickly in the same direction during the following years. By 1984, all network programming and most syndicated programming was being delivered by satellite. Prior to that, the networks had used satellite links over the INTELSAT system to provide coverage of overseas events.

The predominant frequency band employed in North and South America is C band, for the simple reason that more ground antennas and satellites are available than at Ku band. The program distribution satellite shown in Figure 3.11 is used to broadcast the edited program feed on a point-to-multipoint basis. The downlink is received at each TV station by its own RO Earth station and from there it is either transmitted over the local TV channel or stored on video tape. In the case of a live broadcast from the network studio, the signal is connected from the camera to the uplink Earth station and over the program distribution satellite. A video switching capability in the studio and at each TV station allows technicians to insert taped advertising and computer-generated graphics. Even though most programs are played from video tape, it generally is more economical to distribute taped programs by satellite to the TV stations where they are again recorded rather than mailing the tapes (a process called "bicycling") around the country.

Whether the programs are live or taped, the local TV stations are able to insert their own paid advertising in time slots left for that purpose by the network or the syndicator. Stations and networks have automated those functions to improve productivity and to reduce errors. The reason for the latter improvement is that, because commercial stations and networks are paid by advertisers, every lost or degraded commercial results in lost revenue. The technology to do that is an outgrowth of the computer industry, where very high capacity hard disk drives and high-speed data transmission lines have been made relatively cheap and reliable.

Backhaul of Event Coverage

All sports events and much news coverage are brought back to the studio over a separate point-to-point satellite link called a *backhaul*. In the case of football games, for example, stadiums in North America have access via fiber or terrestrial microwave to a local Earth station, which can uplink the telecast to a backhaul satellite (shown on the left in Figure 3.11). The network or stations pay for the use of the satellite and uplinking Earth station by the minute or the hour. Anyone with a receiving Earth station can pick up the backhaul, which does not yet include either the "commentary" or advertising spots that are inserted at the studio prior to re-uplinking to the program distribution satellite. If coverage is of a one-time event such as a natural disaster or an Olympic race, then a truck-mounted transportable Earth station is driven to the site and erected prior to transmission. The use of Ku band (14/12 GHz) is particularly attractive for this type of rapid deployment service and is called satellite news gathering (SNG). A Ku-band SNG vehicle, shown in Figure 3.12, is much more compact and mobile than its C-band equivalent and can be operated almost immediately after it has been parked on location. In addition, the use of nonshared Ku-band frequencies eliminates any need for prior frequency coordination. Whether C band or Ku band, the time demands and economics of event coverage can mean that a backhaul satellite link will be attractive where the distance to the studio is anywhere from 50 to 5,000 km. For example, a backhaul was used during the Los Angeles Olympics in 1984 to reach from Lake Casitas to Hollywood, a distance of approximately 90 km.

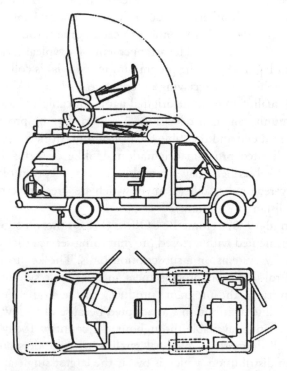

Figure 3.12 An example of an SNG vehicle.

Ground Antenna Utilization

A network-affiliated TV station will use one fixed-mounted Earth station antenna to receive full-time network programming from the point-to-multipoint program distribution satellite. In addition, some "roving" among other satellites can be done with a movable antenna to pick up special programs provided by syndicators and to receive live coverage of sports events of interest to the local community (e.g., when the local baseball team is playing an away game in another city). Antennas used by the networks in backhaul service therefore would need to be movable since events and satellites change form time to time. Some of these movable antennas can be controlled by the network without intervention by the local station. The actual control signals are generated automatically by the broadcast center and transmitted digitally over the program distribution satellite feed in an unused piece of bandwidth.

3.2.1.2 Cable Television

The cable television medium has achieved widespread acceptance in major cities around the world, with over 60% of North American households subscribing to cable service. Originally a means to bring over-the-air broadcasts into remote areas with otherwise poor reception, a local cable TV system uses coaxial and fiber optic cable to connect to each home through a point-to-multipoint distribution network. The arrangement of a typical local cable TV system is show in Figure 3.13. The programming material is collected at the head end, which has high-gain receiving antennas to pick up TV signals with reasonably good quality. In fact, the original name for cable TV was CATV, standing for community antenna television. A studio may be provided at a point between the head end and the cable distribution network for playback of commercials and limited program origination. Unlike over-the-air broadcasting, viewers (called subscribers) pay a monthly fee for reception of the several TV channels delivered by the cable (many of which are "free" advertising supported by local and distant TV stations).

In the mid-1970's, Home Box Office (HBO), now a subsidiary of Time Warner, experimented with a closed programming service that was made available to cable TV systems on a subscription basis. The key to that success has been the control that cable programmers and the cable system operators have over the delivery of the program "product" to the home. As mentioned in Chapter 2, satellite distribution was adopted because of its low cost of making the program channel available throughout the country. Today, approximately one-third of all domestic or regional satellite transponders are employed for cable program distribution, which is by far the largest single application.

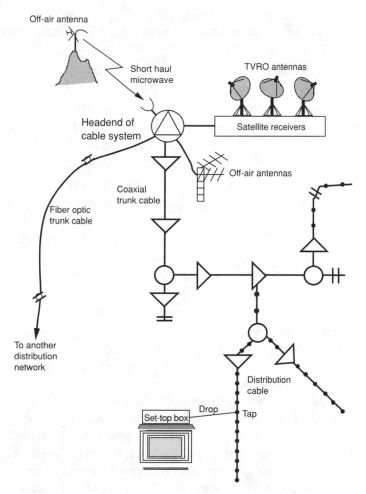

Off-air antenna

Short haul
microwave

TVRO antennas

Headend of
cable system

Satellite receivers

Off-air antennas

Coaxial
trunk cable

Fiber optic
trunk cable

To another
distribution
network

Distribution
cable

Drop

Set-top box

Tap

Figure 3.13 Typical layout of a cable TV system.

Classes of Cable Programming

Several classes of specialized cable programming exist, as summarized in Table 3.3. Many are available internationally over C-band and Ku-band satellites, from which they can be received through DTH and local cable services. The satellite-delivered channels are each focused into particular niches. HBO established the most lucrative niche, that of recent movies. In fact, prior to widespread use of VCRs, HBO and Showtime (a service of Viacom International) provided the only means to watch recent movies uninterrupted by commercials. To continuously receive those services along with the regular channels, subscribers pay an

Table 3.3
A Selection of Cable Television Programming Networks That Originate in the United States

Type of service	Examples
Pay movie (premium service)	HBO, Showtime
Superstation (advertiser supported)	TBS, USA Network
24-hour news	CNN, CNBC, CSPAN
24-hour sports	ESPN, Fox Sports
Music	MTV, The Nashville Network
Children's programming	Nickelodeon, Disney
Cultural programming	Arts & Entertainment
Science and education	Discovery, The Learning Channel
Home shopping	Home Shopping Network, QVC
Adult entertainment	Playboy

additional monthly fee, which is shared between the programmer and the cable system operator. The revenue of the programmer is then used like the advertising revenues of a commercial TV network, paying for the acquisition of programming and for its distribution by satellite.

Because of the marketing strategies of the movie studios that produce most of the product offered by the premium services like HBO and Showtime, there is a time delay (release window) of 6 months or more between when a movie is first shown in movie theaters and when it is made available to cable. Furthermore, the time delay is somewhat shorter with regard to the availability on video tapes for home VCR usage. A PPV service allows the subscriber to place an order to receive a specific movie broadcast at a scheduled time for a fee. Because the administration of this type of service is complicated, the ordering function usually is implemented through the settop box.

Satellite Utilization in Cable TV

The satellite delivery system used by cable essentially is identical to that employed by the networks. As shown in Figure 3.13, program distribution satellites are used to carry the cable programming channels to the cable system head ends. A typical cable system head end in the United States will continuously receive 30 to 50 satellite delivered channels, coming from two to six different satellites. That means multiple receiving antennas are required. The cost of the receiving antenna is relatively low because of the small reflector size required (3–6m). Expense sometimes can be minimized by using a single reflector with several feeds, each aligned for reception of closely-spaced satellites. In actuality, it is the cost of the indoor receivers and considerations as to physical space for antennas that dominate the economics of the cable head end.

To better control access to their product, the cable programmers scramble the programming at the uplink using the analog Video Cipher II system, developed by General Instruments, or a digital system based on the MPEG 2 or digital video broadcasting (DVB) standard. The receiving site at the cable head end then requires a descrambler for each channel being recovered. This approach allows the programmer to certify the particular downlink because the descramblers can be activated (addressed) individually. Scrambling also allows the programmer to reach the backyard dish installation and collect a monthly subscription fee as if it were connected to a cable system.

Some cable programming services cover remote events and therefore employ satellite backhaul. CNN relies heavily on backhaul to provide nearly around-the-world news coverage. Also, Fox Sports channel distributes games of professional sports teams and is a major user of backhaul links. An important requirement in the use of satellites for cable is that the backhaul (which is occasional) be separate from the program distribution (which is full time and continuous). Cable systems themselves generally do not alter the programming coming from the satellite, and many cannot handle a break in transmission the way TV stations are equipped to do.

3.2.2 Direct-to-Home Television

A brief history of DTH services was provided in Chapter 2. From the U.S. perspective, DTH appeared as a viable business in the early 1980s through the manufacture and sale of inexpensive C-band receiving systems to the public. Called backyard dishes, their attraction at the time was the cable programming available for free that the satellites delivered. By 1985, over one million backyard dishes were installed at homes, making the concept of direct broadcasting a reality at least to many U.S. viewers. The pay cable services began to scramble their signals in 1986, and a new phase of industry development was underway. Individuals could buy a Video Cipher II decoder and pay a monthly or annual fee to the cable programmers so their decoders would be authorized to unscramble the signals. The development of this new (but not unexpected) market was important to the success of the Ku-band direct broadcast satellite concept itself.

Nearly all DTH systems employ high power Ku-band satellites that use the BSS channel plan. Under that plan, each orbit position has 32 channels assigned to it. A medium-power satellite might transmit 16 channels at one time, thus requiring that two satellites be operated in the same slot. This

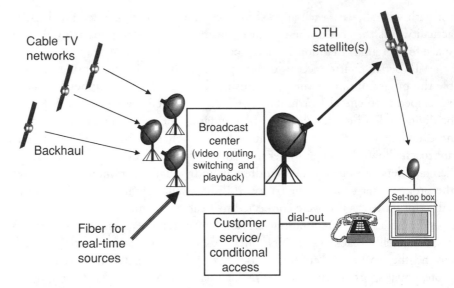

Figure 3.14 A typical DTH satellite broadcasting network.

approach has the added benefit of protecting the service against the failure of a single satellite. The full complement of 32 channels of about 27 MHz each can deliver between 150 and 250 compressed digital video channels. A typical DTH system architecture is shown in Figure 3.14. The broadcast center in this case must be capable of either originating or retransmitting a substantial number of simultaneous video and audio feeds. Other features include subscriber control through conditional access and provision for customer service and billing. Subscribers can select PPV movies at any time, with the record of usage transmitted over a connection from the settop box to the PSTN. In the late 1990s, several digital DTH systems were augmented with data broadcasting to make the service even more attractive to new users. This type of application is covered in the next section.

3.2.3 Data Broadcasting

The most common data communication application implemented by satellite is data broadcasting, illustrated in Figure 3.15. Point-to-multipoint connectivity is used to deliver information in digital form (numbers, characters, and compressed images) to numerous receiving ground antennas. The data typically are transmitted in packets, allowing information to be addressed to individual receivers. Full access to all information in the broadcast is provided in the case of a news wire service such as Associated Press.

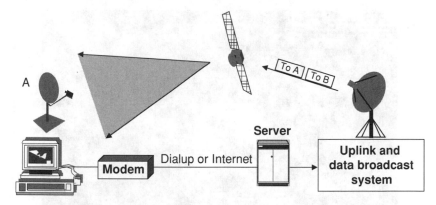

Figure 3.15 One-way data broadcasting to low-cost RO terminals, illustrating terrestrial "dialup" response.

3.2.3.1 Data Broadcast Network Arrangement

A central hub Earth station, shown at the right of Figure 3.15, uplinks a nearly continuous stream of packets to the satellite. The actual content information comes from a database server located at the hub or connected to it by a private line. A data packet assembler is a digital processing device that takes the information to be sent, organizes it into packets, and places appropriate address bits at the beginning of each packet. Because each packet is self-contained with source and destination addresses, it can be routed in any way over the satellite and ultimately through a terrestrial packet-switched network such as the Internet.

The data broadcast reaches RO terminals, shown at the left in Figure 3.15, which are all within the satellite footprint. The RO terminal contains digital processing electronics that can identify the packets addressed to it and recover the data for delivery to the user terminal or storage device. In the illustration, the user at site A has a personal computer to view the data and store them for analysis or later use. Also, the modem and telephone instrument allow the user to reach the hub station or database to request that certain additional data be transmitted over the satellite. With that dialup capability, the broadcast network can perform as if it were a multipoint-to-point interactive network. The DirecPC system from Hughes Network Systems, shown in Figure 3.16, is based on that principle. This system provides medium-speed broadband access to the Internet at up to 400 kbps on the downlink. DirecPC, having received several computer industry awards for innovation and performance over the Internet, has been made available in Europe, Japan, and selected countries in Asia. Similar services are Cyberstar, offered by Loral Space and Communications, and GE-Star, offered by GE Americom.

Figure 3.16   The DirecPC home-based satellite Internet service. (*Source:* Courtesy of Hughes Network Systems.)

3.2.3.2   Data Broadcasting With Video

Data broadcasting also has been applied as an adjunct to video transmission, which is convenient because the video signal requires a lot of power and bandwidth while the data require very little (assuming a usable data rate of 1 Mbps or less). The concept has been applied to DTH systems to increase the attractiveness of the basic service and to add additional revenue streams. The cost of reception is relatively low because the data terminal can be plugged into the video receiver or PC.

In one analog technique, a low-speed data stream is inserted into the vertical blanking interval of the video transmission. (The vertical blanking interval is the horizontal black band that is visible in a gently rolling television picture.) The data are removed from the video signal by a special decoder unit to which the display terminal or data recorder is connected. Another approach employs a separate baseband "subcarrier" onto which is modulated a low- or medium-speed data stream. The receiving Earth station requires a separate subcarrier receiver and decoder to recover the data; however, the subcarrier approach potentially can carry much higher data rates and does not interfere with the video signal. For those reasons, data subcarriers can be found on a large percentage of analog video carriers used for full-time delivery of cable TV programming. Alternatively, broadcast data service is being provided on subcarriers without the video as a means to optimize the link for reception by the smallest possible RO antenna.

In modern digital DTH service, a portion of the high-speed data stream is allocated to data. That provides a flexible "pipe" to deliver broadband data services at rates in excess of 1 Mbps. Both DIRECTV in the United States and B-Sky-B in the United Kingdom introduced those capabilities in 1997. The subscriber adds a special card into the PC to act as the downlink modem. Any return path information is sent through the telephone network using a conventional dialup modem.

3.2.4 Audio Broadcasting

The delivery of high-fidelity audio services by satellite to broadcast radio stations represents an effective, although relatively small, niche in the telecommunications business. Sound quality is excellent because of the wide audio bandwidth (specified anywhere between 5 and 15 kHz, depending on the application) and low noise provided over the satellite link. These links gradually have been converted from analog to digital to improve quality even further. Using point-to-multipoint connectivity, either monaural or stereo sound is uplinked to the satellite typically with the single-channel-per-carrier technique. The transmission is received at a conventional AM or FM radio station, where the desired audio channel is demultiplexed.

The network arrangement is nearly identical to that used for broadcast TV. An affiliated radio station would use one antenna to receive a continuous network feed containing music, news, and national commercials. Local commercials and information are inserted at the radio station in the same way a TV broadcast station adds to its network feed. A second, roving antenna is used for sports events and for syndicated programming such as nationwide talk shows. In one business approach in the western United States, a radio station originates programming from Salt Lake City that is delivered by satellite to stations in several western cities. The main announcer is located in Salt Lake City while the local stations insert only their specific call signs, weather forecasts, and advertising (they could be added at the central uplink site using a submultiplexing technique).

3.3 Point-to-Point Networks

The simplest type of connectivity is point-to-point, illustrated in Figure 3.1 with a pair of Earth stations transmitting RF carriers one to the other. Because both stations simultaneously receive the respective carriers, the link is full duplex. In other words, the users being served can talk or transmit information in both directions at the same time. The uplink section of the satellite repeater receives both transmissions and, after translation to the downlink frequency range, transmits them back toward the ground. If the satellite provides a single footprint covering both Earth stations, then a given station can receive in the downlink its own information as well as that of its communicating partner. That supplementary ability provides a unique way for stations to verify the content and quality of their own satellite transmission. (That would not be the case if the stations were in different beams provided by the same satellite.)

The use of separate frequencies is called frequency division multiple access (FDMA). Alternatively, the Earth stations can transmit their data on the same frequency but separated in time. That technique, in which the transmissions are in terms of limited duration bursts, is TDMA. FDMA and TDMA represent a pair of alternatives that can isolate Earth stations that share a common transponder channel or piece of RF bandwidth. It is even possible to combine the two multiple access modes, for example, TDMA/FDMA, where there are multiple TDMA networks, each operating in a different narrow frequency channel. The reason for doing it that way is to reduce the required uplink power or transponder bandwidth for a small TDMA network. A third technique, called code division multiple access (CDMA), permits the stations to transmit on the same frequency at the same time. All these techniques are reviewed in Chapter 5.

3.3.1 Mesh Networks

A satellite mesh network provides several links to connect multiple Earth station locations. The maximum number of possible links (a full mesh) between N Earth stations is equal to $N(N-1) / 2$. A point-to-point link that is continuously maintained is called dedicated, or preassigned, bandwidth. They typically are used for special applications like heavy telephone trunking between switching centers or high-speed data links in computing networks. A typical telephone network (Figure 3.17) uses a variety of links to achieve mesh connectivity. For example, the preassigned links designated 1, 2, and 3 are wider bandwidth to carry a large quantity of simultaneous conversations. Typical bandwidths of preassigned links range from a low of 64 kbps for a single channel per carrier (SCPC) channel that occupies an RF bandwidth of approximately 32 KHz to a high of 155 Mbps (STM-1 in Table 3.2) on a single-carrier basis in a transponder RF bandwidth of approximately 90 MHz. The need for that bandwidth depends directly on the types and quantities of digital information that the link must transfer. For example, the narrowband channel can be used for high-quality voice communication or for medium-quality video teleconferencing. On the high end, 155 Mbps would provide a broadband ATM service between two nodes in a broadband integrated services digital net-

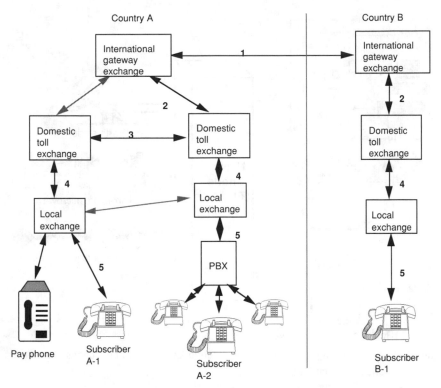

Figure 3.17 Commercial telephone service can be obtained either through public switching facilities or through dedicated private lines.

work (B-ISDN). In between, we find a variety of applications in TDM information for telephone trunking, high-quality video teleconferencing (e.g., full motion video), and bulk file transfer between mainframe computers. The next section covers one particular usage of such broadband links.

3.3.2 Combining Digital Services

Corporate users of point-to-point satellite links can employ multiplexing as a means to integrate digital services, including private telephone, video teleconferencing, and wide area network (WAN) data. As illustrated in Figure 3.18, the appropriate customer equipment provides the digital input to the integrating network node. Devices on the market are based on one of a number of technologies: time division multiplexing, statistical multiplexing, fast packet switching, frame relay, or ATM switching. On the right, high-speed digital streams (typically T-1 or E-1) reach other locations over a combination of satellite and

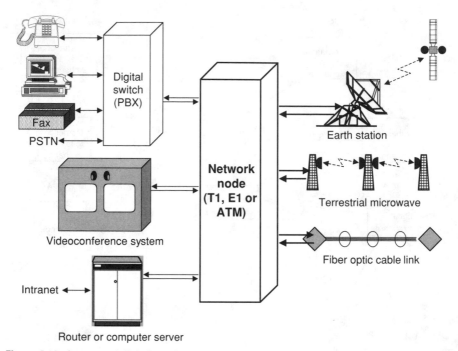

Figure 3.18   Integrated digital services using a network node with connections to satellite and terrestrial T-1 or E-1 transmission facilities.

terrestrial point-to-point links. An equivalent node would be located at each major user location to access the network. Specific traffic patterns between nodes are programmed in a routing matrix that can be altered at any time by an operator using a computer terminal connected to any node in the network. Another important feature of this class of equipment is that it provides statistical multiplexing to efficiently squeeze together smaller streams of data traffic. The following paragraphs describe how this type of network provides services to users.

3.3.2.1   Digital Telephone Services

For telephone service, modern local switches and private branch exchanges (PBXs) digitize the incoming voice frequency information (including voice band data and facsimile) and connect users to each other and to those at distant locations. The trunk side of the switch would be at the T-1 or E-1 level, providing 24 or 30 digital voice channels, respectively. With digital circuit multiplication equipment (DCME), the number of voice channels that can be carried by a T-1 channel is in the range of 44 to 90, depending on the mathematical process (algorithm) employed. The increased telephone capacity shows the effectiveness of digital compression for maximizing the use of the transmission facility.

The digital PBX shown at the top of Figure 3.18 allows many locally connected telephone users to share the bandwidth over the network. Users are connected to the switch by local loops within the same building or by terrestrial tie lines. The interface device at the switch that determines the type of subscriber service is called a port. A typical multibutton telephone instrument digitizes the speech and connects to a digital port on the PBX. Fax machines and voice band data modems must access the switch through two wire analog voice ports. Analog four-wire trunks can be connected to the switch through a tandem port.

To compensate for either satellite transmission delay or the corresponding delay of a long-distance terrestrial link, the multiplexer provides echo cancellation. Switches allow direct digital connections typically in the range of 56 to 384 kbps, depending on the port capacity. The PBX or network node can add the DCME function to increase the number of simultaneous voice and data connections over the fixed bandwidth of the trunk connection. Finally, in the case of ISDN, digital switches are modified to provide greater capability and to accommodate ISDN-compatible terminal devices.

3.3.2.2 Compressed Video Teleconferencing

The video teleconferencing system shown at center left of Figure 3.18 can establish a two-way video link with one or more distant locations. Again, thanks to digital compression, a full motion color TV signal with sound can be transmitted at 128 to 384 kbps. The device that digitizes and compresses the video signal from the camera is called a video codec. The quality of this type of video is adequate for business meetings showing "talking heads" and presentations consisting of color slides or computer graphic images. It is common to operate the video teleconferencing system with a digital multiplexer to introduce other services into the outbound T-1 or E-1 stream. In contrast to video teleconferencing, a broadcast quality picture requires a rate of 2.5 Mbps or greater with currently available technology.

3.3.2.3 Wide Area Network Links

The last type of user access shown in Figure 3.18 is for high-speed data on a point-to-point or mesh basis. This type of service could be required to interconnect two or more large computer systems, each capable of transmitting at a multiple of 64 kbps. Links of this type are to connect local area networks (LANs) together into a WAN. A device called a router is used to transfer network packets from LAN to LAN. Another application for such high speeds is for digital facsimile used to reproduce newspaper or magazine pages. With rates up to E-1, an entire newspaper can be transmitted with absolute clarity in under 1 hour.

The advantage of using digital integration is that such WAN requirements can be accommodated with a conventional network composed of T-1 or

E-1 transmission facilities and state-of-the-art network nodes. The point-to-point links, using satellite and terrestrial transmission, can be arranged in star, mesh, or ring topologies. The selection depends on such considerations as (1) the flow and connectivity patterns of user information, (2) the need for diverse connections, and (3) the cost of those facilities. A common use of the satellite link is as a second or alternative path to act as a backup for a vital terrestrial link. The nodes have the ability of automatically rerouting traffic onto the satellite link in the event of a loss of service over the terrestrial portion of the network. When it is working correctly, users will not notice when the node automatically responds to a failure by making such a transfer. The network operator, on the other hand, is informed of that status by a network management system. At that point, operations personnel take immediate action to restore the lost terrestrial link.

3.3.3 Digital Integration With TDMA

Modern TDMA equipment can perform the function of the sophisticated network node, which is particularly attractive for private digital networks with large and diverse service requirements. Unlike the use of terrestrial nodes discussed in the previous section, TDMA provides a large common bandwidth that all Earth stations share. That means a given Earth station can transmit any amount of data to one or more other stations, up to the maximum bandwidth of the entire network. The flow of information is maintained in a common traffic matrix that defines the connectivity between stations. An important benefit of TDMA is that it exploits the point-to-multipoint capability of the satellite and thereby reduces the number of individual transmission links that otherwise would be required with point-to-point T-1 or E-1 channels. This aspect is covered in greater detail in Chapter 5.

While the first use of TDMA was in a high-speed, full transponder mode, more recent applications are for TDMA/FDMA networks of small VSATs. As discussed in Section 3.4, VSATs can use TDMA to achieve near-continuous communication service even though the actual bandwidth is relatively small. TDMA, therefore, is an effective access technique over satellite channels because of its unique ability to share bandwidth among multiple users.

3.4 VSAT Star Networks

VSAT star networks have helped many businesses solve difficult network problems though the use of communication satellites. The basic principle is to start with data broadcasting to remote sites and to add a return link over the same satellite, bypassing the terrestrial network. As shown in Figure 3.3, multipoint interactive connectivity provides two-way communications because the remote VSATs receive the broadcast from the central hub station and can transmit back typically at a much lower throughput data rate. It is different from a point-to-point network because the remote stations cannot communicate directly with one another but must do so through the hub. The generally accepted principle is that the data broadcast, called the outroute, continuously transmits packetized or TDM information to all VSATs, which in turn utilize burst transmissions, called inroutes, so that several can share a common frequency channel.

The fact that VSAT inroutes are received only by the hub means that their transmit power is quite low. In addition, the outroute is at a high power level to permit reception by the small VSAT antennas. There is, therefore, an economical balance between the high cost of the hub (which is shared among the VSATs) and the low cost per VSAT (installed at each remote served location). The best applications of VSAT star networks are where there is need for remote locations to access a headquarters host computer or LAN and where remote to remote communication is limited.

3.4.1 VSAT Data Networks

The VSAT is relatively inexpensive and easy to operate, justifying its use on the customer premise. Figure 3.19 shows how a VSAT is used in a multipoint-to-point network; Chapter 9 contains a detailed discussion of the typical VSAT and hub equipment configuration. With an antenna size of 1.2m (4 feet) and a purchase price comparable to that of a high-end PC, a VSAT gives full access to a private network involving hundreds or thousands of locations. Such networks would be prohibitively expensive if implemented with dedicated point-to-point private lines, whether terrestrial or satellite.

In a VSAT network, a user can view the satellite as a router or toll switch in the sky, eliminating the need for local loops, public telephone switching, or even PBXs in some cases. The key to the effectiveness of a VSAT network is its interactive nature, allowing two-way communications from remote locations in the same manner as the terrestrial telephone network, provided, however, that a star arrangement is appropriate for the application. If a true mesh connectivity is required, each station must be capable of direct communication with every other.

Star networks are easily implemented using VSATs because the low sensitivity of the small dish is compensated for by the high sensitivity of the large antenna and high-power amplifier at the hub. That is why the first VSAT networks to be introduced in North America all employed hubs in a

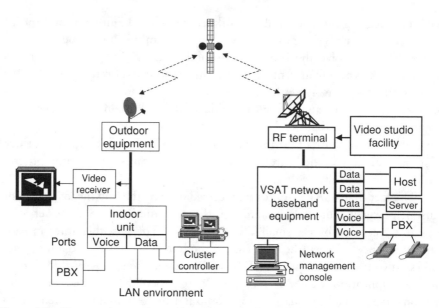

Figure 3.19   The complete end-to-end architecture of a typical VSAT star network.

star architecture, which continues to be the majority of installations around the world. As discussed in the next section, star networks are satisfactory for one-way and interactive data communications.

The VSAT has other dimensions that are outside the current capabilities of the terrestrial voice network. Video teleconferencing has been gaining acceptance as a valuable business communications service, rendered through the TV RO feature of the VSAT. Also, the VSAT is capable of receiving and transmitting high-speed data at transmission rates considerably in excess of what can pass through the 3 kHz bandwidth of the telephone local loop. Many of the features of ISDN, including simultaneous voice and data communications, are accommodated through a VSAT. For many users, those added dimensions provide justification for employing VSAT technology as the centerpiece of their private telecommunication systems.

3.4.2 Interactive Multimedia User Networks

The expansion of high-capacity digital communications systems is blurring the dividing lines between telephone, data, and video transmission. In fact, as discussed previously, it is possible to convert any and all of those services into a digital format and thereafter utilize common terrestrial and satellite links. New standards and equipment are being developed for B-ISDN, by which many applications are digitized at the source and combined (integrated) for efficient routing and transmission. That is the function of network nodes.

The multimedia revolution is approaching for satellite networks at the time of this writing. By convention, multimedia includes all forms of information, such as text, image, audio, and moving picture. The World Wide Web over the Internet provides many of those elements but lacks the broadband feature. Multimedia implies that we have an interactive link, but that can be simulated in a variety of ways. For example, in Figure 3.20, we see a simple picture of a user in front of a PC that is connected by a network to an information server. With a Pentium or Power PC class of personal computer, the user has a high degree of interactivity on the desktop. However, when we connect that same PC to a network, the quality of interactivity degrades quickly. That happens at each point of information transfer, with most of the drop in performance happening within the network and finally within the information server at the right. How can we improve the degree and quality of interactivity for the user on such an information highway? Answers lie in making the network much simpler, with fewer intermediate nodes and points of congestion. That is where advanced satellite networks will come into play.

Advanced multimedia satellites can employ GEO and non-GEO systems, provided that adequate radio spectrum is available. Ka band has drawn a lot of attention as the ideal spectrum for exploiting the multimedia VSAT. There have been a variety of experiments and demonstrations of Ka-band systems and

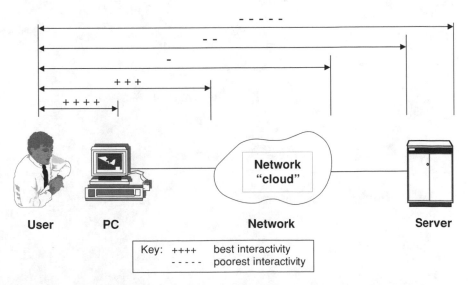

Figure 3.20 Factors that affect interactivity when a PC is connected to a server.

applications, culminating in the ACTS program in the United States. Also, Japan has invested heavily in Ka-band hardware over the past 25 years. The difference today is that for the first time there is a good business opportunity to support the required investment. When properly engineered, Ka-band satellite networks offer high capacity to relatively small fixed Earth stations that could be cheaper than the VSATs currently popular with many corporations around the world.

References

[1] Elbert, B. R., *The Satellite Communication Applications Handbook*, Norwood, MA: Artech House, 1997.

[2] Elbert, B. R., and B. Martyna, *Client/Server Computing: Architecture, Applications and Distributed Systems Management,* Norwood, MA: Artech House, 1994.

4

Microwave Link Engineering

Microwave link engineering is the branch of communication engineering that deals with the analysis, design, implementation, and testing of radio paths in the atmosphere or in space, operating in the range of about 1 to 100 GHz. Satellite and other microwave radio links obey certain rather predictable laws of nature, allowing engineers to design satellite networks on paper (or, more likely, on computer) with good accuracy. An unobstructed line-of-sight path provides the most predictable transmission link; however, various forms of signal absorption (attenuation), blockage, and multipath propagation introduce time-varying fading, which can only be estimated using statistical techniques. Nevertheless, the service obtained from microwave links between the satellites and ground stations can be very acceptable, even when compared to terrestrial cable and radio (wireless) systems.

This chapter is intended for engineers and scientists new to the field and is perhaps the most mathematical chapter in the book. Nontechnical readers may wish to scan it rather than delve into the details. In general, microwave energy travels in straight lines through space in the same manner as a light beam. That should not be a surprise because both light and microwaves are forms of electromagnetic energy, light being at much higher frequencies or, equivalently, shorter wavelengths. Radio waves in general and microwaves in particular can propagate through space like light and heat, spreading out as they move farther and farther from the source. The waves travel through a vacuum at the universal speed of light, the same for all forms of electromagnetic energy. Microwaves can be kept from propagating freely by forcing them to travel in a closed metallic pipe called a *waveguide*. The physical similarity to common household pipes has led to the use of the term *plumbing* when referring to the waveguides found in Earth stations and satellites. At the input end of the waveguide would be found the transmitter, the electronic device that generates the microwave energy on the proper frequency, with the other end connected to an antenna. Waveguides are similarly used on the receiving end.

Microwave power is measured in watts, the same unit we associate with electric power used for lighting and motors. Link effects that either reduce or increase that power almost always are expressed in terms of the decibel (dB), which is why it is important to introduce the concept early in the discussion. Converting a quantity to decibels simply means taking the logarithm to the base 10, or common logarithm, and multiplying the result by 10:

$$A \text{ (in dB)} = 10 \log_{10} (a)$$

where a is a real, nonnegative number. In actuality, a is a ratio, such as the gain of an amplifier, that is,

$$a = P_{out}/P_{in}$$

A sample of decibel values versus their equivalent ratios is provided in Table 4.1. In the table, relative power changes are computed to the closest tenth of a decibel (i.e., 0.1-dB precision) because measurement accuracy on an actual satellite link is about 0.5 dB at best. Since the logarithm of the number 1 is equal to zero, a zero-dB change simply means there has been no change in the power level of the signal. You may recall that logarithms are exponents, and in the case of logarithms to the base 10, they are exponents of the number 10. The inverse of a decibel is expressed as:

$$a = 10^{A/10}$$

which again is a ratio.

A decrease would be represented by a negative decibel value. For example, reducing a power to one-half its value (e.g., multiplying it by 0.5) is the same as subtracting 3 dB (e.g., adding −3.0 dB). Another consequence is that a decrease of 1.0 dB (adding −1.0 dB) represents a decrease of 20%. Table 4.1 indicates that adding 1.0 dB is an increase of 26%.

An inexpensive scientific calculator is the most convenient device for converting back and forth between ratios of powers and decibels. Just be sure to use the common logarithm (log) button and the 10^X button, where appropriate. This is safer than trying to memorize Table 4.1 (although such knowledge is the sign of a "dB artist").

Table 4.1
Example of a Decibel Table, Indicating the Ratio, Percentage Increase,
and Corresponding Decibel Value

Power ratio	% increase	dB
1.0	0	0
1.12	12	0.5
1.26	26	1.0
1.41	41	1.5
1.59	58	2.0
1.78	78	2.5
2.00	100	3.0
2.24	124	3.5
2.51	151	4.0
2.82	182	4.5
3.16	216	5.0
3.55	255	5.5
3.98	298	6.0
4.47	347	6.5
5.01	401	7.0
5.62	462	7.5
6.31	531	8.0
7.08	608	8.5
7.94	694	9.0
8.91	791	9.5
10.0	900	10.0
	20	13.0
	50	17.0
	100	20.0
	200	23.0
	500	27.0
	1000	30.0

The following are a few simple examples of the decibel approach. A 3-dB difference in power level is nearly equal to a factor of 2, while a factor of 10 is exactly equal to 10 dB, that is, $10 \log (10) = 10$. In satellite communications, it is not uncommon to encounter a factor of 10^{-20} decrease in signal power.

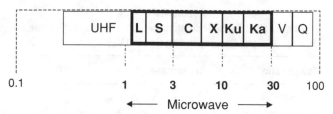

Figure 4.1 The microwave spectrum, indicating the approximate location of satellite bands.

That can more conveniently be expressed in decibels as a 200-dB reduction, called a loss. Gains are added, while losses are subtracted. When dealing with factors in an equation, it is convenient to first convert each factor to decibels. Then the equation can be solved by summing the decibel values instead of having to resort to multiplication. Likewise, division becomes simple subtraction, as in the example of the 200-dB loss. That is the basis for the link power balance calculation and the link budget, reviewed later in this chapter.

The part of the electromagnetic spectrum that is of interest in commercial satellite communications lies between 1 and 60 GHz, as illustrated in Figure 4.1. The relative merits of each band segment were covered in Chapter 1. Using the popular letter designations for the various segments, the spectrum offers a total bandwidth of approximately 60 GHz. The frequencies below 30 GHz are those currently in use or under development; the frequencies above 30 GHz are a still untapped resource. RF bandwidth can be used and reused many times by multiple satellites and the radio beams that they can generate over portions of the Earth. The following sections review the propagation, generation, and reception of microwave signals, in preparation for studying the unique aspects of satellite communications presented in Chapters 5, 6, 7, and 9.

4.1 Propagation on the Earth-Space Link

The process by which radio signals reach the receiving antenna from the transmitting station is called radio wave propagation. As mentioned previously, radio waves represent a part of the electromagnetic spectrum, encompassing radio, infrared, visible light, ultraviolet, and X rays (given in increasing order of frequency or decreasing order of wavelength).

4.1.1 Basic Microwave Propagation

All radio waves behave similarly in free space, but various forms of matter produce interesting and potentially disruptive results when placed in their path.

Table 4.2
Propagation Effects To Introduce Additional Attenuation of Microwave Signals as They
Propagate Between the Transmitting Station and Receiving Station

Propagation mode	Effect	Source
Absorption	Conversion of microwave energy into simple heat	Air, water vapor, precipitation (rain and snow), nonmetallic structures, and foliage
Scattering	Multiple reflections and bending due to minute particles	Water vapor, precipitation, dust, ion plasma (the ionosphere)
Refraction	Bending of microwaves due a change in the refractive index	Air, ion plasma
Reflection	Total reflection of the microwave signal	Temperature inversion layers (ducting), metal structures, bodies of water, buildings, and flat terrain
Diffraction	Bending around physical obstacles	Metal structures, buildings, and terrain blockage

That is because microwave energy can be absorbed, scattered, bent, and reflected, as indicated in Table 4.2. As a result, there is additional loss of signal power, in decibels, which must be accounted for in the link design.

A question that comes to mind is, "How does such a wave, traveling at the speed of light, come into existence in the first place?" Because space consists of nothing, a radio wave does not propagate like the crest of a wave in the ocean. Instead, it usually is the result of the high-frequency vibration of electrons in a piece of wire or other conducting material of appropriate dimensions. Alternating current from a transmitter causes the electrons to vibrate back and forth; by not flowing continuously in one direction, the electrons lose most of their energy by throwing it off into space. That is illustrated in the drawing sequence provided in Figure 4.2.

A high transmitter power, say, 1,000W (1 kW), produces a correspondingly high level of microwave radiation at the point of exit into space. That is the same principle behind a microwave oven, which uses a high-power magnetron amplifier to generate an intense field for cooking purposes. The field causes the molecules in the food to oscillate at the same frequency; that generates the necessary heat to raise the temperature. Such intense radiation

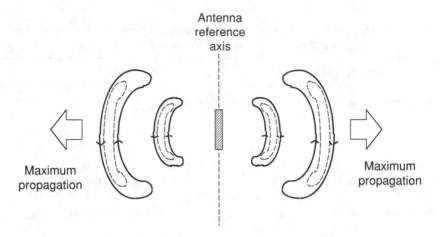

Figure 4.2 The fundamental concept of electromagnetic wave propagation from an elemental antenna

around a high-power microwave transmitting station can "bake" you if you come too close. At sufficient distance from a transmitting antenna, the microwave energy induces less heating. There is a defined area around the antenna, called the *near field region*, where the energy has not quite coalesced into a clearly defined beam. Within the near field, microwave energy varies in intensity, depending where one stands relative to the antenna structure. At sufficient distance from the antenna, the radiation field is formed into the type of beam associated with the particular antenna structure, called the *far field region*. A parabolic reflector antenna has a sharply defined far field beam that is aimed at the satellite. The transition between the two regions is roughly defined by the following equation:

$$R_f \approx 2\,D^2/\lambda$$

where R_f is the approximate distance, D is the diameter of dish reflector in meters, and λ is the wavelength in meters. This equation indicates something unexpected: the distance to the transition increases with frequency rather than decreasing. For example, proper testing of a Ku-band reflector antenna demands that the transmitting (or receiving) source be farther away than if the same reflector were to be used at C band. From the beginning of the far field, the radiated signal keeps on decreasing in intensity by the inverse of the square of the distance from the source (the same property as visible light from a point source such as a star). This particular concept is the basis for link design.

4.1.2 Isotropic Radiator

The most fundamental type of radio antenna is the isotropic source, which is analogous to the lightbulb illustrated in Figure 4.3. At a fixed radius from an isotropic source, which defines a sphere, the energy intensity is constant regardless of direction. The area of the sphere of uniform received energy is equal to the constant π (3.14159...) multiplied by 4 multiplied by the square of the radius, that is, $A = 4\pi R^2$. It is common practice to measure the intensity at the particular radius in units of watts per square meter, calculated by dividing the power of the isotropic source by the area of the sphere in square meters:

$$P/A = P_t/4\pi R^2, \qquad W/m^2$$

The denominator, $4\pi R^2$, is sometimes called the spreading factor.

RF power driving the isotropic source produces a constant power density at a fixed distance. That density decreases by $1/R^2$ as the point of reception moves farther away from the source. Ignoring losses, it is theoretically possible to capture all the transmitted power by collecting it with a closed surface around the source (the physical size of the collector is immaterial).

Receiving antennas, therefore, work by way of the area that they expose to the RF energy radiating from the source. This is illustrated in Figure 4.4 by an isotropic source that radiates energy equally in all directions, shown as equally spaced rays. There are two receiving antenna surfaces of equal area: one

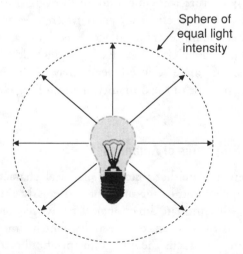

Sphere of
equal light
intensity

Figure 4.3 An isotropic source can be represented by a lightbulb radiating energy in all directions with equal intensity.

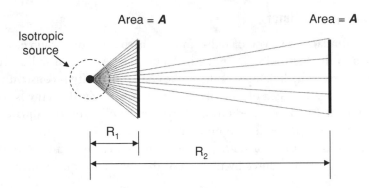

Figure 4.4 The radio energy captured by a fixed antenna area decreases as the distance to the source increases.

at distance R_2 is farther away from the source than the other at distance R_1. Notice how the closer antenna intercepts considerably more power than the more distant area. That illustrates how a radio signal becomes weaker as the receiver is moved farther from the transmitter. It also demonstrates the important concept of capture area, that is, the relationship between the effective area of an antenna and the strength of the signal received by it. The challenge of antenna design is to maximize the fraction of the energy that the antenna actually delivers from the reflector surface to the receiver. The key parameter is the efficiency of the antenna, defined as the ratio of the effective area (i.e., the area that would perfectly capture the same amount of energy) to its physical area. Typical values of dish antenna efficiency are between 55% and 70%. Antennas that do not use reflectors but employ an active receiving area to the incoming wave can achieve up to 90% efficiency. Examples include horns and arrays of elemental antennas. The increase in efficiency may be illusive for frequencies below about 1 GHz because of the difficulty of controlling element beamwidth and interaction.

4.1.3 Directional Properties of Antennas

The isotropic antenna is neither practical (its ideal characteristic cannot be achieved with a simple physical structure) nor particularly useful (due to its low gain and inability to discriminate directions). What we rely on for receiving and transmitting signals is the concept of directivity, which simply means that the antenna has the ability to focus the energy in specific directions. Energy that would have been radiated in unnecessary directions around the sphere is concentrated by the structure of the antenna and redirected to increase the intensity in the desired direction. The directivity in a given direction can be expressed as the ratio of the measured signal to the maximum signal in the peak direction. The gain, on the other hand, is an absolute measure, obtained by comparing the signal from the antenna to that of an isotropic radiator. Our absolute measure of maximum gain could be as little as 10 or greater than a factor of 1 million. When expressed in decibels, these particular values would be 10 dBi and 60 dBi, respectively, where the "i" indicates that the gain is with respect to our fictional isotropic antenna. That implies that an isotropic antenna has a gain of 0 dBi. There must be a fixed offset in decibels between the absolute value of gain and the corresponding relative value of directivity, because our definition of directivity is that it starts at a value of 0 dB at the peak and follows the pattern in a negative-going sense for all other angles.

Another important property of an antenna is called reciprocity, which means that the gain and the directivity are the same at a given frequency whether it is used to receive or to transmit. That allows the antenna to receive with precisely the same directional characteristics as it transmits. Figure 4.5 illustrates a simplified antenna gain pattern for transmission or reception with a main beam (the region of maximum directivity) oriented toward the right. For comparison, the uniform pattern of an isotropic source is superimposed to scale on the directional antenna pattern. In addition to a main beam, every real antenna operates in undesired directions, shown in the figure as a pair of sidelobes and a backlobe. The maximum gain, also called the peak gain, is indicated at the center, and the backward direction is indicated at ±180 degrees.

There are a number of other useful definitions of antenna performance besides the peak gain. The half-power beamwidth (often called simply the beamwidth) is the angular width of the main beam measured between the points where the power intensity is one-half that of the peak. An equally accurate name that is often used is the 3-dB beamwidth, since the half power point is where the directivity is 3 dB down. Assuming that the microwave link can still function with a 3-dB decrease in signal strength, the half-power beamwidth

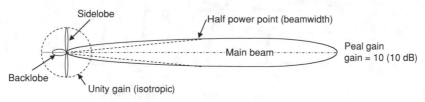

Figure 4.5 Gain of a high-gain antenna expressed as a power ratio and presented as a 360-degree polar plot.

defines the range of antenna pointing (alignment angle) over which the antenna or satellite can move. It is a common practice, however, to allow only a 0.5-dB (12%) drop in signal power, which demands either tighter antenna pointing accuracy or satellite position control.

When installed at an Earth station, the antenna is attached to a mount, which may allow the beam to be repointed. Simple RO antennas used for direct reception usually are fixed to their mounts and need only be pointed toward the satellite when initially installed. Movable antennas usually have motorized mounts to allow remote repointing; that could also be part of an automatic tracking system to keep the beam aligned with the satellite. It is normal to refer to the elevation angle of the direction of pointing, where zero is for pointing at the horizon and 90 degrees corresponds to directly overhead (zenith). The mount also points the antenna in azimuth, that is, along the directions of the compass. Most mounts work directly in terms of elevation and azimuth, but some are polar and behave like an astronomer's telescope. The advantage of a polar mount is that it can repoint along the equatorial orbit plane by adjusting only one axis.

Presentations of antenna performance are called antenna patterns and are illustrated in Figures 4.6 (a hypothetical example) and 4.7 (a real 10m antenna) in terms of gain and directivity, respectively. Almost every antenna has a back-lobe in the opposite direction from the main beam. However, as shown in Figure 4.6, the gain of the backlobe can be made to be less than unity, in this case, producing a negative gain of −2 dBi or lower. Sidelobes and backlogs are important characteristics of Earth station antennas when one is considering transmit or receive interference. Figure 4.7 includes a smooth curve, called a sidelobe envelope, that defines a specification of maximum sidelobe gain. Two

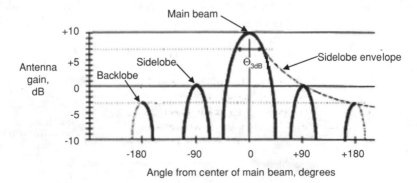

Figure 4.6 Gain of a high-gain antenna in decibels for all angles with respect to the peak of the main beam indicating the worst-case envelope of sidelobes.

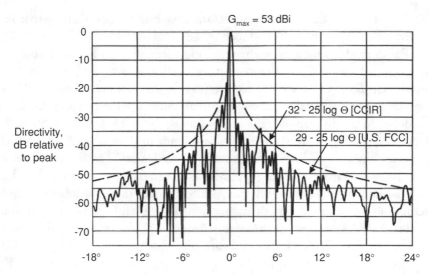

Figure 4.7 Typical Earth station antenna directivity pattern indicating the gain envelope for the CCIR and the U.S. FCC specifications.

such envelopes are shown, the lower being a tighter specification adopted in the United States.

4.1.3.1 Isolation

The directive property of an antenna determines how effective it will be for getting signal power from the source to the receiver. However, any link can be degraded by signals on the same frequency that enter the receiving antenna from a direction other than along the main beam. Likewise, a transmitting station can cause degradation to other systems by sidelobe radiation. Any undesired signal that can potentially degrade reception is RFI. There is a whole field of engineering study that focuses on the identification of sources of RFI, the establishment of criteria for acceptable operation in the presence of RFI, and the development of techniques for countering its effects. Techniques such as beam shaping, cancelation, and shielding are effective in that regard.

The presence of RFI is a consequence of the fact that all radio communications systems reuse frequencies, that is, there will be more than one radio station operating on any particular frequency at any particular time. What keeps the RFI within acceptable limits is isolation that is either natural (e.g., geographical or angular/orbital separation) or artificial (e.g., measures such as RFI shielding or beam cancelation). Satisfactory operation of independent microwave users on the same frequency often requires cooperation (called frequency coordination) because "one user's radio link is another user's RFI." Several satellites can operate in the geostationary arc in the same frequency band because each directional ground antenna can focus on one particular satellite, suppressing the RFI produced by adjacent satellites.

For example, if we assume the receiving user's antenna has the characteristic given in Figure 4.6 and the interfering satellite is located 45 degrees away from the satellite that we wish to receive (the desired satellite), then the RFI is suppressed by at least 20 dB. Actual ground antennas used in FSS and BSS links provide 40 to 60 dB of peak gain with suppression at 45 degrees of 50 to 70 dB. At closer-in angles, the isolation may amount to only 20 to 30 dB, which is adequate to allow satellite spacing as small as 2 degrees. The first sidelobe, in particular, is the strongest, being typically 15 to 20 dB down from and within 1 degree (more or less) of the peak gain. The control of those sidelobes demands careful design and installation of such antennas. A useful specification in that regard is the sidelobe envelope (shown in Figure 4.7), which defines the worst-case potential for RFI. A standardized formula, which was adopted by the ITU, provides a common standard for the larger ground antennas used at C and Ku bands. In it, gain at a particular off-axis angle is specified in the direction of a potentially interfering (or interfered-with) satellite by:

$$G(\Theta) = 32 - 25 \log \Theta, \qquad\qquad \text{dBi}$$

where Θ is the offset angle between the direction of the main beam and that toward the interfering or interfered-with satellite, for angles between ±1 degree and ±36 degrees.

$$G(\Theta) = -10 \text{ dBi, for angles between ±36 degrees and ±180 degrees}$$

Again, this is not really electromagnetic physics but a specification based on practical experience with real parabolic reflector antennas.

4.1.4 Polarization (Linear and Circular)

Up to this point, we have dealt with radio propagation as if it were pure energy. However, there is a property of an electromagnetic wave called polarization, that depends on the orientation (or angle of rotation) of the transmitting antenna. You may be familiar with polarization as it relates to light. For example, true three-dimensional (3-D) movies utilize vertically and horizontally polarized light to simultaneously project appropriate left and right images on the screen. The images are separated by invisible polarizing grids imbedded in the lenses of the viewer's glasses, where one lens is aligned vertically while the other is aligned at a 90-degree angle (horizontally). Vertical and horizontal polarization, therefore, can "reuse" (i.e., use twice) a transmission path, such as the projection of a movie on a screen or, as described below, a radio path.

4.1.4.1 Linear Polarization

The concept of polarization discrimination in radio communications is illustrated in Figure 4.8. Shown at the top of the figure is a type of simple wire or rod antenna called a dipole. Electrical current from the transmitter flows along the rod first upward and then downward, oscillating at the frequency of transmission. At C-band downlink frequencies, the rate of oscillation is 4,000,000,000 times per second (i.e., 4 GHz). The alternating current in the rods produces an electromagnetic wave that propagates off into space (Figure 4.2). A dipole is not a true isotropic source because there is no radiation along the direction of the rods. Instead, what is formed is a doughnut-shaped pattern that is aligned horizontally. The electrical currents in the rods cause the electromagnetic wave to have its electric component to be lined up in the same direction, which is vertical (in the direction of the two arrows). This type of polarization is called linear polarization (LP) because the electric component has a fixed orientation. Horizontal LP is obtained when the dipole is rotated 90 degrees, so that the direction of the electrical current also is horizontal. Reception occurs when the electric component of the incoming wave produces a current in the receiving antenna, which cannot occur if the conductors of the receiving antenna are perpendicular with the incoming polarization. In the lower half of Figure 4.8, horizontally polarized transmitting and receiving antennas provide for the maximum amount of power to be carried (coupled) between them.

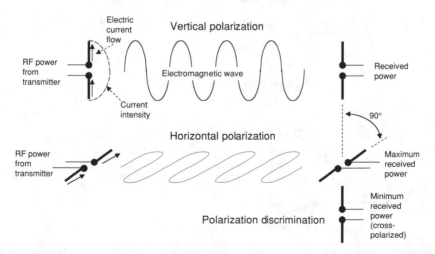

Figure 4.8 Properties of linear polarization as radiated and received by dipole rod antennas.

A vertically polarized receive antenna, which is perpendicular to and therefore cross-polarized with the transmitter, minimizes the amount of coupled energy.

Microwave antennas usually make use of waveguide structures and solid reflecting surfaces because they are much more efficient and predictable than wires and rods. Rectangular waveguide, illustrated in Figure 4.9, is simple in design and efficient in the transmission of microwave energy and hence is very popular in satellite and Earth station design. The electric component is vertical for the orientation shown, extending between the upper and lower side walls (dimension b). The magnetic component lies in the plane of the top and bottom walls and is perpendicular to the electric component at every point along the waveguide. The horizontal dimension of these walls (dimension a) must be greater than one-half wavelength at the operating frequency. At any lower frequency, the wave cannot propagate because its wavelength literally cannot fit. This is termed waveguide beyond cutoff. The narrower dimension, b, is not as critical and is adjusted like a transformer to match the characteristic impedance of the waveguide to the antenna.

The relationship between antenna orientation and wave polarization is shown in Figure 4.9 for a waveguide horn, which basically is a slightly flared piece of rectangular waveguide. Instead of electrical currents, the waveguide carries a tightly focused electromagnetic wave with the electric components extending between the parallel walls. Creation of the guided waves in the first place is accomplished with the small dipole element inserted into the waveguide

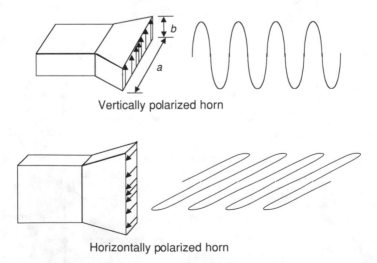

Vertically polarized horn

Horizontally polarized horn

Figure 4.9 Linear polarization of microwave radiation from vertically and horizontally polarized feed horns.

at the input end. Note how the electric components vary in intensity along the width of the horn, being maximum at the center (closely spaced arrows) and dropping to zero at the edges. That particular natural property of a simple horn provides a radiation taper used in the design of horn-fed reflector antennas. In any case, the radiated electromagnetic wave is linearly polarized, just like that from a vertically oriented dipole. The polarization can be changed to horizontal simply by rotating the horn and waveguide by 90 degrees.

As with wire antennas, maximum coupling occurs when the transmitter and receiving horns are co-polarized. If one horn is rotated 90 degrees with respect to the other, minimum energy transfer (coupling) results and they are said to be cross-polarized. The performance of LP at intermediate points follows a simple law: The relative energy coupled is equal to the square of the cosine of the angle. That characteristic is plotted at the left of Figure 4.10 for the co-polarized case, that is, for the level of signal received as the receiving antenna is rotated from maximum coupling to minimum coupling. An ideal LP wave and antenna are assumed, for which the coupling goes from a maximum of 1 to a minimum of zero. Maximum coupling always occurs at some angle, and minimum coupling is as close to zero as physically possible. Properly designed and installed antennas can deliver minimum coupling values of 0.0001, or −40 dB. Larger structures that are highly symmetrical in the direction of radiation can even exceed that value. However, due to slight misalignments and other imperfections, values in the range of 30 to 40 dB are more typical.

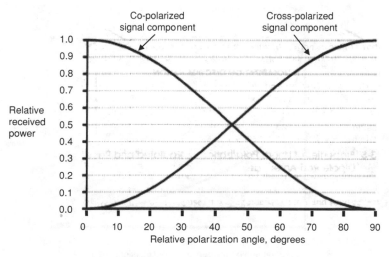

Figure 4.10 Relative receive power as a true ratio for the co-polarized and cross-polarized signals in a linearly polarized link as the polarization angle of the receiving antenna is rotated.

Probably the most important application of polarization is in frequency reuse, where two cross-polarized signals are transmitted at the same time on the same frequency. The right curve in Figure 4.10 shows how the level of the cross-polarized signal increases as the receiving antenna is rotated from zero to 90 degrees. Notice how at 45 degrees both signals are at the same level. Figure 4.11 plots coupling in decibels, termed *polarization isolation,* between the desired and undesired polarizations. Maximum isolation occurs at zero offset angle, that is, where the receiving antenna is aligned in polarization with the transmitting antenna and the undesired polarization is nulled out (minimized). Alignment for maximum coupling is not particularly critical (±5 degrees of error introduces very little loss of signal); however, the cross-polarization isolation is extremely critical, and precise alignment is a necessity. The maximum value of isolation of 40 dB in Figure 4.11 corresponds to the typical value described in the preceding paragraph. All modern satellite communications systems use polarization isolation to increase the capacity of the particular orbit position. Therefore, users must employ antennas on the ground that meet polarization requirements and that can be adjusted from time to time by rotating the feedhorn for minimum RFI. (The only way around the latter requirement is to use circular polarization, discussed in the next section.) Isolation can subsequently degrade from propagation effects, particularly Faraday rotation, which is a twist of linear polarization caused by the ionosphere.

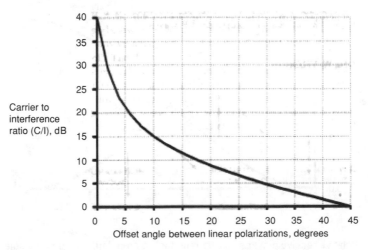

Figure 4.11 Polarization isolation in decibels for a linearly polarized RF signal as a function of the offset angle from maximum isolation.

4.1.4.2 Circular Polarization

Another type of polarization that is used in international and DTH broadcasting satellites is circular polarization (CP). Being circular in polarization, the receiving antenna is not aligned for the transmitting antenna's polarization. Rather, it is the inherent polarization of the signal, either right-hand CP (RCP) or left-hand CP (LCP), that must be supported by both ends of the link. An example of RCP is provided in Figure 4.12.

To understand the composition of CP, first observe that LP can be represented in the form of a vector, where the direction of the vector is in line with the electric component and its length is proportional to the power of the signal. This vector is always perpendicular to the direction of propagation. CP is a particular combination of two equal LP waves that are cross-polarized with respect to each other (horizontal and vertical are two of the four vector states in Figure 4.12). If the two polarizations represent the same signal on the same frequency and they are precisely in phase with each other, the resultant vector lies at a 45-degree angle between the two polarizations. In CP, however, the same two vectors are out of phase with each other by 90 degrees. That is accomplished by first splitting the transmit signal in two at the source antenna and then delaying one component by a quarter period before radiating the components through a dual-polarized antenna element. What is produced is a resultant vector that rotates like a corkscrew (Figure 4.12) as it propagates through space. The sense of polarization is either rotating to the right (RCP) or to the left (LCP), looking in the direction of propagation. That is determined by

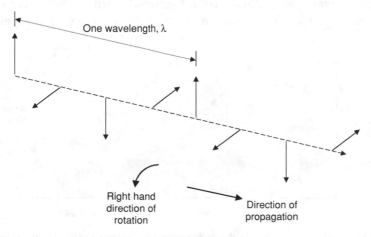

Figure 4.12 Rotation of the electric component during the propagation of a circularly polarized RF signal.

which linear component is delayed with respect to the other. Frequency reuse is possible because LCP and RCP waves are oppositely polarized with respect to each other, just like vertical and horizontal LP waves.

The advantage of CP is that the receiving antenna does not need to be aligned in polarization since the electric component rotates 360 degrees at a rate equal to the frequency of transmission. A receiving antenna would consist of combined vertical and horizontal LP elements, allowing the components to be coupled and recombined with the proper phase relationship. The sense of combining determines whether RCP or LCP will be recovered. Because most waveguide is linearly polarized, a device called a polarizer is used to make the conversion to CP. Also, the waveguide must be square or round to prevent cut-off at the operating microwave frequency. The most common approach is to generate two orthogonal linear polarizations, delay one of them by 90 degrees (one-quarter wavelength), and then add them. In general, the amount of polarization isolation thus obtained is not as high as that achieved by LP because of the imbalance between the two components that often results in practice.

4.1.5 Propagation Losses

The free-space path between transmitter and receiver may pass through a combination of ground obstructions and atmospheric phenomena. They can be divided into those that are constant (and usually predictable) and those that are both time variant and random (which can be estimated only statistically, at best). Taken together, these losses reduce the strength of the received signal and can cause its level to vary over time even to the point where the signal fades away. This is illustrated in Figure 4.13, which shows received signal power (signal level) as a function of time during a period of rapid varia-

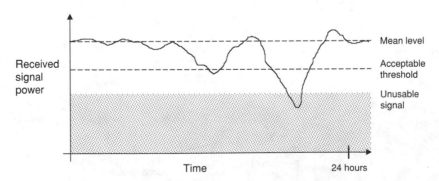

Figure 4.13 Variation of the received carrier power on a microwave link during a period of severe fluctuation (fading).

tion (fading). The mean signal level results from the constant and predictable losses on the link; the random losses cause the level to decrease and occasionally drop below the acceptable level. When the signal dips below that threshold, it could become unusable for communication purposes because of excessive noise and, in the case of digital links, an undesirable high error rate and ultimately a loss of synchronization as well. The difference in decibels between the mean level and the threshold is called the link margin, which is discussed at the end of this chapter.

Table 4.3 summarizes the primary losses that must be determined and accounted for in any microwave link design. The table compares typical downlink values for L, C, Ku, and Ka bands. Certain losses are critical in all satellite systems, while others are of concern only in specific cases.

4.1.5.1 Free-Space Loss

We already have covered in some detail the mechanism for free-space loss, which diminishes signal strength due to distance alone. For 1 m^2 of antenna area, the captured energy is simply

$$P_r = P_t / 4\pi R^2$$

where P_t = the radiated power in watts and R = the distance in meters.

The typical distance from a point on the Earth to a satellite in GEO is 36,000 km (22,300 miles).

This relationship, which should be recognized from the previous discussion of isotropic radiation, does not depend on frequency. It is a common practice to introduce frequency into the equation by splitting apart the propagation loss from the gain of the receiving antenna (here assumed to have an effective area of 1m^2). The free-space path loss, which would be in the denominator of the preceding equation, is

$$a_0 = (4\pi R/\lambda)^2$$

where λ = the wavelength in meters and R = the path length in meters.

Substituting frequency for wavelength (e.g., $f = c/\lambda$) and converting to decibels, we obtain the following basic relationship for free space loss:

$$A_0 = 183.5 + 20 \log(f) + 20 \log(D / 35,788)$$

where f is the frequency in gigahertz and D is the path length in kilometers. (The multiplier of 20 for the log accounts for the squaring of the terms within the brackets.)

Table 4.3
Comparison of Significant Propagation and Fading Modes and Resulting Losses That Affect
Microwave Links Over GEO Paths (Values Are Rough Order of Magnitude, in Decibels)

Propagation mode	Relative importance on satellite links	L band (1.6/1.5 GHz)	C band (6/4 GHz)	Ku band (14/12 GHz)	Ka band (30/20 GHz)
Free space downlink	Dominating factor	187	196	205	210
Atmospheric	Relatively small and nearly constant at high elevation angles	0.1	0.2	0.3	0.5
Rain attenuation	Severe at times at higher frequencies	0.1	0.5	2	6
Refraction	Significant at times at higher frequencies	6	3	2	1
Scattering	Produces local RFI				
Diffraction	Considered for mobile links without line of sight	6 to 12			
Ionospheric scintillation (multipath)	Occasional wide signal variation	3 to 6	1 to 3	<1	
Doppler	Frequency shift for moving vehicles or satellites				

This formula is plotted at three downlink frequencies as a function of altitude in Figure 4.14. The figure illustrates the basic free-space loss that must be countered as a function of the particular orbital configuration. In particular, we see a fundamental difference of 30 dB between GEO and LEO satellite systems.

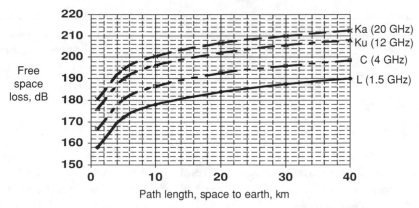

Figure 4.14 Free-space loss, A_0, as a function of orbit altitude and frequency band.

4.1.5.2 Absorption

RF energy is absorbed to some extent (and converted into heat) as it passes through clear air, water vapor, and smog. The impact of frequency of operation is shown in Figure 4.15 for an Earth station at sea level and transmitting or receiving at a high elevation angle. The absorption essentially is constant and will be under 1 dB at C and Ku bands. As frequency increases above 15 GHz,

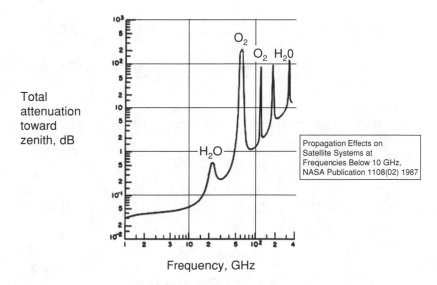

Figure 4.15 Typical relationship between atmospheric absorption and the microwave frequency of transmission, expressed in decibels.

the constituents of the atmosphere reach individual points of resonance, and absorption can become very high, even total. The bands of frequencies around 22 and 60 GHz correspond to resonances for water vapor and oxygen, respectively, and are not employed for either uplinks or downlinks. Direct links between satellites, called intersatellite links (ISLs) or, alternatively, cross-links, bypass the atmosphere and hence may utilize the absorptive bands.

4.1.5.3   Rain Attenuation

After free-space loss, the most detrimental effect on commercial satellite links above C band is rain attenuation, which results from absorption and scattering of microwave energy by rain drops. That loss, which increases with frequency, was discussed in Chapter 1 in the comparison of frequency bands. Rain attenuation is not predictable with great accuracy, but estimates can be made that allow links to be designed. Obviously, dry seasons and regions of the world with low rainfall would not suffer greatly from this phenomenon. However, links in regions with heavy thunderstorm activity—and hence rainfall—should be provided with greater link margin, or service might not be maintained with sufficient availability to satisfy commercial requirements

Intense rain is contained in rain cells, which have somewhat limited geographic size. As indicated in Figure 4.16, the lower the elevation angle from the ground, the greater is the amount of attenuation from a given amount of rainfall. Heavy rainfall also alters the polarization of the signal because atmospheric

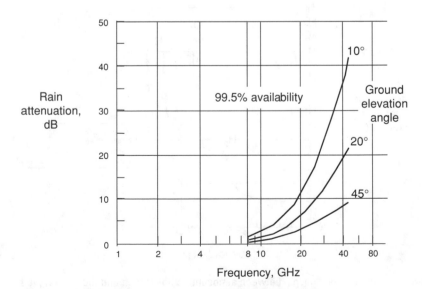

Figure 4.16 Rain attenuation versus frequency and elevation angle.

drag causes raindrops to flatten and not be perfectly spherical. That reduces cross-polarization isolation between linear polarized transmissions. For example, the maximum isolation in Figure 4.11 decreases at Ku band in heavy rain from 40 dB to approximately 20 dB. That would not be detrimental to most transmissions if appropriate adjustments are made in the link design. Depolarization in rain is not particularly harmful at C band and the lower parts of Ku band, but it can greatly reduce isolation at frequencies above 15 GHz. It also turns out that depolarization of CP transmissions is more severe than for LP.

The technique used to estimate the worst-case rain attenuation is based on the work of NASA and the ITU. The Crane and Commitée Consultatif Internationale de Radio (CCIR) rain models, developed under NASA and ITU sponsorship, respectively, have been employed for several years with satisfactory accuracy. The first step in using either model is to identify the particular rain region where the service must be provided. The map in Figure 4.17 shows different regions, each characterized by a type of climate and associated

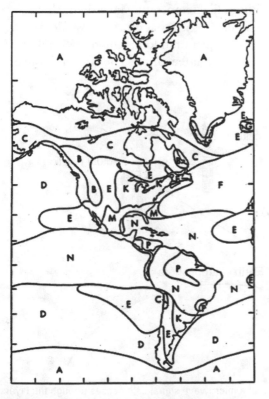

Rain region	Intensity mm/hr	Margin at 99.8%
A	6	0.3
B	12	0.5
C	15	0.7
D	19	0.9
E	22	1.1
F	28	1.4
G	30	1.5
H	32	1.7
J	35	1.8
K	42	2.2
L	60	3.2
M	63	3.4
N	98	4.8
P	145	5.8

Figure 4.17 Rain-rate regions of the Americas [1].

rain properties. Tropical regions are the worst, due to a higher incidence of thunderstorms, which contain the most intense rain cells. The letters designate the relative importance of rain in the service design. From that, the required amount of link margin is projected using the relationship between rain rate (in millimeters per hour) and decibels of attenuation. The results are displayed for different levels of link availability; for example, an availability of 99.8% requires at least 2.2 dB of rain fade in region K when using Ku band (12 GHz).

4.1.5.4 Refraction

The lower portion of the atmosphere, called the troposphere, decreases in density upward from the Earth's surface. As illustrated in Figure 4.18, electromagnetic waves are bent as they pass through the medium; as a result, the satellite has a virtual position slightly above that of its true position. That characteristic is taken into account in terrestrial microwave system design by increasing the Earth's radius by a factor of 4/3 in the profile map used to plot the line-of-sight path. In satellite links, the bending is less significant because of the higher angle toward the satellite.

Unstable atmospheric conditions like temperature inversions, clouds, and fog produce discontinuities and fluctuations in what otherwise would be a uniform air density distribution. The consequent random, temporary bending can cause significant signal fading akin to rain attenuation. The effect is more pro-

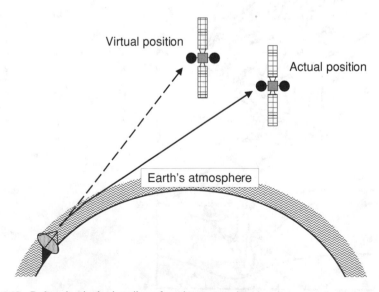

Figure 4.18 Refraction is the bending of a microwave signal as it passes through the layers of the atmosphere.

nounced for paths that are nearly parallel to the Earth (line-of-sight microwave and satellite links at low elevation angles). That is also a worry for making precise gain measurements of an Earth station antenna when the reference transmitting source is on a ground-based tower. Generally, refraction by itself does not impair typical satellite links because the amount of bending is small relative to the beamwidth of the satellite and the Earth station antennas.

The ionosphere, the layer of ionized particles at around 150 km, completely reflects frequencies between 0.1 and 30 MHz under certain conditions, producing the "sky wave" effect, which allows short-wave transmissions to cover long distances. In addition to refraction, the ionosphere rotates the polarization of microwave transmissions (Faraday effect), although that can be compensated for by appropriate adjustment of feed-horn angle at the receiving Earth station. Faraday rotation increases during periods of high sunspot activity, according to the 11- and 22-year solar cycles. One of the reasons why CP was first used in satellite communication was to avoid the need for adjusting feed polarization angles to compensate for Faraday effect. Faraday effect decreases with frequency and can be ignored at Ku band and above.

4.1.5.5 Scattering and Diffraction

It was mentioned previously that water droplets scatter microwave signals. That would reduce the direct-path power level, spraying some of it back toward the source. At an Earth station site, the occasional scatter can place RFI in the direction of terrestrial microwave receivers that otherwise would have been adequately protected.

Diffraction, on the other hand, occurs when microwaves encounter and bend over a physical obstacle such as a building or mountain. The principle behind diffraction is illustrated in Figure 4.19, which shows how an Earth station might be shielded from local microwaves by an intervening hill or mountain. The heights, H_1 and H_2, introduce shielding but also provide the possibly diffraction path for microwave signals in both directions. If the hilltop just happens to touch the line-of-sight path between the Earth station and local microwave antennas (e.g., H_1 and H_2 are zero), a diffraction loss of precisely 6 dB is introduced. As the obstacle begins to block the path, the amount of diffraction loss increases until there essentially is total blockage.

Frequency coordination of Earth stations takes rain scatter and diffraction into account by requiring additional protection margin against such random RFI. Diffraction also can help mobile satellite links by allowing signals to bend around obstacles as a vehicle travels down a highway. Lower microwave frequencies, between 1 and 3 GHz, can exploit diffraction, so they are preferred for those applications. Even so, this particular mode of propagation requires considerable link margin.

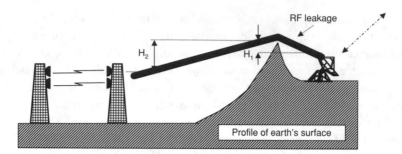

Figure 4.19 The use of terrain shielding to block RFI between a terrestrial microwave link and an Earth station.

4.1.5.6 Ionospheric Scintillation and Multipath

Ionospheric scintillation and multipath are the result of the same RF signal taking both a direct path and a slightly longer refracted path, the latter arriving at the receiving antenna delayed in time from the former. It is illustrated in Figure 4.20 for refraction caused by discontinuities in the ionosphere. It will be shown that both signal cancellation and enhancement can occur. The twinkling of a star is a multipath phenomenon, where two light rays, one direct and the other bent back in the troposphere, reach the eye and combine to produce a variation in light intensity. The twinkling occurs when the variability in the air causes the refracted path to change over time.

Due to the fact that rays are wave phenomena, the direct and refracted paths combine using vector addition, as illustrated in Figure 4.21. Shown at the

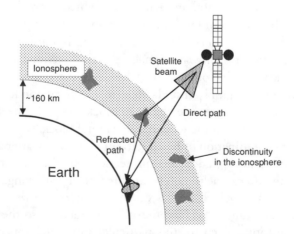

Figure 4.20 Multipath propagation caused by ionospheric scintillation.

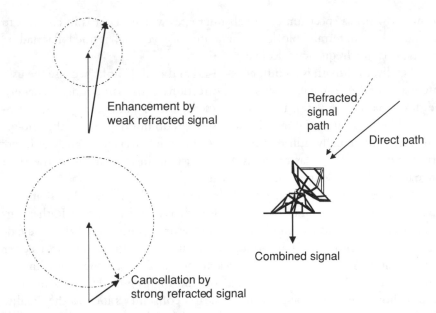

Enhancement by
weak refracted signal

Refracted
signal
path

Direct path

Combined signal

Cancellation by
strong refracted signal

Figure 4.21 Multipath signals produce enhancement or cancellation of the direct path signal when received through a common antenna, as illustrated by vector diagrams.

right are the two signal paths reaching the same antenna, wherein they combine to form a resultant signal (that is because both paths contain the same signal, although the refracted path, being longer, introduces a delay and consequent phase shift). The two drawings at the left of Figure 4.21 represent the extremes: enhancement at the top and cancellation at the bottom. The direct path is shown as the vertical signal vector of constant length. Added to that on a vector basis is the refracted signal (the dotted vector), which can arrive at any random relative phase angle, as indicated by the circle about the tip of the direct path vector. The resultant received signal vector, shown as a heavy arrow, is only slightly affected by the weak refracted signal. An increase or a decrease in strength occurs, depending on the relative phase angles.

If the refracted signal is comparable in strength to the direct path, cancellation occurs when the relative phase is 180 degrees. There also can be significant amplification (up to 6 dB) when the two vectors instantaneously combine in phase. Fortunately, the unstable conditions that produce this in the ionosphere only occur during two periods annually, around the equinoxes. Earth stations near the geomagnetic equator experience the most severe scintillation fades during those periods of activity. Ionospheric scintillation is not constant, with the result that the received signal both decreases and increases rapidly. The fact that the combining is frequency dependent produces selective fading, in which a spectrum of signals fluctuates widely. That is in contrast to variation due to terrain blockage for a moving vehicle, termed flat fading because it is not frequency selective.

Finally, multipath is an influence on L-band mobile links because the user antenna can receive signals from several directions at the same time. Generally, the direct path dominates, but the reflected path or paths can produce pronounced fading. Terrestrial mobile radio services do not depend on the line-of-sight path and actually utilize multipath to provide coverage in areas of heavy terrain blockage. There, as much as 30 dB of margin is afforded by the close range between mobile user and radio tower.

Satellite links, on the other hand, do not normally generate 30 dB of margin and so cannot depend heavily on multipath as a positive factor. Rather, signals reflected off buildings and local terrain features more likely will cause fading. The basic problem is illustrated in Figure 4.22 for a mobile receiver traveling along a highway. Two paths reach the mobile user: the direct line-of-sight path and a path that is reflected off the highway itself. They produce either fading or enhancement, which is nearly constant as long as the satellite remains in the same position relative to the vehicle. The amount of loss depends on the reflection coefficient experienced by the reflected signal. Alternatively, reflection off a fixed feature, such as a hill or a building, produces time-varying fading as the vehicle moves relative to the reflection point. The rapidity of the fading depends on the speed of the vehicle. The design of MSS links is, therefore, the most challenging, because they are subject to greater variation and uncertainty.

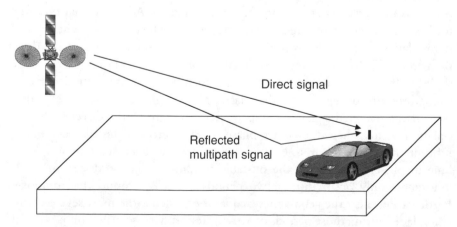

Direct signal

Reflected
multipath signal

Figure 4.22 Generation of mobile multipath on flat, smooth terrain.

4.2 Microwave Transmitters and Receivers

The basic elements in an end-to-end satellite communication link are illustrated in Figure 4.23. In the figure are the transmitting Earth station, which establishes the uplink path, a simplified satellite and its communication payload, the downlink path, and a receiving Earth station.

The entry and exit points to the propagation medium are provided by the transmitting and receiving antennas. As previously discussed, antennas convert electrical energy at microwave frequencies into electromagnetic waves, and vice versa. A transmitting Earth station consists of equipment that impresses (modulates) the information to be sent on an RF signal called the carrier, translates it to the appropriate frequency and amplifies it to a high-enough power level to provide an adequate uplink. A receiving Earth station works in the exact opposite direction. Deferred to Chapters 6 and 9 are detailed descriptions of the electronics of the satellite and the Earth station, respectively.

4.2.1 Transmitting Station

Figure 4.24 portrays a single transmitting chain of a microwave station, with the signal input at the left and the RF output from the transmitting antenna at the right. The signal to be transmitted consists of information in electrical form, such as one or more voice channels for telephone service, digital data in the form of a high-speed bit stream, or a composite video signal such as that delivered from a video tape recorder. In modern satellite systems, analog information forms like voice and TV are first digitized and then compressed to reduce the required bandwidth. From that point, the link is digital in nature.

4.2.1.1 Signal Processing and Modulation

In the signal processing element of the station (the first element to be encountered), the signal format is changed in some manner to better prepare it for transmission through the link. Analog processing in the form of preemphasis is used in FM systems to improve the signal-to-noise ratio (S/N). Dolby™ noise reduction (DNR) is an audio technique, used in cassette tapes and on FM radio broadcasts. Both preemphasis and DNR processing must be reversed at the receiving station, or the distant user will be delivered a distorted signal. In the last several years, digital signal processing has become widespread because of its ability both to improve quality and to compress the bandwidth of transmission. If the signal input is analog in form, such digital signal processing must include the conversion from analog to digital (A/D conversion), wherein the signal becomes a bit stream.

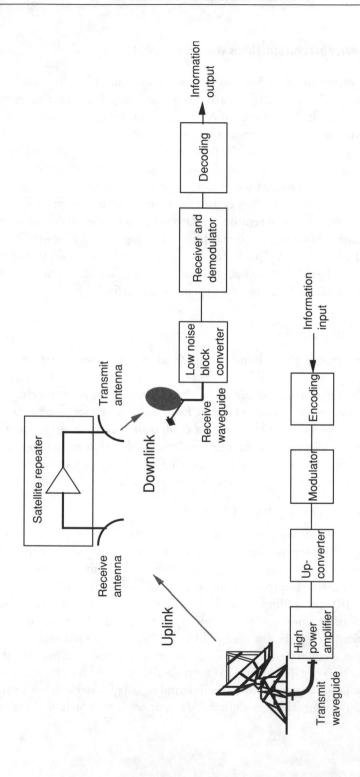

Figure 4.23 The main elements of a satellite link on an end-to-end basis.

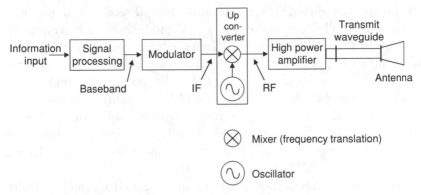

Figure 4.24 Simplified block diagram of a microwave transmitting station capable of baseband signal processing, modulation, frequency translation, and high power
amplification.

The analog or digital output of the signal processing stage is referred to as
the baseband. It is the purpose of the modulator to take the baseband and apply
it to an RF carrier. The process of modulation is also what goes on in a dialup
telephone modem used in a personal computer. However, the type of modulator used in a microwave station often handles a wideband baseband input such
as that obtained from a high-capacity data stream measured in megabits per second. In SCPC service, the output of the modem is kept on to allow a continuous stream of data to be uplinked to the satellite. TDMA operation, on the
other hand, requires that the modem transmit in noncontinuous bursts. That
is because the RF channel is being shared by multiple Earth stations that transmit in different time slots.

The opposite of modulation is simply demodulation, which is the process
whereby the baseband is removed from the carrier. The demodulator initially
must acquire the incoming carrier, demodulate the bit stream, and then produce the baseband in a form that can be used by the receiving application. The
process is repeated for each received transmission, particularly for TDMA operation. Within the demodulator is typically the decoding circuitry used to correct a majority of the errors produced by noise and interference on the uplink
and the downlink. The modulator and the demodulator for two-way communication often are packaged into a bidirectional modem, again comparable in
function to a standard PC modem.

4.2.1.2 Frequency Conversion and RF Amplification

The RF carrier coming from the modulator typically is not at microwave frequencies but rather is centered within a standard frequency channel, the

intermediate frequency (IF). Most transmitting and receiving stations use 70 MHz as the IF, allowing modulators and demodulators to be conveniently interchanged and interconnected by patch cords and coaxial switches. In low-cost consumer equipment, the IF is internal to the terminal unit and hence need not follow any particular standard. Another point is that there are cases where the RF bandwidth is larger than 140 MHz, making it unfeasible to use 70 MHz as the IF frequency (since the bandwidth would extend below zero frequency). This can be overcome by directly modulating a carrier at the microwave frequency of transmission or, more likely, by using an adequately high IF, such as 140 MHz.

The transmitted RF bandwidth is proportional to baseband bandwidth in linear analog modulation like AM and a digital modulation system like binary phase shift keying (BPSK). In a nonlinear modulation system like FM, the RF bandwidth is not simply proportional to baseband bandwidth, but rather is determined by a complex relationship between the baseband bandwidth and the amplitude of the input signal. A somewhat gross simplification called Carson's Rule allows us to estimate the RF bandwidth from the FM frequency deviation and baseband bandwidth, for example:

$$B = 2(\Delta f + f_m)$$

where Δf is the frequency deviation and f_m is the highest frequency in the input baseband.

The frequency deviation generally is set to produce the desired baseband S/N, considering the particular television signal standard and type of preemphasis.

For digital modulation, the bandwidth is directly proportional to the input data rate:

$$B = k \cdot R$$

where R is the data rate in bits per second and k is a constant determined by the particular type of digital modulation. For example, BPSK is a popular form of digital modulation used on satellites, in which case k is nominally equal to 1.2. The proportionality constant, k, also depends on the order of the modulation. For example, going from BPSK to quadraphase shift keying (QPSK) cuts k in half, to 0.6.

The function of the upconverter is to translate the carrier without modification from IF to the desired microwave frequency of transmission. Within the upconverter are a microwave mixer and a local oscillator (LO). Translation is governed by a simple mathematical relationship that states that the output frequency equals the sum of the input IF and the frequency of the LO. For example, if the IF is at 70 MHz and the LO is at 6030 MHz, the output RF is 6100 MHz. From a practical standpoint, the RF usually is assigned by someone else and the IF is fixed. Therefore, the LO frequency must be selected properly to put the RF carrier in the right place (e.g., on the assigned frequency, which is 6110 MHz in the example). Modern upconverters are frequency agile, which means that the LO can be tuned for a different RF channel from the front panel in much the same way as a TV set or car radio. The particular type of LO is called a frequency synthesizer because the frequencies are generated digitally using computational techniques. A nonsynthesized type of agile LO could cover the frequency range but would not be adequately stable over time and temperature. The desired frequency would be selected either from the front panel using an input keypad or remotely by computer over a data line.

4.2.1.3 High Power Amplification

The last active element of the transmitting station is the high-power amplifier (HPA). Because all processing and frequency translation have been accomplished in prior stages, the only function of the HPA is to increase the power of the microwave carrier from the low output of the upconverter to the power level needed to achieve satisfactory uplink operation. The HPA must have sufficient bandwidth to operate at the assigned microwave frequency and cover the active RF bandwidth of any anticipated carrier type. Examples of typical microwave HPA devices and their respective power capabilities are shown in Figure 4.25. Practical commercial HPAs, which do not require water cooling, have been chosen for this presentation. They include the klystron power amplifier (KPA) and the traveling wave tube amplifier (TWTA). Lately, there is great interest in low-power and low-cost solid state power amplifiers (SSPAs), which provide only enough transmit power to sustain relatively narrowband transmissions from coming generations of VSATs and other inexpensive user terminals.

The KPA has a capability measured in kilowatts and is very popular for video uplinks. Within the klystron microwave tube there is a resonant waveguide cavity that is tuned to the specific frequency of operation. The operating bandwidth of a KPA is in the range of 50 to 100 MHz, making it necessary to retune the internal structure to change transponders. That difficulty is overcome with the TWTA, another class of microwave amplifier that happens to be very popular on satellites as well (see Chapter 6). TWTAs have a higher-percentage bandwidth, which is the ratio of usable bandwidth to the center operating frequency, than KPAs and so can transmit over a total bandwidth of 500 MHz at C band and as much as 1000 MHz at Ku and Ka bands. Practical TWTAs can be found with power outputs of from 50W to as much as 800W, although 10-kW water-cooled TWTAs were used in early INTELSAT Earth

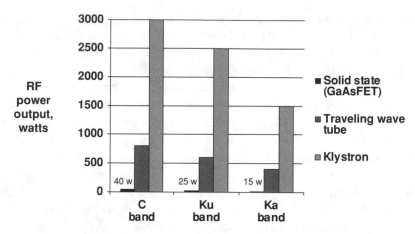

Figure 4.25 RF output power capability at C, Ku, and Ka bands of standard microwave power amplifiers.

stations. Both the TWTA and the KPA are vacuum tubes that require sophisticated high-voltage power supplies, and both employ heated cathodes to emit electrons for use in the process of amplification. In the case of HPAs used in Earth stations, it is unavoidable that both types of high-power tubes will wear out and need to be replaced after a few years of operation. TWTAs used in satellites must have longer lifetimes, lower mass, and greater dc-to-RF efficiencies and so have to be designed differently. That also results in lower power outputs, generally under 250W at the time of this writing.

In low-power applications, such as the VSATs mentioned above, SSPAs can be employed for stable, long-term operation without maintenance. A detailed discussion of SSPAs can be found in Chapter 6. Figure 4.25 presents SSPA power levels available in 1998: up to 40W at C band, 15W at Ku band, and 5W at Ka band. The basic building block of the SSPA is the gallium arsenide field effect transistor (GaAsFET), and the overall power capability is determined by the number of individual transistor stages that can be effectively paralleled. Bandwidth of SSPAs is somewhere between those of the KPA and the TWTA, based on the biasing and tuning of the transistor stages. That generally cannot be altered after the SSPA is manufactured.

4.2.2 Receiving Station

The reverse process found in the receiving station is illustrated in Figure 4.26. Because the microwave signal collected by the receiving antenna is weak, it first is necessary to raise the power to a level that can be accommodated by the pro-

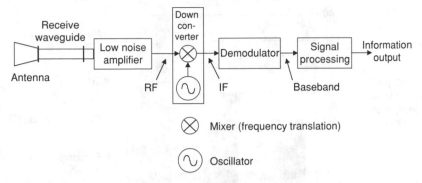

Figure 4.26 Simplified block diagram of a microwave receiving station capable of low-noise amplification, frequency conversion, demodulation, and baseband signal processing.

cessing elements. That is performed by the low-noise amplifier (LNA), whose gain must meet the requirements described above. However, the internal noise contribution of the LNA must be held small enough, or the weak signal input literally can be buried in noise. The rest of the elements perform functions that are inverse to those of the transmitting station.

4.2.2.1 Low-Noise Amplifiers

It is possible to express noise in terms of an equivalent noise temperature in Kelvin (K). The scale begins at the noiseless state of absolute zero and measures the average random energy of motion of electrons within the receiver electronics. According to the theory, the random energy of the electrons is proportional to the noise power that overlays the desired signals within the passband of the amplifier. A super-cooled device, therefore, would not contribute noise to the system.

Figure 4.27 compares the internal noise level of three classes of common LNAs, indicating how noise temperature increases with frequency. The amplifier with the lowest internal noise and, unfortunately, the highest price, is the parametric amplifier (paramp). In addition to being costly, the paramp is also considerably more complicated electronically and tends to have the greatest chance for failure. Nearly all ground stations and satellites now utilize solid state LNAs containing GaAsFET amplifiers. Some noise reduction can be accomplished, as shown in the figure, by lowering the physical temperature of the FETs using a device called a Peltier cooler. This confirms that noise temperature is not just a measure of noise power but also is directly related to how physically hot the amplifier is.

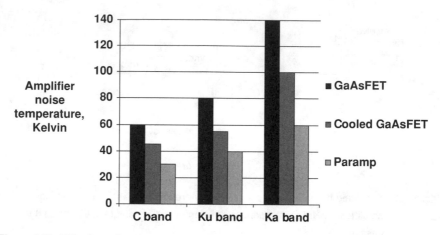

Figure 4.27   Effective noise temperature in Kelvins at C, Ku, and Ka bands of standard microwave LNAs.

It is common practice in Earth station design to place the LNA as close as possible to the feed and thereby minimize the input loss. That has two benefits: (1) It improves the effective gain of the receiving system because any loss on the input must be subtracted from that of the antenna, and (2) reducing this input loss also decreases the effective noise of the receiving system. The latter effect is discussed in later chapters. There are three common configurations of low-noise front ends:

1. LNA is the defining type of amplifier and contains only amplifier stages. The output frequency range is the same as the input range. The same RF channels must then be carried by an interfacility link (IFL) that must deliver the downlink range to the indoor systems. At C and X bands, for example, the IFL could be composed of low-loss coaxial cable, but at Ku and Ka bands, waveguide may be required.

2. Low-noise block converter (LNB) includes a wideband downconverter after the LNA stages. Its function is to translate the entire receive band, typically 500 MHz, to a relatively high IF centered around 1 GHz. At that output frequency, low-cost coaxial cable can be used to carry the RF bandwidth a reasonable distance with relatively low loss. The operating RF (e.g., Ku or Ka band) is immaterial because the cable carries the lower IF. The downconverter in the rest of the receiving station still must select one particular channel and translate it to a low IF, such as 70 MHz, for demodulation.

3. Low noise converter (LNC) is a combination of an LNB and a true downconverter. The output contains a single transponder bandwidth centered at the nominal IF of, typically, 70 MHz. The type of IFL used in this case is the least costly because the frequency range is so low. LNCs are not as common as they once were because of the limited output bandwidth. However, they could be used at L band and other low-frequency applications where the total bandwidth is supportable by the 70 MHz IF.

4.2.2.2 RF-to-Baseband Chain

Assuming that all signals first pass though an LNA, frequency translation from microwave to IF is accomplished by the downconverter, with the LO appropriately tuned again to the difference frequency. As discussed in Chapter 1, the satellite's downlink frequency usually is offset from its uplink frequency (the exception is for single-band systems like Iridium that transmit and receive packets on the same frequency). To receive the carrier, the frequency of the LO of the downconverter would differ from that of the upconverter by the fixed frequency offset (2225 MHz for standard C band) introduced by the satellite repeater. To complete a previous example, the downconverter LO should be set to the (6100 − 2225) − 70 MHz, which is equal to 3085 MHz.

The demodulator produces the baseband from the received carrier and typically provides baseband signal processing or decoding. Along with the information are versions of noise and interference picked up over the link. In analog modulation, the demodulate output quality is measured by the ratio of baseband signal power to the noise power that accompanies it (e.g., the S/N). In FM systems, the process of modulation and demodulation yields S/N that is significantly greater than the carrier-to-noise ratio (C/N). The increase in quality is called the modulation improvement factor (MIF), measured in decibels. The MIF is a direct consequence of the particular value of frequency deviation, with the amount of benefit proportional to Δf^2, that is,

$$MIF = 10 \log [3\beta^2(1 + \beta)]$$

where β, the modulation index, equals $\Delta f / f_m$.

A digital carrier on the downlink is demodulated to deliver a digital bit stream with the digitized version of the desired information. The demodulated performance is measured in terms of bit error rate. The bits are processed further to reduce the error rate, decompress the underlying information, and, if appropriate, convert the data back into the original analog form. Greater detail on digital modulation characteristics is contained in Chapter 5.

4.2.3 Definition of a Transponder

The satellite communication industry has long used the term *transponder* in reference to a defined RF channel of communication within the communication payload. The term itself is a contraction of *transmitter-responder*, originally referring to a single-frequency repeating device found on aircraft. The purpose of the aircraft transponder is to actively enhance the power to be reflected back to the radar transmitter. A satellite transponder is entirely different because it is more of a microwave relay channel, also taking into account the need to translate the frequency from the uplink range to the downlink range.

We can better define a transponder by examining the two different payload configurations shown in Figure 4.28. The single-channel repeater (shown in A) does just what its name implies: provide a single channel of transmission within the satellite. As shown at the top of the figure, the entire uplink band is translated in the downconverter and applied to a single power amplifier. As a consequence, this amplifier must accommodate the entire uplink bandwidth, which would amount to 500 MHz at C band. That tends to limit the power-handling capability, particularly if multiple carriers are amplified by the common output stage. The first generation of LEO and GEO satellites like Telstar and Syncom, respectively, had repeaters of this type, capable of carrying one TV channel each.

The transponderized design (shown in B in Figure 4.28) breaks up the downlink range into individual frequency channels. Figure 4.29 suggests how eight transponders would be divided from the uplink bandwidth. Note that the

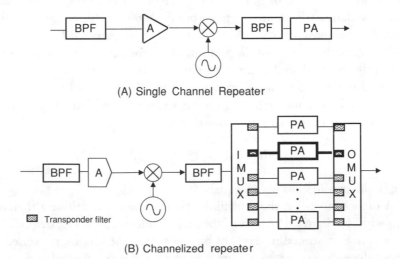

(A) Single Channel Repeater

(B) Channelized repeater

Figure 4.28 General arrangement of a single-channel repeater and a channelized repeater with multiple transponders.

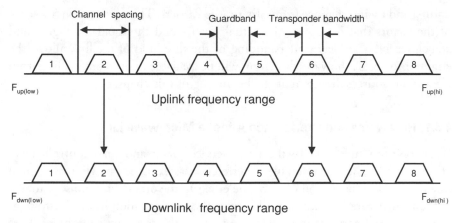

Figure 4.29 Hypothetical frequency plan for eight transponders.

uplink section still carries the entire bandwidth of 500 MHz and is shared by all transponders. After the downconverter, there is a bank of RF filters, called the input multiplexer (IMUX), that separates the transponder channels on a frequency basis. The output of each IMUX filter contains one transponder bandwidth, to be amplified in an individual power amplifier (TWTA or SSPA, depending on the power required). On the output side, the RF power of the amplifiers is summed in a reactive power combiner called the output multiplexer (OMUX); it is composed of specially designed low-loss waveguide filters. A single waveguide transfers all the power from the OMUX to the antenna system.

The transponder, then, is a combination of elements within the payload. On the input side, it represents a share of the common uplink and receive equipment within the repeater. We are able to identify specific equipment for each transponder on the downlink side, consisting of the input filter, power amplifier, and output filter. Not shown in the figure are the necessary spare active elements (redundancy) to ensure continuity of service in the event of amplifier or receiver failure. The fact that a transponder can be assigned to a particular user application network has caused them to be rented or sold like condominium flats. In actuality, it is the microwave channel of communication bandwidth that the transponder lessor or purchaser acquires.

4.3 Overall Link Quality

One of the more complex problems in microwave link engineering is predicting how a particular signal will be affected by the noise, which is random in nature, and by interference from other radio carriers. This section reviews some of the factors that determine the actual quality and threshold for digital and analog receivers. A clear understanding of the threshold of the link allows the engineer to predict how much margin will be available to overcome fading from the various sources discussed at the beginning of this chapter.

4.3.1 How Noise and Interference Affect a Microwave Link

Both noise and interference will degrade service quality and if not controlled will at times render the link unusable. Interference, as explained previously, is often due to the RF transmission of someone else. On the other hand, noise is totally random in nature because it results from the random motion of electrons or other elemental particles in the environment or the receiving equipment. Pure white noise is most common on satellite links and produces random voltage fluctuations whose probability follows a normal (Gaussian) distribution. The term *white* refers to the fact that the frequency spectrum does not have discrete components at specific frequencies but rather is a continuum of frequencies, like white light. Ideal white noise, running from zero to infinite frequency, is physically impossible since it would represent infinite power. Noise is sufficiently white if it is constant over the bandwidth of the signal in question.

The resulting density of the noise power (noise power density) in watts per hertz is proportional to the equivalent noise temperature in Kelvins, that is,

$$N_0 = k \cdot T$$

where k is Boltzmann's constant (i.e., $1.3803 \cdot 10^{-23}$ W-sec/K or -228.6 dBW-sec/K when expressed in decibels).

This simple relationship demonstrates why the noise performance of an LNA of a receiving Earth station is rated in terms of an equivalent noise temperature. The actual noise power that affects a given signal is that which lies with the signal's bandwidth, B, that is,

$$N = k \cdot T \cdot B$$

The effect of white noise on a modulated carrier containing binary data can be shown using time waveforms, presented in Figure 4.30. The noise-free case is shown in Figure 4.30(a) in the form of digital information impressed on a carrier. The transition from binary 1 to 0 occurs where the phase of the sinewave reverses. That type of modulation is called phase shift keying (PSK) because the shifting (flipping) of phase by 180 degrees is the means by which information is transmitted. It is the job of the demodulator to detect the phase

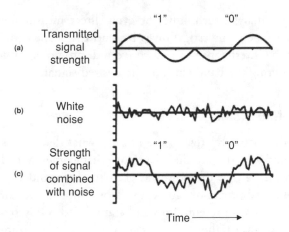

Figure 4.30 A digital phase-shift keyed signal combines with white noise to produce a distorted signal waveform that must be detected by the receiver.

reversal to convey to the user the proper bit sense. The PSK signal is sent through the link and enters the receiver along with white noise inherent in the electronics. That noise is depicted in Figure 4.30(b), where the mean voltage is zero and the standard deviation of the noise is approximately one-half the carrier amplitude. Because power is proportional to the square of the voltage, the true ratio of carrier power to noise power is 4, or equivalently 6 dB. The rather high relative noise (low C/N) level yields the sum of carrier plus noise shown in Figure 4.30(c), where the carrier appears to be somewhat obliterated by noise spikes.

A well-designed demodulator and digital signal processor can detect the transition from the 1 to 0 state in most cases. The noise can occasionally reach a voltage greater than the signal amplitude; if it has the opposite sense to the transmitted signal, it can cause the demodulator to make the wrong choice. The assumed binary digit will then be incorrect (i.e., a 1 instead of a 0 or vice versa, depending on what was sent). It should now be possible to visualize that the strength of the signal relative to the noise will determine the rate at which errors in detection are made. This is a significant area of engineering study in the communications field, and some of the most important applications of very large scale integration (VLSI) and digital signal processing (DSP) allow low-cost sending and receiving equipment to identify errors and reverse them prior to delivery to the user. Special codes that provide such forward error correction (FEC) accomplish that in practice.

Wideband interfering signals can be treated as white noise and often are. However, if the interference is narrower in bandwidth than the desired signal, more sophisticated analysis techniques or even direct measurement must be employed. As that type of investigation is beyond the scope of this book, we will assume that the interference can be converted into an equivalent amount of white noise covering the bandwidth of the desired signal.

4.3.2 Carrier-to-Noise Ratio

The strength of the carrier relative to the noise is what determines the quality of transmission and not the absolute level of what comes out. That can be understood by realizing that the power level of the baseband signal from the receiving station's signal processor can be raised by simple power amplification. However, the noise that is present also will be amplified. Only by suppressing the noise at the input to the receiver can the quality of transmission be improved. That is why the true performance of the link is measured by the ratio of RF carrier power to noise power. A variant used in digital transmission is the ratio of the energy per bit to the noise power density, E_b/N_0. It will be shown that there is a simple constant adjustment factor between C/N and E_b/N_0.

A receiving system showing a carrier, white noise, and a single source of interference is presented in Figure 4.31. The frequency spectrum is similar to the display on a microwave spectrum analyzer, a useful piece of test equipment. The figure shows a constant spectrum of white noise, providing what is referred to as a noise floor. The vertical scale is linear, measuring power density in terms

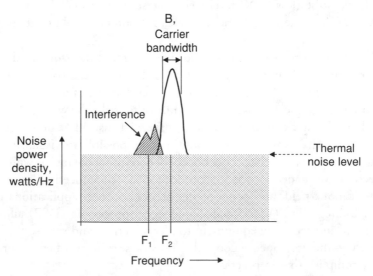

Figure 4.31 Frequency spectrum showing the desired carrier and the interfering carrier and white noise, as seen on a spectrum analyzer.

of watts per hertz. Piercing through the floor are the desired carrier occupying bandwidth B and centered at frequency F_2, along with an interfering carrier at frequency F_1. The total noise power, N, over the bandwidth of the carrier is the product N_0B. An actual spectrum analyzer would have a vertical scale measured in decibels to simplify the measurement of C/N. In that case, the C/N in decibels is equal to the difference between the measured decibel levels of total carrier power and noise power within the bandwidth B. While bandwidth can be taken into account by subtracting the quantity 10 log B from the spectrum analyzer reading, it is also possible to preset the spectrum analyzer's measurement bandwidth to eliminate the need for manual correction.

The type of interference shown in Figure 4.31 is not noiselike and therefore cannot be analyzed in a straightforward manner. One consideration is that only the upper corner of the interference actually affects the carrier. The ratio of carrier power to interference power can be taken from the spectrum analyzer display, but its effect on the reception process also depends on the frequency offset ($F_2 - F_1$) between the two carriers and their relative bandwidths. In real-world situations, the interference may rove through the bandwidth of the carrier, affecting it as well as other carriers on adjacent frequencies. Spectrum analysis is vital to microwave communications because it provides one of the few windows for viewing the reception process.

4.3.3 Link Budget Analysis

The key station elements and propagation phenomena having been described, it is now possible to review the analysis and prediction of link performance. Figure 4.32 shows a simplified microwave link with a key parameter indicated

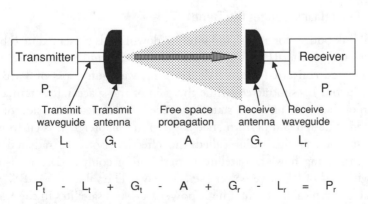

Figure 4.32 The relationship between the critical elements of a microwave link and the power balance equation.

for each element. The transmitter can be characterized by the HPA power output, with the transmitting waveguide introducing some loss as it carries the power to the transmitting antenna. The electromagnetic wave propagates outward from the antenna into the medium, where it is subjected to various losses, the free-space loss being dominant. The small amount of signal power gathered by the receive antenna is carried through waveguide to the LNA of the receiver. It is possible to characterize the receiver by the minimum acceptable threshold, which takes into account the RF noise as well as the quality desired by the user.

4.3.3.1 Power Balance Equation

The simple mathematical relationship shown at the bottom of Figure 4.32 is the power balance equation. The parameters of the link are actually factors; however, it is a common practice to express all of them in decibels because that reduces the analysis to addition and subtraction. The power balance in decibels can be stated simply: The power received equals the power transmitted plus all gains and minus all losses. Transmitter power is expressed in decibels relative to 1W (dBW). For example, 1W is by definition 0 dBW; 2W is 3 dBW; 10W is 10 dBW; 100W is 20 dBW; and so on. The power balance equation usually is arranged in a table called a link budget, in which each parameter of the link is provided with its own explanation and quantity. We review a simple example to demonstrate the process. In practice, there are as many possible formats for link budgets as there are engineers who use them. It appears to work best if the problem is set up for each case to be analyzed, using the facilities of a spreadsheet program like Microsoft Excel or Lotus 1-2-3. There are specialized link budget programs on the market, but they tend to constrain the analysis to the particular approach of the commercial software developer.

4.3.3.2 Typical Link Budget (Ku Band)

A typical link budget for a Ku-band satellite downlink is presented in Table 4.4. The first entry is the RF power output of the transmitter, expressed first in watts and then converted into dBW. Transmit waveguide loss of 1 dB (26% power reduction) is subtracted, while the gain of the spacecraft antenna in the direction of the receiving Earth station is added. A detailed discussion of spacecraft antenna design and performance is given in Chapter 7. It is customary to show a subtotal at this point called the effective isotropic radiated power (EIRP), indicating how the satellite is performing compared to an isotropic source with 1W of RF drive power (i.e., 0 dBW). The value of 53 dBW shown in the table is typical for current high-power Ku-band satellites in the BSS. For a large country such as the United States, China, Indonesia, or Australia, this can be delivered by two 125W TWTAs operated in parallel.

With the satellite in GEO at an altitude of 36,000 km, the link budget contains a single entry of –205.6 dB of free-space loss. In contrast to spreading loss (e.g., $4\pi R^2$), path loss includes λ as a component, having the form:

$$a= (4\pi R^2)(4\pi/\lambda^2)$$
$$= (4\pi R)^2/\lambda^2$$

Other propagation losses, such as rain attenuation, that are random in nature are evaluated separately and compared against the overall link margin. The next two items relate to the receiving Earth station: the peak gain of an assumed 45-cm antenna and 0.5 dB of waveguide loss. The combined power balance yields a received power of –120.4 dBW at the input to the Earth station LNA and receive electronics. For that hypothetical link, assume we have a receive system noise temperature of 140K and a signal bandwidth of 27 MHz. For the corresponding received noise power of –132.8 dBW (e.g., kTB), the downlink C/N result is 12.4 dB.

Uplink noise within the spacecraft receiver contributes a lesser amount to the total link noise and is ignored in the present example. However, the standard approach is to prepare a separate link budget for the uplink, following approximately the same format as Table 4.4. Then the overall link C/N is determined using the following formula:

$$C / N_t = (N_d/C + N_u/C)^{-1}$$

calculated as ratios and not decibels. The final step is to convert back to decibels.

4.3.4 Link Margin

The final and perhaps key concept in microwave link engineering is that of link margin, which is nothing more than the excess power in the carrier relative to the threshold value. It is possible to measure the threshold in the laboratory under controlled conditions. First, the link performance without fading can be determined using the power balance equation and spectrum analysis. The margin, then, is the difference in decibels between the minimum value and the expected value. For example, if the minimum acceptable (threshold) C/N is 8 dB, the link margin is 12.4 – 8 = 4.4 dB. After the link is operational, the margin can be verified by manually decreasing the transmit power to the point where the received baseband signal is barely acceptable. The recorded change in transmit power in decibels is essentially the link margin.

Adequate link margin allows the link to deliver exceptionally good quality (high S/N or low bit error rate) under mean signal, clear weather conditions.

Table 4.4
Typical Ku-Band Link Budget for the Downlink to Small DTH Receivers

Parameter	Value
Transmit power	(250W)
	24.0 dBW
Transmit waveguide loss	−1.0 dB
Transmit antenna gain	30.0 dBi
EIRP	53.0 dBW
Free-space loss	−205.6 dB
Receive antenna gain (45-cm diameter)	32.7 dBi
Receive waveguide loss	−0.5 dB
Received power (clear sky)	−120.4 dBW
Received noise power*	−132.8 dBW
C/N	12.4 dB

* Assumes T_{sys} = 140 K and B = 27 MHz

However, when either the uplink or the downlink is experiencing heavy rain or multipath fading, the available link margin determines how often and for how long the link will drop below threshold. The criterion often used is the availability, that is, the percentage of time the link is above threshold. Typical satellite links operate in the range of 99% to 99.95%, demonstrating the high reliability of line-of-sight paths between the satellite and its associated Earth stations.

Reference

[1] Final Acts of WRC-77, 2. Radio Propagation Factors, International Telecommunication Union, Geneva, 1997.

5

Modulation and Multiple Access

Digital communication formats and transmission systems have long been used over commercial satellites. Analog formats have been retained in some of the older applications, many of which are still viable in economic terms. Perhaps a few years after 2000, even those systems will have been retired in favor of more efficient digital technologies. To understand how a satellite link actually transfers information between Earth stations, we must delve into the twin topics of modulation and multiple access. We will quickly see that these relate directly to the analog or digital format being applied.

The discussion that follows is reasonably technical and detailed, but the topics of modulation and multiple access are extremely important to the cost-effective use of communications satellites. In most networks, the efficiency of transmission is increased by combining several channels of information with a process called *multiplexing*, creating a *baseband* spectrum in electrical form. *Modulation* is the process by which the baseband is then impressed on a carrier. The arrangement and the sequence of these functions for analog and digital formats are shown in Figure 5.1. In Chapter 4, we touched on the role of modulation in transmitting user information with adequate circuit quality on a microwave link. The type of modulation as well as the specific parameters employed also determine how much traffic the satellite repeater can carry in aggregate. Also, the types of connectivity and flexibility achieved depend to some degree on the modulation system. Those aspects of RF transmission and link connectivity are referred to as *multiple access*.

In Section 5.1, we discuss the role of analog modulation and its impact on the satellite link. The analog input signals are usually continuous in time

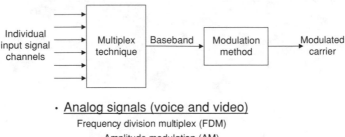

- <u>Analog signals (voice and video)</u>
  Frequency division multiplex (FDM)
    - Amplitude modulation (AM)
    - Phase modulation (PM)
    - Frequency modulation (FM)
- <u>Digital signals (binary coded information)</u>
  Time division multiplex (TDM)
    - Frequency shift keying (FSK)
    - Phase shift keying (PSK)
    - Spread spectrum modulation

Figure 5.1 Multiplex and modulation methods for use with analog and digital information.

and employ frequency modulation over the satellite. The amount of bandwidth consumed is not constant but varies with the amplitude and baseband bandwidth of the input signal. The overall quality can be predicted with reasonable accuracy and can be held at high levels of user acceptance. However, the amount of transponder bandwidth consumed generally will be larger than what is possible with modern digital transmission systems. That will be evident as we move into the discussion of digital signals in the later part of the chapter. Readers who are not interested in analog formats can skip to Section 5.2.

5.1 Analog Signals

Information in analog form is continuous in time and represents, as closely as possible, the detailed nature of the source waveform. For example, analog voice information, or simply voice, contains all the true frequency components along with the unvoiced sounds associated with speech. Analog information is generated by and recognized by us humans, but it also can be used to precisely control machines and vehicles. One difficulty for analog transmission is that the channel must be maintained in its operating state even when user information is not present. For example, when a person pauses during conversation, there is no signal being emitted. The channel may still have to be up and running, thus consuming transponder power and bandwidth. A way around that is to use talker activation, in which the carrier is disabled during an idle period. However, that may introduce another impairment in that some low-level speech may not be properly detected and the resulting speech train would be clipped. The way around the problem is to digitize and delay the speech and then, through mathematical techniques, detect the idle periods with greater precision.

5.1.1 Analog Signal Conditioning

A device called a *transducer* converts some type of sound, light change, or motion into analog electrical impulses, which are carried to the radio facility by wire. The electrical impulses can be converted back into the original physical form with reasonable fidelity by a transducer at the distant end. In telephony, the pair of transducers are contained in the standard telephone handset (e.g., the ear piece and the speaking end). Perfect reproduction, however, is neither practical nor, in most cases, necessary. Diminished quality of reproduction results from restrictions on the bandwidth of transmission as well as the distortion of the amplitude and phase characteristics of the analog information.

Analog signal conditioning minimizes the demands on a satellite link by reducing the bandwidth and/or the amplitude range of the analog input. The resulting signal will still be understandable (intelligible), although some of its subtler aspects may be lost. Bandwidth is reduced by filtering, effectively chopping off the higher frequency components as well as the lowest. In voice communications, human speech can be sent with nearly perfect fidelity over a frequency range of 100 to 10,000 Hz, a bandwidth of approximately 10 kHz. In the public switched telephone network (PSTN), that bandwidth is filtered down to a bandwidth of approximately 3 kHz, covering 300 to 3,400 Hz and offered to homes and offices on a dialup basis. Another popular term for this telecommunication facility is *POTS* (for "plain old telephone service"). A similar process applies to audio broadcasting, television, and scientific measurement using telemetry. The important reason for filtering is that it reduces bandwidth occupancy, permitting the efficient stacking of many channels together using the process of multiple access and multiplexing.

In addition to simple filtering, the signal can be prepared by altering its voltage level or volume (in the case of audio). The preemphasis technique, plotted in Figure 5.2, introduces an upward tilt to the input signal spectrum as a way of compensating for a similar tilt of the receiver noise coming out of the distant demodulator. Another type of analog signal conditioning is *companding* (a contraction of compressing and expanding), which, like preemphasis, is a way to overcome noise. The characteristics of compression and expansion are plotted in Figure 5.3. In this case, the compressor at the sending end automat-

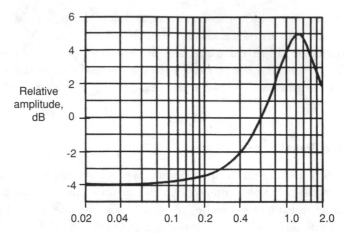

Figure 5.2 Preemphasis characteristic used at the FM modulator to improve baseband S/N at the demodulator output.

ically amplifies the input signal when it is weak and attenuates it when it is strong, responding in real time to the rapid variation of the audio level. The expander at the receive end works in an opposite fashion to restore the proper audio response. The benefit comes when the expander automatically attenuates the noise in compensating for what was originally a weak audio input at the

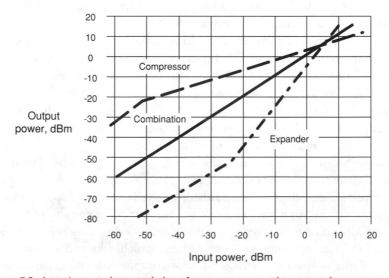

Figure 5.3 Input/output characteristics of a compressor and an expander.

transmitting end. The operation of the "compander" is undetectable to most listeners, indicated by the straight line at the center of Figure 5.3, and the effect is to enhance the S/N by 15 to 20 dB. All modern telephone transmission systems, be they analog or digital, now employ companding in one form or another. The DNR system mentioned in Chapter 4 is particularly effective because it incorporates features of both preemphasis and companding. Readers familiar with high fidelity recording are aware that DNR is applied to audio cassettes and FM radio broadcasting (but not to audio CDs, which are inherently free of background noise).

5.1.2 Frequency Division Multiplex

The most common analog multiplexing scheme is frequency division multiplex (FDM), which was developed in the 1930s to reduce the number of wires needed to connect two telephone switching offices. Later, FDM found application in microwave radio links because without it a separate transmitter and receiver pair would be needed for each end of a voice path. Each individual information channel is assigned its own unique frequency, much like TV channels and AM radio broadcast frequencies. The bundle of FDM frequency channels occupies the baseband and follows a single transmission path between terminal stations. The incoming channels themselves can either originate from the same signal source or be tied together at a communication node.

A simplified block diagram of the FDM transmit equipment chain is shown in Figure 5.4; frequency spectra corresponding to the intermediary points are shown in Figure 5.5. Each channel is taken from its normal frequency range

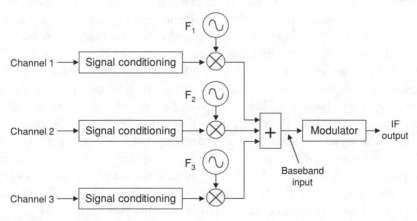

Figure 5.4 Block diagram of equipment used to generate a three-channel IF carrier using FDM/FM.

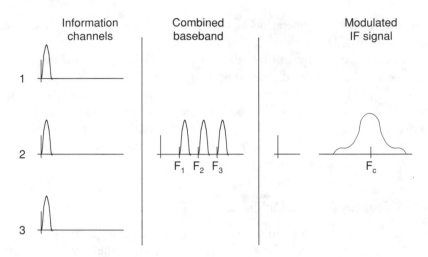

Figure 5.5   Assembly of a baseband using FDM and generation of an IF signal with frequency modulation.

(300 to 3,400 Hz in the case of voice, or 20 Hz to about 5 MHz in the case of analog video) and translated to an assigned frequency channel by a mixer-LO combination. That is the same technique as that used in an upconverter, discussed in Chapter 4. The mixing process in Figure 5.4 causes channel 1 to appear just above frequency F_1, channel 2 just above F_2, and so on, as illustrated at the center of Figure 5.5. Being on separate frequencies, the channels can be safely added in the summing amplifier (shown as a box with a plus sign) without causing interference among them. This FDM baseband spectrum would then be modulated on the IF carrier (indicated at frequency F_c in Figure 5.5).

In previous generations of FDM telephone equipment, the translation was done in multiple stages so the baseband could be made from building blocks. In North America, the basic group consisted of 12 channels spaced 4 kHz apart, lying between 12 kHz and 60 kHz. Using successive frequency translations, five groups can be multiplexed to obtain a 60-channel supergroup. A mastergroup consists of 10 supergroups and covers a baseband bandwidth of approximately 2.5 MHz. FDM equipment also includes facilities for maintaining constant signal levels and for detecting any fades or equipment failures. Different standards apply around the world to the specific frequency ranges and groupings of channels, so the numbers cited here should be considered only as examples. Equipment of this type was largely retired in the United States when the local and long-distance carriers digitized their networks. That same trend is underway throughout the world, so it is unlikely that new FDM telephone equipment will be installed in public networks again. One remaining application is the use of wave division multiplex (WDM) to add lightwave channels onto existing fiber optic links.

For analog TV, audio channels usually are multiplexed with a video channel using FDM. In that case, the video is carried across in its nominal range between 20 Hz and approximately 5 MHz. The audio is not directly translated but rather is modulated on a distinct baseband frequency called a *subcarrier.* In the case of stereo, the two audio channels normally would be sent on separate subcarriers at different baseband frequencies. Other subcarriers containing additional audio channels or data can be added to the baseband for narrowcast applications. The modulator shown at the right in Figure 5.4 operates on a baseband consisting of the video with audio and data subcarriers (if present).

5.1.3 Frequency Modulation in Analog Transmission

Phase modulation (PM) and FM are used in satellite communications systems because of their ability to deal with nonlinear distortion, noise, and interference. The typical bent-pipe transponder is operated for maximum power efficiency and is therefore highly nonlinear, which means that signals that have time-varying amplitudes will be seriously distorted. To control that effect, the amplitude of the carrier should be held constant so there is no change in the power level. The following basic equation for FM and PM indicates how both of these modulation methods meet that criterion:

$$a(t) = A \cos\{2\pi [f_c + f_m(t)] t + \varphi(\tau)\}$$

where

$a(t)$ is the modulated carrier signal;

A is the constant amplitude (e.g., for a nonamplitude modulated signal);

f_c is the carrier frequency;

$f_m(t)$ is the frequency modulation function or information;

$\varphi(t)$ is the phase modulation function or information.

The equation can be explained as follows. The carrier has a constant amplitude, A, and is centered at carrier frequency, f_c. It could have one of two forms of angle modulation: FM, $f_m(t)$, and PM, $\varphi(t)$. To introduce PM, the unmodulated carrier is delayed in time (plus and minus) in proportion to the amplitude variations of the information being sent. Delaying of a carrier causes the angle of its phase to change, which can be detected at the receiving end and converted back to the original form. FM is easier to comprehend because it is produced by varying the frequency of the carrier in proportion to the amplitude of the information. The carrier effectively swings back and forth in frequency; the amount of swing is determined by the loudness of the information (in the case of audio), while the maximum range of the swing (i.e., the peak-to-peak deviation) establishes the bandwidth of the modulated carrier. FM is directly related to PM; it turns out that the change in frequency is equal to the rate of change of phase (the first derivative of phase).

An electronic circuit called a phased lock loop (PLL) is very handy for demodulating PM and can also demodulate FM with additional components to perform the necessary mathematical transformation. These devices can track the carrier as it traverses the occupied bandwidth, recovering the modulated information with maximum efficiency. The PLL is also described as a threshold extension demodulator (TED) because it can perform adequately even as the C/N drops below the point where earlier FM discriminators could no longer function adequately. (The threshold effect is reviewed below.)

The bandwidth of an FM carrier is not a constant due to its dependence on the amplitude variations of the information. Generally speaking, RF bandwidths for these modulation types are many multiples of the baseband bandwidth. As introduced in Chapter 4, Carson's Rule provides an estimate of the bandwidth of a wideband FM carrier: It is approximately equal to the peak-to-peak frequency deviation plus twice the highest baseband frequency. That approximation generally gives a somewhat larger bandwidth than actually would be used. The benefit of bandwidth expansion in FM comes about after demodulation, when the narrower baseband bandwidth is restored with only a corresponding fraction of the RF noise accompanying it. That is because the PLL-type demodulator has an effective bandwidth closer to that of the modulating signal rather than that of the RF carrier. Stated another way, the PLL tracks the deviation of the carrier, recovering only a portion of the RF noise and interference.

A familiar concept in FM radio communications is the threshold effect, illustrated in Figure 5.6. As discussed in Chapter 4, the S/N from the demodulator follows the change in the link C/N decibel for decibel as long as the power level is above the threshold of the demodulator. If the C/N is at or below threshold, the demodulator delivers a degraded signal because, at times, random noise spikes will be stronger than the carrier. As C/N decreases below threshold, noise spikes will become more rapid until the point is reached where the signal is totally useless. For as little as 1 dB of margin above threshold, the recovered information is remarkably clean and stable in quality (referred to as "quieting").

Another benefit comes when there is a weaker interfering carrier present. The desired carrier will suppress the interference, which is called the capture

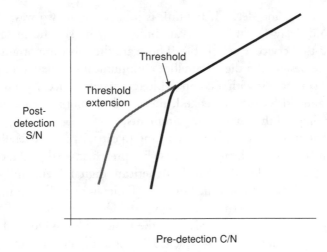

Post-detection S/N

Threshold

Threshold extension

Pre-detection C/N

Figure 5.6 The threshold effect in FM reception provides a substantial quieting of baseband noise as long as the threshold is exceeded.

effect, as long as there is a 6- to 10-dB difference in carrier power. Capture effect is what allows an automobile FM receiver to lock onto a stronger station in favor of a weaker one on the same frequency. With capture effect, a strong-enough FM carrier can overcome low levels of interference that are present in all operating satellite networks.

Typical threshold values and C/N for FM are in the range of 6 to 10 dB, where the lower end of the range is obtained with the more modern PLL type of demodulator. With sufficient margin above threshold, the baseband noise level will be very low and noise spikes extremely infrequent. That is a particularly desirable situation for voice channels carrying low-speed data since the noise spikes almost always will introduce bit errors. In the case of analog color video transmission, noise spikes near threshold appear as short horizontal lines called "sparklies." In summary, audio or visual impulse noise effects will not be observed as long as there is adequate margin above threshold, a condition that also provides good S/N performance.

5.2 Digital Baseband Signals and Hierarchies

Information in digital form can be represented as a stream of bits, that is, 1's and 0's. A digital baseband, therefore, is a digital stream that is to be transferred by a communication link or network. Generally speaking, it does not matter if the link is terrestrial or satellite, provided that the baseband is delivered with an acceptable error rate and delay. If the link is interactive or two way, the error rate quality and delay (including its variability) can be significant factors in determining if the service is acceptable. Those are the most important technical issues in the design of a digital satellite communications network, particularly when we compare it with a comparable terrestrial service. Also of importance is the bandwidth of the baseband, measured in kilobits or megabits per second, depending on the capacity requirements.

Digital bandwidths are defined according to capacity and generally follow one of the standard digital hierarchies. Table 5.1 presents two digital hierarchies in use around the world: the North American digital hierarchy and the European digital hierarchy, also called the CEPT hierarchy (CEPT is the abbreviation for the French term for European Commission of Post and Telecommunication). Both hierarchies use the basic building block of the 64-kbps channel. Each is referred to as a pleisiochronous digital hierarchy (PDH), meaning that they are nearly synchronous. That can be seen from the fact the digital rates above 64 kbps are not integer multiples but include some additional bit rate capacity inserted at each level. For example, the T1 rate of 1.544 Mbps is higher than the precise value of 1.532 Mbps, which is exactly 24 times 64 kbps. The extra bits are inserted to provide synchronization information and to allow for maladjustments in clock rates, which were endemic in earlier networks.

Because of two developments, such poor synchronization is largely a thing of the past. The first development is the production of low-cost atomic clocks that are used to synchronize the local and national networks within a region and globally as well. Complementing that is the second development: the increasing use of fiber optic communications with its inherent stability and quality of transmission. That means highly synchronized digital transmission systems are both feasible and attractive. That reduces wasted bandwidth and

Table 5.1
The Plesiochronous Digital Hierarchy (PDH) as Employed in North America and Japan (1.5 Mbps) and Europe, South America, Asia, and Africa (2 Mbps) (Entries Expressed in kbps)

Level	North America (1.5 Mbps)	Japan (1.5 Mbps)	Europe, South America, Asia, and Africa (2 Mbps)
1	1,544	1,544	2,028
2	6,312	6,312	8,448
3	44,736	32,064	34,368
4	97,728	139,264	

provides the means to increase the bit rates that users can employ. Today, it is the synchronous digital hierarchy (SDH) which is becoming the digital hierarchy of choice, initially in public networks, but more and more in private installations as well.

In North America, where the standard was first introduced, we have the synchronous optical network (SONET, pronounced "sonnet"); in Europe the standard is simply called SDH. A comparison of the standards is presented in Table 5.2. The principle differences between SONET and SDH relate to terminology, not technology. They use different naming conventions and start at different basic line rates (which are multiplied up). The basic rate for SONET is 51.84 Mbps; we see that SDH's basic rate of 155.52 is exactly three times 51.84 Mbps. As it turns out, all practical SONET equipment begins at the OC-3 level, which, at 155.52 Mbps, happens to be precisely the same as STM-1, the basic line rate of SDH.

As of the time of this writing, there are vast quantities of PDH equipment and links in operation around the world. They interface with user devices such as multiplexers, video teleconferencing equipment, various types of high-performance computers and computer networks, and digital PBXs. In all likelihood, the demand for PDH will last well after the year 2000. At the same time that the inventory is being expanded to meet current application networks, we see an increasing demand for even higher speed end-user equipment and for advanced public networks that employ high-bandwidth fiber optic links. SONET and SDH interfaces already have begun to appear on the U.S. PC market. Also, SONET is fast becoming the preferred digital transport medium of the Internet backbone in the United States. Satellite systems, therefore, will

Table 5.2
SONET (North America) and SDH (Worldwide) (Entries Are Expressed in Mbps)

SONET level	SONET designation	Bit rate (Mbps)	SDH level	SDH designation
STS-1	OC-1	51.840		
STS-3	OC-3	155.52	1	STM-1
STS-9	OC-9	466.560		
STS-12	OC-12	622.080	4	STM-4
STS-18	OC-18	933.120		
STS-24	OC-24	1,244.160		
STS-36	OC-36	1,866.240		
STS-48	OC-48	2,488.320	16	STM-16

have to accommodate both forms of digital baseband transmission to properly interface with the multitude of application systems and networks.

5.2.1 Digital Information Sources

The range of digital information sources that provide input to satellite communication networks has grown rapidly over recent years. It begins with the digitized voice from the PSTN and low-speed computer data and extends through the more exotic and promising applications of the coming millennium, notably broadband and interactive multimedia. There is even likely to be a place for HDTV, a technology that experienced a very slow start until a digital implementation could be proposed. In this section, we merely wish to identify the more familiar and promising of what can only be described as an opening cornucopia.

The basic digital information inputs are classified according to the following criteria:

- Bandwidth or data rate requirement at the user interface, usually a peak or maximum value in kilobits, megabits, or even gigabits per second;
- Variability of this data rate, that is, whether it is constant, time varying, or bursty;
- Acceptable end-to-end delay (latency), which is more important for interactive services and less important for one-way information distribution on a broadcast or narrowcast basis;
- Communication connection paradigm, namely, dedicated bandwidth, circuit-switched or packet-switched.

It is interesting to note that, from a historical perspective, services having different requirements in terms of those criteria had to be carried over totally different and unique networks. Table 5.3 presents some of those "traditional" telecommunication networks and services, measured against the parameters listed above. Those four criteria are probably the most important, but there are many more that can be critical to the success of a particular network or service. The secondary factors, such as call setup time, ease of use, availability, and cost per bit or minute of use, are important differentiators in the competitive commercial marketplace.

We note from this analysis that there are perhaps only two widely available networks that address multiple information requirements: the PSTN and some form of public data network, particularly the Internet. Neither has much chance of addressing all or most of those needs, as well as the forthcoming move to broadband multimedia, where all those elements are seen to converge.

Table 5.3
Comparison of Information Requirements for Traditional Telecommunication Networks

Type of information or service	Bandwidth requirement (kbps)	Variability of data rate	Latency requirement (one way)	Connection paradigm	Traditional network
Telephone (POTS)	16 to 64	40% duty cycle for voice	Between 150 and 500 ms	Circuit switched	PSTN
Fax	9.6 to 64	Continuous but time varying	Between 150 and 500 ms	Circuit switched	PSTN
Internet access	14.4 to 2,048	Highly variable to bursty	Between 500 ms and 2 sec, depending on content	Circuit switched	PSTN; dedicated bandwith T1/E1
Video teleconferencing	128 to 2,048	Continuous but time varying	Between 500 ms and 1 sec	Circuit switched	Narrowband ISDN; switched T1/E1
Data transactions	4.8 to 64	Time varying to bursty	Between 500 ms and 1 sec, depending on application	Packet switched	PDN; private packet switched network (intranet)
Data file transfers	64 to 100,000	Continuous during transmission	Between 500 ms and 10 sec, depending on application	Any	PDN; intranet
TDM baseband	1,544 and up	Continuous and constant	Between 150 and 300 ms	Dedicated bandwidth	Dedicated bandwidth, PDH, and SDH
Digital TV	Approximately 5,000	Continuous but time varying	Up to 10 sec for broadcast or narrowcast	Dedicated bandwidth	Satellite area coverage transmission; local radio broadcast and cable systems
HDTV	Approximately 20,000	Continuous but time varying	Up to 10 sec for broadcast or narrowcast	Dedicated bandwidth	Satellite area or spot beam coverage; local radio broadcast or cable systems

The entry in Table 5.3 for TDM reflects an old technology for combining many channels of information into a wideband baseband. The channels can be identical, such as multiple 64-kbps voice circuits within the telephone network, as well as other services listed in the table. Using conventional TDM multiplexers, public network operators and private networks operated by companies and governmental organizations gain an economy of scale by either expanding the bandwidth on a common link or sharing the bandwidth by including such features as circuit switching or statistical multiplexing (discussed in Chapter 4). We treat TDM as an information type because many organizations and public network operators use the technology to integrate multiple information sources. This is a tradeoff that amounts to a decision of whether to own and operate the multiplexers or to outsource the function to the network operator.

There is an evolving telecommunications network standard that at least promises to perform this broadband integration miracle: ATM. It is beyond the scope of this book to review the technical approach and the implementation of ATM on private and public networks, but readers can find a wealth of information [1]. In summary form, ATM provides what is called ATM adaptation, in which a packet protocol is used to address the same set of applications and information types that are covered in Table 5.3. What the developers and potential operators of public ATM networks could have to offer is a "one network serves all" proposition. At the time of this writing, only ATM stands out as an integrated solution to address multiple/simultaneous information transfer needs. What we are going to do in this book is discuss how satellite communication also can be used as a means of data transfer and integration. The reason that is possible is that satellites deliver broadband links today on a nearly universal basis, something that could take terrestrial networks many years to implement.

5.2.2 Analog to Digital Conversion

The process by which a continuous information signal is converted into a stream of digital data is called A/D conversion. Quite obviously, the reversal of digital data to analog information is D/A conversion. It has been mentioned that A/D conversion is necessary to convert analog signals, which generally are associated with some natural form of activity or communication (e.g., voice or image), into a bit stream that can be applied to a digital network. Many user devices such as digital PBXs and video codecs perform that function prior to delivery of the information to the Earth station. For example, the network node shown in Figure 3.18 contains A/D and D/A conversion circuits in its analog interface port cards. In applications such as MSS telephony and DTH reception, the user terminal must perform the appropriate direction of conversion.

The first step in A/D conversion is to sample the analog signal at equally spaced points in time. Electronic devices that perform that measurement and conversion operate at high rates, corresponding to twice the highest baseband frequency. The voltage range being sampled is divided into narrow bands called quantization levels, illustrated in Figure 5.7. (This linear quantization scheme is not representative of real-world A/D systems which use some form of companding.) The intelligence of the A/D converter puts out a numerical code, typically in the binary system.

A/D converters have evolved over the years, first appearing in the 1960s as large shelf-sized units using discrete transistors. In the 1980s, the device was reduced to individual integrated circuit (IC) chips. Today, they perform multiple functions as elements within VLSI circuits, many of which are application-specific ICs (ASICs). We find A/D and D/A converters in digital cell phones, PC sound boards, and some settop boxes used in home entertainment.

The A/D conversion process introduces some inaccuracy into the end-to-end link. Called quantization error, it is the small amount of information loss due to the use of steps (as opposed to transferring a continuous range of

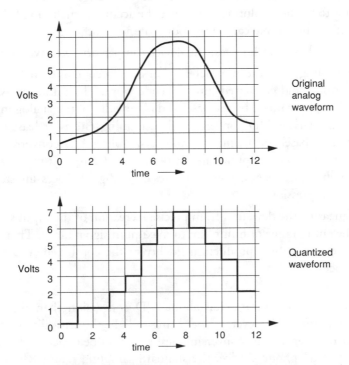

Figure 5.7 Comparison of the original analog waveform with the quantized version, based on 8 levels (or 3 bits).

voltage). The basic relationship for quantization error, measured by the ratio of signal power to quantization noise, is:

$$S/N_q = 2^{2M}$$

where M is the number of bits per sample. The same formula [2] can be expressed in decibels as follows:

$$S/N_q = 2M \cdot 10 \log 2 = 6.02 \cdot M$$

The importance of having enough bits per analog sample is evident in this relationship. The main features of the A/D and D/A converter that affect end-to-end performance include the following:

- Resolution, specified in terms of the number of bits in the digital code used to represent each sample. An 8-bit code, for example, can represent $2^8 - 1$ (i.e., 255) different levels. The more bits in the code, the more steps and hence, the smaller the range of each level. That improves the resolution and hence the accuracy (including S/N_q) with which the sample can be reproduced on the other end.

- Speed, based on how fast the device can perform a conversion.

- Sampling rate, the number of times per second the analog signal is sampled and converted into the code word. The minimum sampling rate is determined by the baseband bandwidth of the analog input. For Nyquist-type A/D converters, the sampling rate is at least twice the highest baseband frequency. An oversampling A/D converter operates at an elevated sampling rate, in the range of 2 to 64 times the highest baseband frequency. That facilitates other digital processing steps, such as compression.

- Linearity, the deviation from precise recreation of the input signal over the entire range of minimum to maximum signal range. That, in turn, determines the reproduction fidelity of the combined A/D and D/A process.

This is possibly more detail about A/D conversion than most readers would care to know. However, it does bring to light the criticality of choosing the right parameters for the conversion, lest the result be unacceptable to users. In the case of telephone service, the standard sampling rate is 8 kHz (8,000 times per second), which is slightly more than twice the highest signal frequency (i.e., 3,400 Hz). With regard to quantization, good telephone quality is obtainable with 128 levels (seven bits when coded in binary). Standard pulse code modulation (PCM) adds an eighth bit for synchronization, yielding a transmission speed of 64 kbps per telephone channel. Companding in the form of nonequal levels is used to help overcome some of the quantization error and channel noise.

Another important consideration in satellite network design is the number of times that an analog information source might be exposed to the A/D and D/A process. The example of an end-to-end communication link in Figure 5.8 shows how an analog input can experience four conversions on the way to the distant end user. Each pair of conversions introduces distortion in the form of quantization error (due to the selected number of bits per sample) and nonlinearity. Therefore, the final product may not provide the desired quality. What sounded or looked acceptable after a pair of conversions (e.g., A/D followed on the other end by D/A) might indeed be considered of poor reproduction after two or more such pairs. While it is desirable to preclude more than two pairs of conversions, that may end up being outside the control of the satellite network designer.

5.2.3 Compression

The number of bits per sample can be reduced by taking advantage of predictable aspects of a particular signal type such as voice or video. Removing

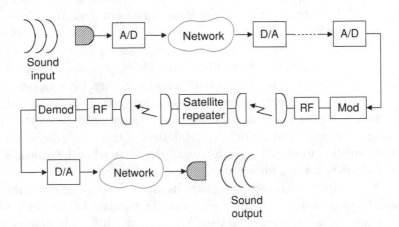

Figure 5.8 Example of an end-to-end communication link, indicating two pairs of A/D-D/A conversion.

excess bits in the sample effectively compresses the signal bandwidth by allowing the resulting data to be sent at a lower speed. That requires that the compression equipment have intelligence in the form of a microcomputer or DSP and an algorithm for controlling it. In the following discussion, we assume that the information already has been digitized at the source.

There basically are two classes of compression: lossless and lossy. In lossless compression, the reduced bit stream still contains enough of the original information content to allow the decompressor to recreate the input without distortion. It is said to be a reversible process because the output is, for all intents and purposes, exactly the same as the input. This is the one case in which compression and decompression can be done over and over without any additional distortion. To make that work, the input signal usually must be constrained to fit within a certain bandwidth and amplitude range, such as is common in the analog telephone network.

An example of lossless compression is adaptive-differential PCM (ADPCM), a standard voice-encoding technique that cuts in half the number of bits per sample while maintaining the quality of voice. The speech is reproduced without distortion even though only half the bits are transferred. On the other hand, there does appear to be a tradeoff in that the speed with which inband data can be sent is slightly reduced. In PC graphics, an uncompressed image (e.g., a file of type .BMP) contain a megabyte (MB) quantity of data, while one compressed without loss (type .PCX) would be in the 100 kB to 1 MB range.

That brings us to the more interesting topic of lossy compression, which, as the name implies, does not precisely reproduce the input information. Why would anyone accept lossy compression for a commercial service, satellite or otherwise? Well, it turns out that every means of communication is lossy, and we are all accustomed to "filling in" missing details. Our brains are very good at doing that, which explains why we believe that the typical picture post card looks like a photograph. However, if one looks closely at a typical printed post card, the actual image is not at all of photographic quality. Even greater loss of quality can be seen in newspaper printed pictures, especially ones in color. The same principle applies to sound. We certainly can enjoy a conversation over the telephone or a music broadcast on AM radio. Both of these limit the bandwidth of the information to something substantially less than what the source generates and what the average human is capable of hearing.

Lossy compression is used in perhaps its most effective manner in the typical telephone fax machine. This is the Group 3 standard of the ITU, which takes a black and white image that would require, say, 300 kB in graphic form (not as character data, as it is stored in a computer) and transmits it in only 20 sec at the typical modem speed of 9.6 kbps. That works out to only 24 kB of information, representing a compression factor of 12.5. The resulting image is noticeably poorer than the original and can be recognized easily as a fax. However, the value of the technique and the wide availability of fax machines around the world have overcome its lossy properties. Similar comments can be made about the JPEG compression standard (type .JPG) in wide use over the Internet. JPEG lossiness is apparent in highly compressed images. Images that were compressed with JPEG by a factor of, say, 40 are definitely not perfect but nevertheless provide utility and even enjoyment. Of course, JPEG includes the option of lossless compression, but that would not be as effective in reducing the quantity of data in the compressed image.

Compression of motion pictures combines the properties of lossless and lossy compression and must deal with images, motion, and sound all at the same time. As discussed in our previous work [3], the MPEG gave users and service providers a workable approach to information compression and integration. The amount of compression typically is adjusted to the type of image and the needs of the user. With more and more compression, the picture quality can look choppy, particularly when there is rapid motion. In video teleconferencing applications, compression down to 384 kbps usually is acceptable. However, commercial TV is demanding, and reduction to about 2.5 Mbps is considered acceptable using currently available processing algorithms. That was not always the case; in the first edition of this book, we indicated that the industry standard for digitized commercial TV was at least 45 Mbps. Colors can be reproduced with excellent quality no matter how much compression is used, simply because hue information occupies very little bandwidth to start with. Advances in the compression of full-motion video are anticipated as faster digital processors and more sophisticated algorithms are developed.

5.2.4 Error Detection and Correction

Noise, interference, and distortion on a digital communications link will on occasion confuse the receiving data processing equipment, causing bits to be misread. We measure the result through the bit error rate (BER), which is the ratio of error bits divided by the total bits during a particular time interval. The effect of BER can be relatively minor, as in the case of voice communications, or catastrophic if the bits represent numerical data such as computer programs or important statistics.

To control errors in the communications link, there are techniques for preparing the data prior to transmission. In FEC, data are coded to expand the number of bits, and a sophisticated decoder is used by the receiver to reverse (correct) errors that occur along the link. This is a difficult concept to explain or understand without resorting to the mathematics of statistical communication theory and information theory [4]. Nevertheless, the following paragraph reviews a simple example that would not actually be used in practice but that identifies some of the concepts.

Suppose the transmitting end codes the binary 1 into the 4-bit sequence 1111 and the binary 0 into the sequence 0000. The 4-bit sequence is called a code word. This simple code increases the data transmission rate by a factor of 4 since the same information goes through the channel in four times the time. The receiver is designed to be able to recognize that any of the following sequences with no errors or one error started out as 1: 1111, 1110, 1101, 1011, or 0111. It would then deliver a 1 to the user. A 0 bit being coded and sent would work in a complementary manner (e.g., by reversing the 1s to 0s and vice versa in the example). This coding scheme can handle one bit error per code word. If a received code word has two 1s and two 0s, there were at least two errors in reception. In that case, the coder cannot determine the correct bit sense. The way to handle that in the decoder would be to make a guess, recognizing that the choice would have a 50% chance of being correct. The code is a simple example of error-correcting codes that are more powerful in their ability to correct errors caused by the link. Also, actual codes are more efficient in their use of the channel, increasing the data rate by a factor of 2 or less.

Two classes of FEC codes are applied in the majority of cases: block codes and convolutional codes. In the block code, a fixed length of input bits from the user side (say, k bits in length) is translated into a somewhat longer fixed length of a code word (n bits in length). Thus, there is a one-to-one mapping between any possible k-length input word and the corresponding n-length output word. Information theory, credited to Claud Shannon, shows that the error correction performance of the block code improves as the code length increase. That brings with it increasing complexity on the sending and receiving ends, so practical code lengths generally are not particularly long (e.g., with k = 5 and n = 8). A popular example of a block code is the Reed Solomon code (RS code), first applied to digital TV over satellites and later to terrestrial DTV networks as well.

The convolutional class of FEC codes is also heavily applied to satellites and other wireless media. In convolutional coding, the input bits are not grouped in blocks but rather are run through a computation device that, in turn, outputs another stream of bits at an elevated bit rate. The increase in data rate is again related to the ratio of n / k < 1, causing the bandwidth on the link to be increased by the inverse ratio. An example of the type of computational device is a tapped shift register such as that shown in Figure 5.9. An important design parameter in convolution coding is the constraint length, which is the quantity of input bits that influence the encoder output of a single bit.

For data communications, a typical objective is one bit error per 100 million bits transmitted, expressed as 10^{-8}. The channel efficiency of error-cor-

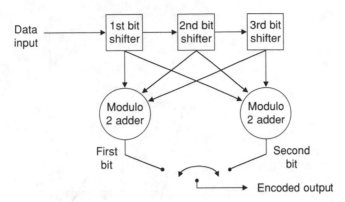

Figure 5.9 Convolutional coder block diagram.

recting codecs (often included in the high-speed modem in the Earth station) is specified by the coding rate (R), which is equal to the ratio of information bits to coded bits on the channel, or simply k / n. Also, the sophistication of the codec determines how much the error rate on the channel is improved (reduced). For example, a rate R = 1/4 convolutional codec can reduce the error rate on a typical satellite link by two orders of magnitude (e.g., from 10^{-2} to 10^{-5}), which is equivalent to a 2-dB to 3-dB reduction in the demodulator threshold for a constant error rate (Figure 5.10). Even greater benefit was obtained by applying RS and convolutional coding in the same link (concatenated coding). The only penalties with using that type of coding are the increase in bandwidth on the RF channel (because the data rate on the satellite is increased by a factor equal to the inverse of the coding rate) and the complexity and cost of including a codec in the RF modem. State-of-the-art microelectronics technology essentially has eliminated the latter penalty. Other more advanced FEC techniques, including turbo codes and nested codes, have been developed by researches and will reach the commercial market.

5.2.5 Multiplexing and Switching

We introduced TDM as a popular technique for combining multiple data streams to form a baseband. Figure 5.11 contains a simplified block diagram of a transmit terminal capable of digitizing and multiplexing four independent analog information channels (sources). Each source can be continuous in nature and must occupy a specified baseband bandwidth (e.g., 300 to 3,400 Hz for telephone and 0 to 5 MHz for video). In the figure, the information is first converted to digital data in a PCM encoder, whose output would run at a continuous data rate (64 kbps for telephone, 384 kbps for compressed video). Both

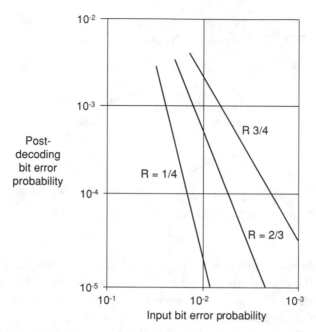

Figure 5.10   Digital link error rate performance versus E_b/N_0, for coded binary data channels.

the PDH and SDH standards are intended to facilitate TDM applications in terrestrial and satellite networks. In PDH, the 64-kbps channel is multiplied up to achieve the higher data rates. The same principle applies to SDH, but the basic building block is much higher, at 155.52 Mbps.

Direct digital inputs can be accommodated by bypassing the A/D conversion elements. The TDM multiplexer at the center of Figure 5.11 compresses the rates in time (retaining the same number of bits per sample) and outputs them sequentially as a single stream of data at a rate equal to or greater than the sum of the individual channel data rates. Some equipment designs reverse the sequence of encoding and multiplexing by using one high-speed coder to convert the TDM stream of analog voltage pulses directly into binary data.

The signal conditions at the various intermediate points of a simple TDM terminal are shown in Figure 5.12. Time waveforms are indicated for each analog input: They are sampled and digitized into 4 bits per timeslot, as shown at the center of the figure. The baseband spectrum of the TDM data stream starts with a maximum near zero frequency, decreases to zero, and then rises again in the form of a sidelobe. There is a similarity to the sidelobe of an antenna

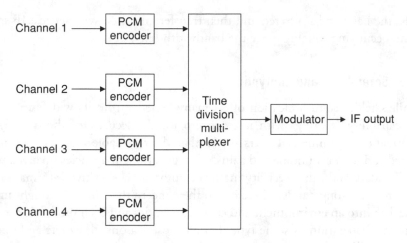

Figure 5.11 Block diagram of equipment used to generate a four-channel IF carrier using PCM-TDM/PSK.

because both are governed by similar mathematical laws. Sidelobes occur because the input has limits (a fixed-bit time period for data and a fixed reflector size for an antenna). In theory, the sidelobes continue to infinity, decreasing in strength more or less exponentially. When modulated on a carrier using

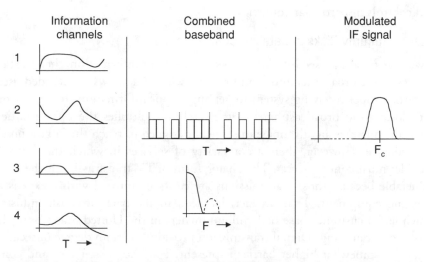

Figure 5.12 Assembly of a baseband using PCM-TDM and generation of an IF signal with PSK.

PSK, the baseband is filtered and then transferred into a symmetrical IF spectrum, occupying roughly twice the bandwidth (e.g., 2F).

5.2.6 Scrambling and Encryption

While satellite communication offers many benefits from its wide-area coverage capability, it brings with it a concern for the security of the underlying information. Commercial users worry that their valuable data can be intercepted and used for undeserved gains. Also, government agencies involved with public safety and other security matters cannot allow sensitive information to fall into the wrong hands. There is also the concern that someone might inject false data into an environment and cause difficulties and even total failure of an important operation of some type. For all those reasons, designers and operators of satellite networks must make proper provision for security. As with any type of countermeasure, the steps taken should be in proportion to the threat, to minimize the spent resources and to not unnecessarily interfere with normal day-to-day use of the system. The following paragraphs provide some guidelines and technological options for addressing those concerns. However, this subject is extremely complex and considers also that security must deal with intentional misuse and intrusion, for which clever humans are ultimately to blame. For those reasons, security technology must constantly be reviewed to counter threats that can become more sophisticated as the opportunities (and rewards) for abuse tend to increase. Another issue is that security technology and encrypted information may be subject to import and/or export restrictions and controls in particular countries.

5.2.6.1 Security Risks in Satellite Communication

If we were to take no security measures to protect our information, then the fact that a satellite broadcasts information over a wide area allows unintended users with the proper receiving system to get any of our information. It is the same with local radio broadcasting, but the satellite multiplies the opportunities. That, of course, is an advantage in services designed to reach the largest possible audience. However, there are a variety of services in which the network provider must control access. The analog form of TV transmission is the most vulnerable because those transmissions easily monitored with inexpensive receiving equipment. That, in fact, has lead to the backyard dish industry, which had a customer base of 1 million owners in the United States until the advent of scrambling. Digital transmission provides a certain degree of security because a somewhat higher barrier is presented by the greater cost and complexity of the receiver. However, it is not an impossible barrier to overcome, particularly as most services become digitized and the requisite equipment becomes a consumer item. We cover many of the commercial solutions offered on the market at the time of this writing. This is not exhaustive and, as discussed below, the nature of the threat escalates as our ability to counter it improves.

The following are representative of the types of abuse with which network operators are concerned:

- Preventing piracy of the information content. This is the most fundamental issue in broadcast application, leading first to the use of scrambling and subsequently to encryption.

- Controlling the customer base, also called conditional access. The network operator has the power to turn individual service on and off through the network itself.

- Protecting intellectual property (IP) rights, derived from the fact that satellites are used to distribute information such as TV and computer programs that are the property of someone else. The rights of the IP holder must be protected, for which conditional access and either scrambling or encryption are appropriate.

- Simplifying pay-per-use services to the end user. For example, PPV TV and information file transfer are services that are becoming a popular application of satellites but demand the type of control cited above.

- Providing levels of service and privacy for sensitive information. This is not a case of requiring payment but requiring that the information be blocked for reasons of competitive sensitivity or even national security.

- Preventing disruption of the normal operation of the network. This is a concern for all telecommunication services and will depend on the nature of the satellite network (e.g., how sensitive it is to disruption).

5.2.6.2 Analog Scrambling

For some commercial applications, enough of the security requirements given in Section 5.2.6.1 can be provided by analog scrambling. The first condition, of course, is that the information be in analog form, particular TV or sound. Scrambling is achieved by altering the normal waveform so that it cannot be used unless the receiver can restore the original waveform through foreknowledge of the scrambling pattern (Figure 5.13). Basically there are two ways to do that. The cheapest is to simply add an interfering or jamming signal into the baseband and then to filter it out at the receiving end. Alternatively, the signal itself can be altered in time or frequency according to some pattern, such that only authorized receivers can reverse the process to produce a proper signal.

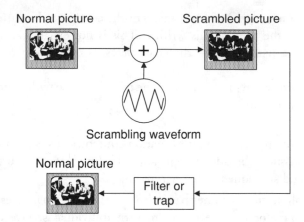

Figure 5.13 The principle of video scrambling as applied to TV security.

The additive technique is found in older analog cable TV systems, where costs are held to the lowest and people are already paying something for cable access. The jammer can be a simple sinewave or a triangular waveform, either of which is easily filtered out or suppressed with an active circuit within the home set-top box. That approach is rarely used over satellites because access cannot so easily be controlled and pirates tend to be more resourceful than the typical rogue cable installer. True scrambling, which is the second technique, is the preferred control technique for analog TV services.

The first scrambling systems were simple in concept, often doing no more than inverting the video signal, by reversing the white and the black of the picture. The result is something that looks like a photographic negative. As pirates developed the necessary technology to thwart that approach, scrambling went to the more sophisticated cut-and-roll technique. Each line of the picture (there are about 500 lines in the NTSC standard and about 600 in the PAL and SECAM standards) is chopped into segments, and the segments are shuffled like the cards in a deck. The shuffling is done systematically according to a code programmed into the receiver as well. The resulting picture cannot even be discerned and definitely offers little to the viewer. While it may be relatively easy to unscramble the picture (frame by frame) like a jigsaw puzzle, it is doubtful that anyone would consider doing that for practical gain or, much less, enjoyment.

The cut-and-roll technique is the forerunner of modern digital encryption, in which there is a seemingly random rearrangement of the information, rendering it unusable to unsynchronized receivers. It operates on the principle that there are two elements: the technique for scrambling the information, using appropriate sending and receiving devices, and a key to the sequencing of the scrambled bits and pieces. The devices can be offered through the commercial market, but they will not descramble unless the proper key is inserted into the receiver. The key mechanism is in the form of a digital code that, in turn, represents the scrambling sequence. If the scrambling sequence repeats itself frequently (which is determined by the technique and the length of the key), then one can analyze enough segments to obtain the key. That can be avoided if the key is long enough so that the repeat happens perhaps only once a week or year. Then it is simply a matter of changing and then redistributing the key to all authorized receivers.

Good scrambling systems are in use around the world, and the technology will be available as long as analog services are in use. In time, all types of information will be digitized, so the more powerful technique of encryption will be the standard for security. That provides the highest protection against piracy, but it cannot replace good security policies, which protect the keys from compromise (e.g., communication of the key to unauthorized users).

5.2.6.3 Digital Encryption

The technologies associated with the various forms of digital encryption have been with us for centuries. Encryption is, by definition, the use of ciphers, which are nothing more than codes. However, simple codes, in which a given letter in the alphabet or a number from 0 to 9 is converted into another on a one-by-one basis, are not particularly secure because they can be easily broken through analysis of the text. The better approach was invented in Nazi Germany in the form of the Enigma machine, a device that gave a rolling code, in which the conversion from letter to letter changed with each input. Thus, the letter A might be coded as a 6 the first time and as a Y the next. The decoding process would have to follow the path of the rolling code to recover the information. That is exactly the same principle as the cut-and-roll technique.

Over the years, more sophisticated and efficient encryption techniques were developed and reduced to standardized computer code or electronic circuitry. As with cut and roll, there is the encryption technique and the key. There are really two classes of techniques: private key encryption, of which the Digital Encryption Standard (DES) is the most common example, and public key encryption, which is embodied in the RSA standard (named after its inventors, Ronald L. Rivest, Adi Shamir, and Leonard Adelman). Additional information on those standards can be found in [5]. For most satellite communication applications, private key encryption is most popular. The DES standard can be utilized in the United States, where it was developed under government sponsorship. However, there were export restrictions at the time of this writing, so substitute public key encryption techniques have become available from other sources.

For all intents and purposes in commercial satellite communications, public and private key encryption are highly adequate to protect user data from intrusion and unauthorized interception. Even without encryption, it has been argued that the requirement to use a specialized receiver provides a low level of security. That argument does not hold up in attractive consumer services with low-cost receivers, such as DTH TV, so encryption becomes the norm. The techniques and the equipment do their jobs very well, but the problem comes from the need to adequately protect the keys. Because of the repeating nature of the encryption techniques and the tendency of pirates to be able to find out what they want, keys are changed routinely and often. They usually are distributed over the same satellite link that is to be protected. If pirates have been able to overcome the key once, they are in the position to acquire the new keys when those keys subsequently are broadcast over the satellite.

Auxiliary systems like smart cards and special chips add another level of security on top of encryption. An authorized user needs three things: the decryption device, the key, and the smart card. If the system is compromised, the operator distributes new cards (which is cheaper than distributing new encryption devices). Another defense is to require users to hook their equipment to the telephone network so the operator can dial in and transfer information such as the new key or an encryption technique upgrade. As is evident, security is a process of implementing a technique that is adequate until broken, then strengthening the system in response to the increasing capability of the pirate. It is, after all, a game of cat and mouse, in which the mouse has all the motivation to break through and obtain the information (TV programs, computer programs, and the like) without paying the originator. Of course, the pirate must pay to play the game, so there is a cost nevertheless.

5.3 Digital Modulation

Digital modulation involves transferring the bit stream to an RF carrier, for which techniques like BPSK and QPSK are popular. This discussion is an introduction to the theory and the practice of digital modulation design, with emphasis on the typical satellite link. The steps in the process include generating a pulse that meets certain criteria for bandwidth and shaping; transferring those shaped pulses onto a carrier, typically using some form of PSK; allowing for impairments, noise, and interference along the link; and providing a suitable demodulator capable of recovering the pulses from the carrier and cleaning them up to reproduce a digital bit stream with minimum errors due to the previous elements. Readers wishing a more detailed discussion suitable for making specific design decisions should review [6,7]. Nontechnical readers not needing an understanding of modulation theory can skip this section entirely.

Nearly all digital modulation relies on the basic characteristic of a pulse, shown in the time and frequency plots in Figure 5.14. The pulse has a fixed duration of T seconds, which means a continuous stream of such pulses would have a rate 1/T bits per second. If T = 0.001 sec (1 ms), then 1/T = 1,000 bps (1 kbps). The frequency spectrum of the pulse, shown in the figure, is a maximum at zero frequency and passes through the X-axis at a frequency equal to the bit rate (1/T, or 1,000 Hz for the previous example). Successive sidelobes have zero values at multiples of 1/T. The bandwidth of the signal, therefore, is proportional to 1/T, the bit rate. The formula for this spectral shape is simply the ratio

$$\sin(\pi\ F\ T)/(\pi\ F\ T)$$

That formula is plotted more precisely in Figure 5.15 and converted in relative power in decibels in Figure 5.16. The shorthand way of referring to the formula is sin(x)/x. The value at zero frequency, which is actually referenced to the carrier center frequency, is not zero even though sin(0) = 0. In fact, the value of sin(0)/0 is 1, because for small angles sin(x) = x. The important relationship here is that a fixed pulse length produces maximum energy at zero (e.g., carrier center frequency) and an extended frequency spectrum. In practice, the carrier at center frequency can be suppressed to save power.

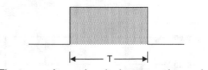

Time waveform of a single rectangluar pulse

Baseband frequency spectrum of a rectangular pulse

Figure 5.14 Characteristics of a rectangular pulse in the time and frequency domain.

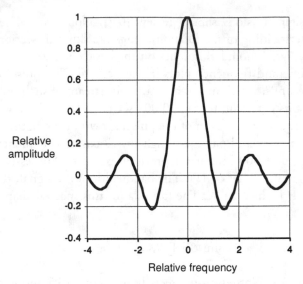

Figure 5.15 Amplitude versus frequency spectrum of a PSK carrier with rectangular pulse modulation.

The baseband and IF bandwidths of signals follow those direct relationships whether frequency shift keying (FSK) or PSK is employed. It seems logical, then, that the spectrum of the pulse should be filtered to minimize the occupied bandwidth. That is illustrated in Figure 5.17, where the attenuation characteristic of a low-pass filter is shown superimposed on the spectrum of a rectangular pulse. After filtering (the heavy line), the output spectrum is con-

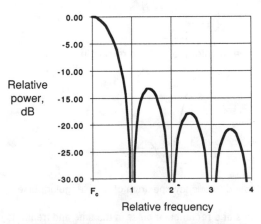

Figure 5.16 Power spectrum of rectangular pulse modulation shown on a decibel scale.

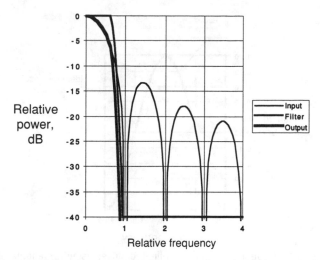

Figure 5.17 Baseband frequency spectra of a rectangular pulse before and after filtering to reduce the transmission bandwidth.

fined within the bandwidth of the pulse spectrum's main lobe. Due to the same mathematical relationship that causes a rectangular pulse to have an infinite spectrum, a limited spectrum alters the pulse so that it has infinite duration in time. Instead of a rectangular shape, a filtered pulse has a time waveform with an oscillating tail. A stream of filtered pulses contains residual tails, which increase the chance for errors. That is comparable to increasing the link noise level or adding interference; hence, the presence of tails is called intersymbol interference (ISI). If too little of the pulse spectrum is passed by the link, the ISI can so degrade reception that satisfactory operation is impossible. That, of course, is why adequate modulation bandwidth must be allowed.

The error-causing effect of ISI can be reduced by aligning the pulse tails so their zero crossings occur at times when the pulse main lobes are at their peaks. That can be done by proper filter design and demodulator timing. Figure 5.18 shows an ideal rectangular spectrum and a spectrum shaped like a cosine curve (called the raised cosine spectrum). Both occupy approximately the same bandwidth. The corresponding pulse waveforms in time are plotted in Figure 5.19, clearly showing the reduction in the oscillating tail produced by spectral filtering. Both pulse waveforms have zero crossings at the times corresponding to main lobe peaks (–3, –1, –1, 1, 2, 3, ...), thus precluding ISI. In addition, the raised cosine spectrum pulse has a much weaker tail, making it less sensitive to timing errors in the demodulator. This type of shaping is one example of the techniques used to efficiently reduce the modulated bandwidth of a digital carrier. Typically,

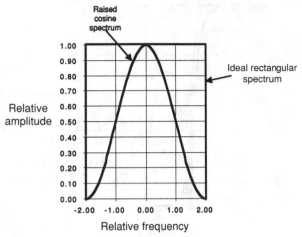

Figure 5.18 Frequency spectra of bandwidth-limited pulses using rectangular and raised cosine shaping.

the necessary RF bandwidth of a properly filtered pulse is approximately equal to 1.2 times the bit rate (expressed in hertz).

5.3.1 Frequency Shift Keying

The earliest form of digital modulation to be used in radio communications was FSK. Illustrated in Figure 5.20, FSK is obtained by switching between two discrete frequencies: F_1 and F_2, which correspond to the "1" and "0"

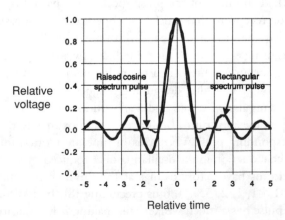

Figure 5.19 Comparison of pulse waveforms with limited bandwidths for rectangular spectrum and raised cosine spectrum shaping.

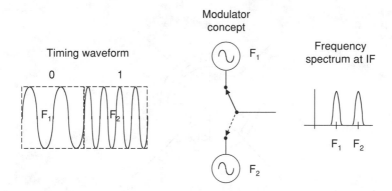

Figure 5.20 FSK as shown in time and frequency can be generated from two oscillators of different frequencies.

states, respectively (or, vice versa). That aspect forces FSK to consume twice the bandwidth of PSK. The principal advantage of FSK historically was that simple hardware could be used in the modulator and the demodulator. That lack of sophistication also means that FSK requires more signal power for the same performance in terms of error rate, which would have to be reflected in the link budget analysis. FSK was used in the first telephone modems, which were capable of between 600 and 1,200 bps. To support high voice band data rates, FSK had to be abandoned in favor of bandwidth-efficient modulation techniques.

5.3.2 Phase Shift Keying

The technique of PSK, discussed in Chapter 4, is the most common system for transmitting and receiving high-speed digital information over satellites. PSK modems operate at rates between 64 kbps and 1,000 Mbps. BPSK uses two phase states: 0 and 180 degrees; QPSK uses four states split into pairs: 0 degrees and 180 degrees (the in-phase, or "I," component) and 90 degrees and 270 degrees (the quadrature, or "Q," component). Four possible phase states are illustrated in Figure 5.21 using vector diagrams, each representing a two-bit combination. A property called orthogonality is used in QPSK, wherein the I and Q components are separated by 90 degrees from each other, shown in Figure 5.22 by orthogonal vector pairs. That is similar in effect to the use of cross-polarization isolation to achieve frequency reuse (see Chapter 4). QPSK transmits 2 bits per signal, meaning that the vector position in Figure 5.21 represents 2 bits of digital information. Both BPSK and QPSK are very efficient in the way bandwidth and power are used. Adding another bit per symbol

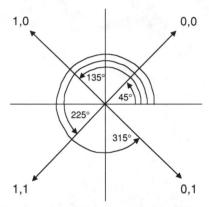

Figure 5.21 QPSK modulation vector diagram, indicating the four vector (phase) states and the corresponding encoded bit patterns.

forms eight-phase PSK, a technique that increases capacity but subjects the signal to greater degradation from the repeater. In addition, state-of-the-art advancements in digital hardware and software have pushed the performance very close to what is theoretically possible. That minimizes the amount of power needed to obtain a satisfactory operating link.

5.3.3 Hybrid Modulation

A modulation method that combines PSK with another modulation technique, notably PM or AM, is called a hybrid modulation method. The pur-

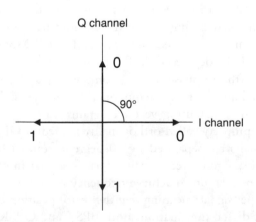

Figure 5.22 Orthogonal properties of the I and Q channels in QPSK.

pose is to increase the transmission speed in a limited bandwidth. For example, the I channel could be used in BPSK to represent the course quantization level and can be transmitted with relatively few bits per sample. The Q channel would then carry the quantization error in analog PM form. Since there is no quantization error in the resulting modulated signal, reproduction at the distant end is nearly exact. However, noise and interference on the channel are still a factor, particularly with regard to the Q channel, which carries an analog signal.

A higher capacity digital modulation system uses the I and Q channels as in QPSK and adds a third dimension by shifting the amplitude of the carrier (amplitude shift keying, ASK). The ASK channel thereby adds more bits to the possible sequence of binary data. A version of that, Trellis modulation, is popular in high-speed telephone modems.

Because of the higher data rate per unit of bandwidth, the S/N on the channel must be significantly higher than would be acceptable for PSK or FSK. In addition, the RF channel must be linear to allow the complex modulation format to pass unimpaired. Most applications are in telephone channels and in high-frequency radio rather than on commercial satellite links. That is because satellite transponders tend to be nonlinear and thereby would distort hybrid modulation to the point where the extra power needed to overcome the impairment would not be economically feasible. However, future improvements in satellite channel performance could make hybrid modulation more popular. An example would be in a digital processing payload, in which the spacecraft receives a hybrid modulated carrier, recovers the bits, and retransmits the signal in a linear fashion.

5.4 Multiple Access Methods

The basic concept of multiple access is to permit Earth stations to transmit to the same satellite without interfering with one another. RF carriers can be maintained separate in frequency, time, or code, as indicated in Figure 5.23 by three perpendicular axes. Hence, the use of FDMA, TDMA, and CDMA. In the first edition, we also identified space division multiple access (SDMA) as a unique scheme, based on using multiple spot beams from the same satellite. SDMA could be viewed as a multiple access method because the beams allow independent access to the satellite from Earth stations located in different beam footprints. The industry now recognizes, however, that SDMA is a standard feature of all satellites being launched, so multiple access is viewed as the manner in which bandwidth is shared within one beam or piece of spectrum.

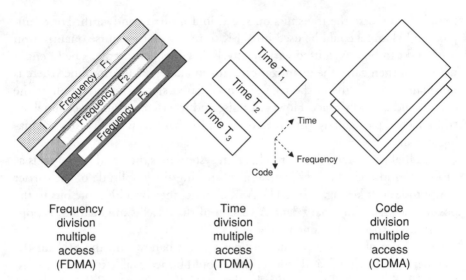

Frequency   Time   Code
division   division   division
multiple   multiple   multiple
access   access   access
(FDMA)   (TDMA)   (CDMA)

Figure 5.23 Generic multiple access methods using the three dimensions of frequency, time, and code.

5.4.1 Frequency Division Multiple Access

The FDMA technique is traditional in radio communications, since it relies on frequency separation between carriers. All that is required is that the Earth stations transmit their traffic on different microwave frequencies and that the modulation not cause the carrier bandwidths to overlap. Three such independent transmissions are indicated at the left of Figure 5.23 by long rectangles extending along the same "time" dimension but on different frequencies (indicated by different shading). A constraint in FDMA is that the sum of the bandwidths of the individual carriers cannot exceed the satellite's available bandwidth. Consequently, all three carriers in the uplink pass cleanly through the satellite repeater and are radiated toward the area of coverage on the ground.

The principle behind FDMA is that each Earth station or user terminal is assigned a separate frequency on which to transmit. That assignment can be fixed for time (permanently assigned), possibly changeable only by a manual command at either the uplink and the downlink or from a central control point. Alternatively, the assignment of the frequency can be dynamic (demand assigned, or demand assignment multiple access, DAMA), responding to user requests for service. Permanently assigned FDMA channels are useful for dedicated bandwidth services, which also are known as leased lines. Demand assigned channels are suitable for circuit-switched services, notably for telephony. Special signaling channels, also provided on an FDMA basis, are added to allow stations to request connections and to alert stations to incoming calls. In either case, the bandwidth of the channel must match the information bandwidth of the signal and allow for some guardband to prevent adjacent channel interference. The guardband may be as small as 5% or 10% of the channel signal (occupied) bandwidth, to allow for proper channel filtering and frequency errors.

The overall efficiency of FDMA is affected by intermodulation distortion (IMD), which results from multiple carriers in a common nonlinear amplifier like a TWTA or an SSPA. Figure 5.24 provides an example of how C/N varies, depending on the total input power of a common spacecraft TWT power amplifier. Two variables are indicated: the TWT input power with respect to saturation and the C/N. The Y axis also represents the carrier to intermodulation (C/IM) ratio, which is the negative-sloping curve. We see that there is an optimum operating point in terms of TWT input, for which the overall C/N is maximized. As discussed at the end of this chapter, that type of analysis is relatively complex and requires detailed knowledge of the particular signal and link characteristics.

Over the years, FDMA became viewed as antiquated compared to TDMA and CDMA. However, the fact that FDMA carriers can be transmitted

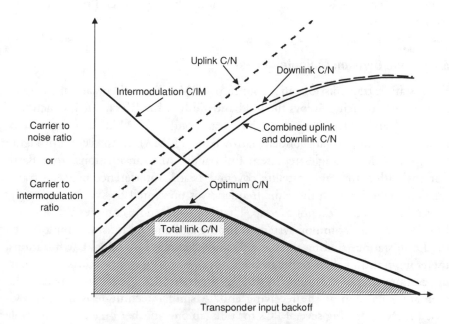

Figure 5.24 FDMA service optimization of the TWTA operating point for maximum C/N total (all curves are representative but not accurate).

without coordination or synchronization gives it longevity in telecommunications in general and satellite communications in particular. Many mesh network applications still rely on the simplicity and the efficiency of FDMA. Due to a variety of practical limitations on equipment power and cost, FDMA often is combined with the other multiple access methods to provide smaller networks that can operate more or less independently of one another. The basic principle of the channelized repeater and the transponder itself reflect the principle of FDMA.

There are some possible constraints to the effective use of FDMA, so satellite communication engineers have to do their homework before making a selection (that is true when doing the tradeoffs, because there are so many factors to consider in that type of decision). Some applications use a very low data rate per user, such as 4-kbps speech or 1,200-bps packet data transmission. Using the SCPC FDMA technique, the RF bandwidth per carrier is comparable to those data rates. At only 4 kHz of bandwidth, the carrier is subject to two sources of carrier frequency error: LO drift, and Doppler frequency shift. Both could cause carriers to run over each other if not controlled. LO drift can be improved through the use of accurate and stable frequency sources, something that gradually is being introduced both in space and on the ground. Doppler shift depends on the relative motion of the satellite and the user. Therefore, systems in which both are in motion (e.g., non-GEO MSS) tend not to use narrowband FDMA.

5.4.2 Time Division Multiple Access

Earth station transmissions in a common TDMA network are all on the same frequency, and each employs the full bandwidth of the RF channel, which may consist of an entire transponder (full transponder TDMA) or a segment of bandwidth within a transponder (narrowband TDMA). In the center of Figure 5.23, the wide rectangles represent full-bandwidth transmissions from Earth stations within the same satellite coverage beam. Interference between transmissions, which are on the same frequency, is prevented by synchronizing the transmission so they do not overlap. That is a much more complex process than FDMA because a common system of timing and control must be employed by the Earth stations sharing the same satellite channel. Individual Earth stations, therefore, transmit their traffic in the form of bursts of information, necessitating compression of the traffic in time at the transmitting end and the complementary expansion at the receiving end. A similar technique is used in PC-based LANs, allowing several PCs to "talk" to one another on a common cable loop. In TDMA, the most appropriate modulation is digital in nature, typically

QPSK, since that is compatible with the compression and timing requirements of burst transmission.

Full-transponder TDMA was developed at COMSAT to achieve the maximum throughput from each transponder within the communication payload [8]. That is possible because the transponder is operated at saturation, providing maximum power and bandwidth utilization. An example of a TDMA frame assignment is provided in Figure 5.25. The only inefficiencies are due to the need for guardtimes between bursts, analogous to the guardbands used in FDMA, "preamble" frame overhead bits used for synchronization and network control. Saturated operation allows the bit rate to be run up to the theoretical maximum, with no significant loss due to IMD. The latter decreases FDMA channel capacity by 50% or more. There are a few degradations that apply to full transponder TDMA. The first is that the bursts themselves produce pulses of dc current demand by the transponder output amplifier; some form of compensation typically is needed in the spacecraft power subsystem to prevent the pulses from affecting operation of other equipment on the same power line. Another concern is sidebands that can be generated in the transponder, potentially causing adjacent transponder interference. However, both factors can be dealt with effectively in the design of the modulation structure and the repeater itself.

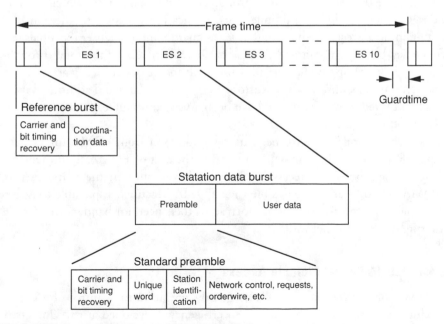

Figure 5.25   An example of a typical TDMA timeframe.

Full-transponder TDMA networks were first introduced in the INTEL-SAT system and provided a maximum throughput of about 60 Mbps. That is based on a transponder bandwidth of 36 MHz, QPSK modulation, and no FEC. In a real system, link performance can be improved with FEC at the expense of throughput. Higher bandwidth systems at 120 Mbps were introduced later to match the transponder bandwidth of 72 MHz. Experimental full-transponder TDMA networks were developed at 250 Mbps and 1 Gbps, but they have not been put into active service at the time of this writing.

The alternative of narrowband TDMA has found wide application in satellite networks as well as popular terrestrial mobile radio systems (notably the European GSM and North American IS-41 TDMA standards). It is a blend of TDMA and FDMA, designed to reduce the power and bandwidth requirements for the Earth stations and user terminals. That is because the power required is directly proportional to the data rate; narrowband TDMA is much less demanding on the user terminal, which can be designed for very low power operation. That being the case, why not simply use FDMA and forget about having to provide accurate timing and synchronization? Even without considering the throughput efficiency of TDMA, there are a number of benefits to using the hybrid technique. First, the bandwidth of each carrier is proportional to the number of stations that share the same channel multiplied by the average data rate that each station is afforded. The wider bandwidth, which can still be well under 100 kHz for mobile voice applications, reduces the demand for maintaining a precise frequency; that saves guardband. Also, user terminals can frequency-hop to grab unused time slots, thereby increasing the total throughput. Frequency hopping has the additional advantage of decreasing access delay and setup time. Thus, new applications (e.g., mobile data) and additional users can be inserted into what would have been a wasted resource. That principle is illustrated in Figure 5.26.

Narrowband TDMA is beginning to take over from FDMA for DAMA types of networks. That can happen because the cost of TDMA equipment has dropped rapidly as the necessary chip sets have come on the market. Also, TDMA networks are more flexible than FDMA because it is possible to assign multiple time slots to users in proportion to their need for bandwidth without altering the RF channel width.

5.4.3 ALOHA Packet Multiple Access

A variant of TDMA called ALOHA greatly simplifies the control of a digital satellite network, although at a cost in efficiency. Developed at the University of Hawaii, it is basically the technique illustrated in Figure 3.3 for multipoint-to-point connectivity. Stations transmit packets of data only when they need to

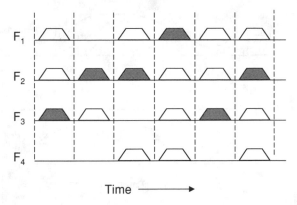

Figure 5.26 Illustration of the principle of frequency hopping applied to TDMA. The shaded bursts are transmitted by the same station, utilizing empty time slots on different frequency channels.

communicate; therefore, there is little need for end-to-end synchronization, and the packets arrive at the satellite at random times. On occasion, two packets will overlap in time in what is called a collision; when that occurs, neither packet will reach the destination properly. The destination station returns an acknowledgment packet if the transfer is successful. The basic structure of the ALOHA timeframe is illustrated in Figure 5.27 for the specific technique called slotted ALOHA, in which packets can be transmitted only in the defined time slots. A reference station transmits the timing reference to all remote ALOHA terminals.

ALOHA is used to transmit from small VSATs to a large antenna at a hub; therefore, individual VSATs cannot properly receive their own transmissions, whether a collision occurs or not. The hub acknowledges receipt of good ALOHA packets and is silent in the event that packets experience collisions. The remote VSAT waits the appropriate amount of time for the acknowledgment; if acknowledgment is not received, it assumes a collision has occurred. When that is the case, the two stations involved in the collision automatically retransmit their packets but only after they randomly select different time delays to preclude a second incident of interference.

The throughput efficiency of ALOHA is illustrated in Figure 5.28. An unslotted Aloha network, which employs very simple Earth stations, cannot provide more than about 15% throughput efficiency due to the time delays introduced by the automatic retransmission process. We see that as that point is reached, the retransmissions increase the average end-to-end delay by many times over the basic packet transmission delay over the single-hop satellite link. In contrast, no such automatic retransmission is needed in TDMA, because the

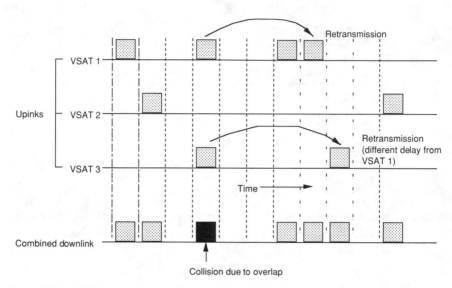

Figure 5.27 ALOHA protocol operation (slotted ALOHA).

timing is fully controlled by a reference station. The slotted ALOHA technique essentially doubles throughput for the same delay.

The efficiency of ALOHA has been improved by allowing stations that have large blocks of data to pass the data in a preassigned block of time using a reservation mode. The more advanced TDMA architectures for VSAT net-

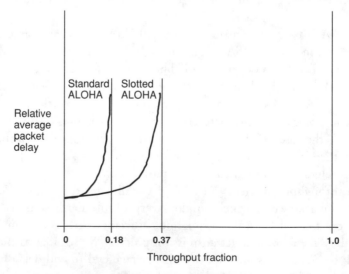

Figure 5.28 Performance of the ALOHA channel for the standard (unslotted) and slotted cases.

works incorporate features of standard TDMA, ALOHA, and the reservation mode to provide the greatest flexibility and efficiency of bandwidth utilization.

5.4.4 Code Division Multiple Access

CDMA combines modulation and multiple access to achieve a certain degree of information efficiency and protection through the technique of spread spectrum communications. First applied to guarantee security for military transmissions, it evolved into a commercial system that promises better bandwidth utilization and service quality in an environment of spectral congestion and interference. The basic concept is to separate or filter different signals from different users, not by using frequency or time, but by the particular code that scrambles each transmission. The code is a long sequence of bits, running at many times the bit rate of the original information. That expands the bandwidth by the ratio of bit rates; for example, if the scrambling sequence is 1,000 times the original bit rate, the resulting bandwidth is 1,000 times as well. On the surface, that seems very inefficient in terms of spectrum use, but it allows nearly as many signals to be transmitted one on top of the other [9, 10].

The basic approach is to use a pseudo-random noise (PN) binary sequence, which is nothing more than a computer-generated randomized sequence of bits designed to guarantee noncoherence with any delayed version of itself or any other PN sequence used in the same CDMA network. Figure 5.29 illustrates the basic process of sending and receiving. Each CDMA signal consists of the original data, protected by FEC, which are digitally multiplied by the PN sequence and then modulated onto a carrier using either BPSK or QPSK. At the same time the bandwidth is multiplied, the power density is reduced substantially by the inverse of the expansion of bandwidth. Multiple spread spectrum signals can then be transmitted one on top on each other as long as the PN codes are not synchronized. The signals can be separated in the receiver by using a correlator that accepts only signal energy from the selected PN sequence and despreads its spectrum. For that to work, the despreading must be perfectly synchronized with the incoming PN sequence, locking onto the correct start bit and maintaining tight timing during reception. At the same time the desired information sequence is reproduced in the receiver, any undesired spread spectrum carriers appear like noise at the output of the PN despreader. Capacity of the RF channel is determined by how much the noise floor from the extra spread spectrum carriers can increase and still provide adequate link margin. That depends on the bandwidth ratio and the relative power levels of the different CDMA signals sharing the same frequency bandwidth.

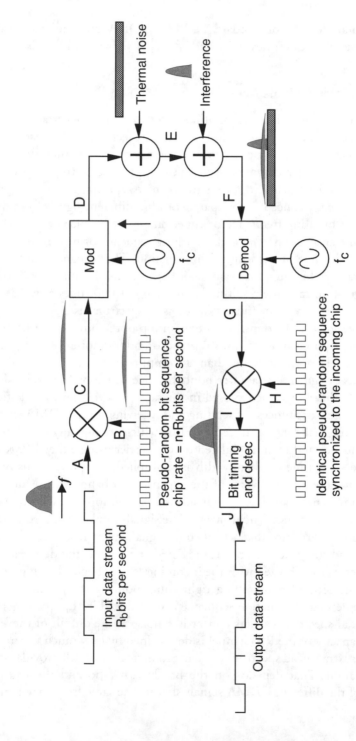

Figure 5.29 CDMA system block diagram, indicating how interference can be suppressed at the receiver.

Processing of a digital bit stream using the spread spectrum technique can make the carrier more tolerant of RFI. Spread spectrum, which could be classified as either a coding technique or a modulation technique, provides two important benefits for commercial satellite communications. First, the high-speed PN code spreads the carrier over a relatively wide bandwidth, reducing the potential of RFI caused by the Earth station transmission. That is important in siting the Earth station and in dealing with uplink interference to closely spaced satellites. Second, the recovery process at the receiving Earth station can suppress almost all external RFI located in the RF bandwidth of the spread carrier. The important data are obtained by despreading the carrier, but that causes any nonsynchronized RF signal to undergo spreading after mixing in of the PN code at the receive end. Modern CDMA receivers also include the RAKE function, so called because it can despread the direct carrier as well as other delayed copies of itself that arrive through multipath propagation.

The first commercial satellite communication application of CDMA on record was by Equatorial Communications, formerly located in Mountain View, California. Equatorial developed a one-way data broadcasting system to provide various kinds of financial market information and wire services with conventional C-band transponders transmitting to small antennas only about 40 cm in diameter. Spread spectrum modulation was chosen, not to allow multiple transmissions, but rather to overcome terrestrial interference, which is rather pronounced in the United States at that frequency. One transponder supported two to four of those carriers, each at a bandwidth of about 1 MHz. The information rate was rather low, being only 19.2 kbps for each carrier. However, the service was found to be very effective at the time since doing the same job over the analog terrestrial telephone network was much more costly and less reliable.

More recently, CDMA has gained acclaim due to the pioneering efforts of Qualcomm, another California-based company, this time in San Diego. Qualcomm successfully introduced two-way communication using CDMA in its highly acclaimed IS-95 digital cellular telephone standard. The same technique was applied to the Globalstar LEO satellite network, using a bandwidth of approximately 1.2 MHz. Unlike Equatorial's data broadcasting system, both Globalstar and IS-95 presume the existence of hundreds of users sharing the same channel at the same time. Furthermore, the technique provides improved frequency reuse, because cells that operate on the same frequency can be placed much closer to each other than is possible with either TDMA or FDMA. The actual improvement on an overall basis is difficult to assess, but the inherent benefits of CDMA still remain.

5.4.5 RF Bandwidth Utilization in Multiple Access

Comparison of multiple access methods is a complex and often daunting task. To be true to this type of evaluation, one must consider a wide range of variables, including technical properties, cost, risk, and operational convenience. However, in this instance, we consider only the general issue of bandwidth utilization in the satellite transponder. The manner in which the three generic multiple access methods use that bandwidth is illustrated in Figure 5.30. Across the top are shown eight transponder channels, with the three in the center expanded to display occupancy. The channel bandwidth and spacing most likely will differ by frequency band (36 MHz at C band, 27 MHz at Ku band, etc.) and also may be arranged differently for a specific satellite design. A detailed description of the design of the repeater and transponders is covered in Chapter 6. We can assume that, for the purpose of this example, the channel bandwidth and the spacing are 36 MHz and 40 MHz, respectively, typical of C-band domestic satellites operating in the North American and East Asian arcs.

The example of a transponder used for FDMA shows two groups of RF carriers, each of equal power and bandwidth. Each carrier would come from a different Earth station and could contain multiple information channels in either analog or digital form. There is no reason to maintain constant bandwidth and power, and most FDMA transponders contain a variety of carrier bandwidths and power levels. One important constraint is that total power of these carriers cannot exceed the available transponder power, minus a margin to account for RF IMD (discussed later in this chapter). Between each carrier is a small guardband to allow the Earth stations to separate the carriers effectively. The satellite downlink contains all the carriers, and any particular one

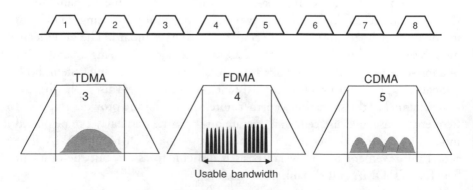

Figure 5.30 Sharing the transponder resource in TDMA, FDMA, and CDMA.

could be selected for reception at any Earth station located in the coverage beam.

In the case of full-transponder TDMA, all stations transmit bursts in the uplink, producing equal power as received at the satellite. A single wideband carrier transmitted by one station exists in the transponder at any instant in time. There would be brief blank periods of time (not visible in a static spectrum display) between station transmissions because full utilization is nearly impossible to achieve in practice. The guardtime intervals are needed to prevent interburst interference from the overlap of transmissions. (The criteria that determine the shape of the digital carrier were covered earlier in this chapter.) It also is possible to use TDMA with a lower data rate that uses less RF bandwidth to permit the sharing of a transponder on an FDMA basis. Networks of inexpensive Earth stations with limited uplink power can utilize narrowband TDMA because the power required is proportional to the bandwidth of the carrier, all other things being equal.

The display for CDMA in Figure 5.30 indicates multiple carriers, which ride one on top of each other (which is difficult to illustrate on paper because the signals appear as one image). Individual carriers are centered on the same frequency and are separated only by their code modulation. An alternative is to divide the transponder on an FDMA basis and to have multiple CDMA channels operating at the same time. That also facilitates frequency reuse, in which the frequencies are also assigned to particular beams in a cellular-type of pattern. The same total power constraint of FDMA applies to CDMA as well.

To summarize, multiple access is the process by which several Earth stations transmit to a common satellite and thereby share its communications capacity. The capacity and connectivity achieved depend on both the multiplexing and modulation formats of the individual transmissions as well as the multiple access method employed. In Section 5.5, we discuss how multiple access signals interact with one another and with the transmission systems used to uplink, repeat, and downlink the associated RF carriers. (Nontechnical readers may want to skip Section 5.5.)

5.5 Distortion and Impairments

Generally, impairments are the distorting effects (exclusive of noise and external interference) that a communication channel has on the information. This section reviews the impairments that involve the transmission of one carrier through the channel at a time. (Multiple carriers such as are used in FDMA are covered later.) An example of typical single-carrier impairments is presented in Table 5.4 from the standpoint of both the cause and the effect. In column 1 of

Table 5.4
Matrix of Transmission Impairments Due to the Uplink, Transponder, and Downlink, From the
Standpoint of Cause and Effect on Analog FDM/FM, FM-TV, and Digital Signals

Cause within the channel	Effect on analog FDM/FM signals	Effect on analog FM/TV signals	Effect on digital signals
Amplitude versus frequency	Intermodulation distortion	Modulation component degradation	I to Q coupling
Phase versus frequency	Intermodulation distortion	Chroma distortion	Intersymbol interference
Amplitude nonlinear distortion (AM to AM)	No effect by itself	No effect by itself	Spectral sideband regrowth
Phase nonlinear distortion (AM to PM)	Intelligible cross-talk	No effect by itself	Signal constellation rotation (I to Q coupling)

the table are the impairing characteristics of the empty channel, that is, those that can be measured without the desired information present. Among those causes are variations in the amplitude versus frequency response (i.e., bandwidth limiting) and the phase versus frequency response (also called delay distortion: the rate of change of phase is equal to time delay). Also of interest are two forms of gain nonlinearity: amplitude nonlinearity (AM-to-AM distortion) and phase nonlinearity (AM-to-PM distortion). The distortion properties are illustrated for a standard C-band transponder and TWTA in Figures 5.31 and 5.32, respectively.

Columns 2 and 3 of Table 5.4 apply to analog transmission systems; column 4 applies to digital. The listed impairments are effects that are actually visible in an information signal or modulated carrier after passing through the channel. For an FM carrier containing an FDM baseband (FDM/FM), amplitude versus frequency variation (also called gain slope) causes interference frequencies to appear in the baseband. Baseband IMD is the process by which those interference frequencies are caused in the FDM baseband. A different process called RF IMD occurs in the transponder bandwidth due to amplifier nonlinearity. The baseband variety, however, is the result of an FDM/FM carrier being swept back and forth across the gain slope and/or delay distortion characteristic of a transponder.

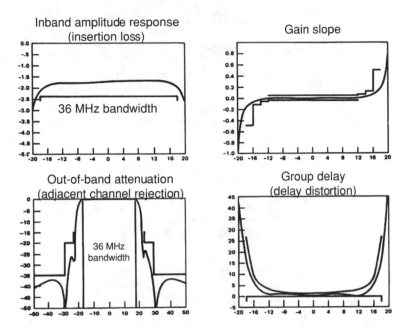

Figure 5.31 Distortion characteristics of a typical C-band transponder (IMUX).

5.5.1 Digital Signal Impairments

Impairment of digital data modulated on a carrier with QPSK (TDM/QPSK) typically results in an increase in the BER. While that can be compensated for by an increase in overall E_b/N_0, there can be a point where increased power is not available as a solution. Predicting those impairments and designing around them involve the use of computer simulation techniques. Since the first edition was published, several excellent design tools have appeared on the market. The process of using a tool involves setting up a simulation model of the end-to-end system within the software. An example of such a model is illustrated in Figure 5.33 for a hypothetical three-transponder system such as would be found in a bend pipe satellite link. The desired channel is transmitted through the center transponder, and potentially interfering channels exist on either side in the adjacent transponders. That allows engineers to evaluate many different types of signals, bandwidths, power levels, and transponder characteristics, all without having to build and test any real hardware. Done correctly, the simulation can get to within 0.25 dB of the performance to be expected in a real system. Trying to do that by direct mathematical analysis is much more difficult and, due to the necessary approximations, would not prove as accurate.

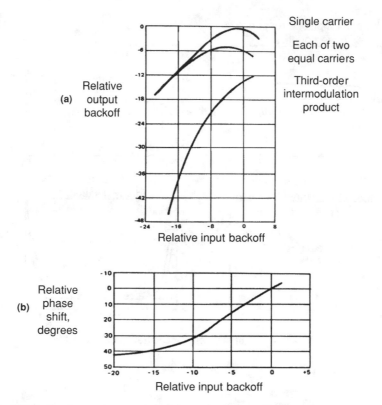

Figure 5.32 Nonlinear characteristics of a typical space TWTA.

The preceding discussion is meant to give the reader an idea of the types of factors that are taken into consideration when a channel is analyzed in detail on a single-carrier basis. Most of the impairments, while important, affect the performance to a lesser degree than power levels and antenna gains, which are taken into direct account in a link budget. However, there are many situations when detailed transponder characteristics can have a meaningful impact on the system, particularly if the designer needs to squeeze the last 0.5 dB out of the design.

The gain nonlinearities (AM to AM and AM to PM) of satellite power amplifiers (shown in Figure 5.32) tend to produce unwanted interference to the output carriers by the process of RF IMD, which appears as unwanted interfering carriers. A single unmodulated carrier with a constant amplitude will not cause RF IMD, no matter how nonlinear the amplifier may be. Any harmonics, which appear at integer multiples of the carrier frequency in the output of a nonlinear amplifier, are located way outside the downlink band and are eas-

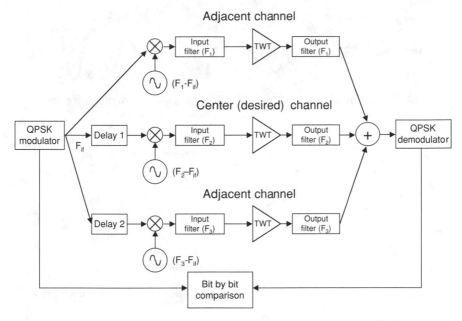

Figure 5.33 Typical computer simulation block diagram for a three-channel QPSK system.

ily removed by the output filters of a satellite transponder. The following paragraphs review the considerations for multiple-carrier RF IMD in FDMA and single-carrier RF IMD for TDMA.

5.5.2 Transponder Impairments

The following is a brief discussion of transponder RF impairments that affect both analog and digital signals. Such impairments produce additional forms of interference noise that need to be accounted for in the link budget.

5.5.2.1 Multiple-Carrier RF IMD

Multiple-carrier RF IMD (abbreviated IMD for the remainder of this section) results from FDMA operation of a common transponder amplifier. The distortion actually looks like additional RFI in the same transponder and can overlap into adjacent transponders as well. The most basic type of IMD comes from two or more carriers that share the same nonlinear amplifier. To see how two unmodulated carriers produce the amplitude variations needed to generate IMD, examine the time waveforms in Figure 5.34. Illustrated at the top of the figure are a sinewave of amplitude 1.0 at frequency F_1 and another sinewave of one-quarter that amplitude at a slightly lower frequency, F_2. The algebraic sum

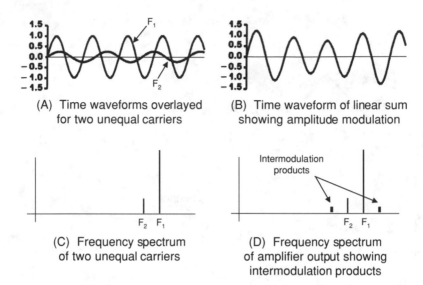

Figure 5.34 AM from two carriers results in intermodulation products in the output of a nonlinear amplifier.

of the two waveforms added point by point is shown in Figure 5.34(B). It turns out that the AM is periodic with a frequency equal to the difference $(F_2 - F_1)$.

The gain nonlinearity of the amplifier responds both to the degree of the AM and to its frequency. If the nonlinear characteristics and input signal parameters are known, the IMD can be predicted with accuracy sufficient to design the link. The input spectrum composed of the two carriers is shown in Figure 5.34(C). The IMD products at the amplifier output occur at frequencies that are arithmetic sums and differences of the input carrier frequencies. For example, the two carriers will generate the strongest intermodulation products, located at frequencies $(2F_1 - F_2)$ and $(2F_2 - F_1)$, illustrated in Figure 5.34(D). The power level of those products will be determined by the relative input power of each carrier into the amplifier, the total output power relative to the maximum capability of the amplifier (the operating point), and the particular nonlinear characteristic of the amplifier. With two equal amplitude carriers operating at the maximum power point of a typical TWTA (the point of saturation), each product will be approximately 10 dB down from saturation. Figure 5.32(a) contains a plot of TWT IMD performance, where the level of the IMD product is shown in the lower curve. At −10 dB, the interference is unacceptable for most transmission modes, although the products do not fall directly on top of the carriers in that example. However, as the total input power is reduced (i.e., as the input is backed off), the level of the products also decreases but at a somewhat greater rate. As illustrated in Figure 5.32(a), a total input power backoff of 10 dB results in an IMD product that is approximately 20 dB below the carrier level. The fact that downlink carrier power drops at the same time means that an optimum point of link operation is reached where the sum of link noise and IMD is minimized (illustrated in Figure 5.24).

The principle of how nonlinearity acts on FDMA is shown at the left of Figure 5.35. The FDMA system illustrated occupies a single transponder and consists of four equal amplitude carriers with rectangular-shaped modulation each transmitted from a different Earth station. The amplitude and phase non-linearity of the satellite power amplifier respond to the rapid amplitude fluctuations of the composite signal. The output spectrum, shown at the lower left of the figure, contains IMD products in the form of small carriers with triangular-shaped modulation. Each such product is the result of the interaction of combinations of either two or three of the input carriers, according to the following algebraic relationships:

$$F_{(2)} = 2F_a - F_b$$
$$F_{(3)} = F_a + F_b - F_c$$

where F_a, F_b, and F_c are the frequencies of sets of three different carriers in the input.

The combinations on the right side of the equations are frequencies in the same band as the carriers themselves and can overlap useful information. The degree of overlap depends on the number of carriers and their placement in the

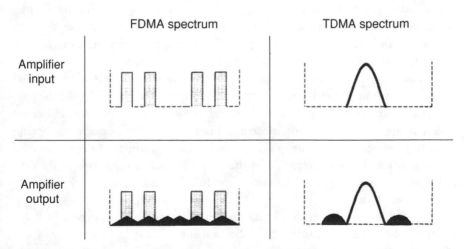

Figure 5.35 IMD products generated in a nonlinear amplifier by FDMA and TDMA.

RF channel, as governed by the formulas. It is important to note that the modulation on the IMD products is a combination of the original carrier modulations; IMD products, therefore, occupy a greater bandwidth than any one carrier. With enough input carriers and wide enough modulation, the spectrum of the IMD appears as a noise floor almost like white noise. Most of the time, IMD can be treated like white noise over the bandwidth of the carrier in question and analyzed in that manner in the link budget calculation. However, IMD products that fall back on one of the generating carriers include the original modulation. That is only a consideration in analog FM transmission and can be ignored for BPSK and QPSK.

The particular IMD products described in the previous example are called third-order products, indicated by the sum of the absolute values of the frequency terms (i.e., $2 + |-1|$ and $1 + 1 + |-1|$, both of which are equal to 3). Fifth-order terms, such as $3F_a - 2F_b$, also fall in band and would need to be included in the analysis since they can add a significant although lesser amount of interference. Generally speaking, it is the odd-numbered orders of IMD that fall in band; as the order increases, their impact on performance decreases.

5.5.2.2 Single-Carrier IMD

A single carrier operating at or near saturation usually is not expected to experience detrimental effects from IMD. However, it will be shown that digital transmission using QPSK often is subject to IMD, particularly when pulse shaping such as that described previously in this chapter is employed. A PSK carrier (or either the I or Q components of a QPSK carrier) has a reversal of phase at the transition from one binary state to the other. Rectangular pulses would cause the modulated carrier to always maintain a constant amplitude, which eliminates the possibility of either AM-to-AM or AM-to-PM conversion in a nonlinear amplifier. As explained previously, rectangular pulses have infinite bandwidth of the form shown at the bottom of Figure 5.14.

The input and output spectra for a TDMA system using PSK with raised cosine spectrum pulse shaping is shown at the right in Figure 5.35. At the top, the ideal raised cosine shaping is evident and no sidelobes are present. However, after passing through the nonlinear amplifier, the sidelobes reappear, having been regenerated in the form of intermodulation products. Another way of viewing this is that the saturation of the amplifier flattens the tops of the AM pulses and the resulting resquared pulse modulation has an infinite spectrum of the form $\sin(x)/x$.

The new sidelobes of the digital carrier defeat part of the purpose of shaping the pulses in the first place. Sidelobes are approximately 14 dB down from the carrier and can cause interference to carriers operating in adjacent transponders. By backing off the amplifier input by 1 dB or more, it is possible to reduce greatly the sidelobe level and make adjacent transponders more compatible with one another. Once again, there is a tradeoff in total link performance, since backing off the input reduces the downlink power. Fortunately, the appropriate amount of backoff may reduce the downlink only slightly.

When several narrowband TDMA carriers are operated in the same transponder, sideband regeneration is nearly insignificant; hence, the carriers can be packed tightly together in frequency. Multicarrier IMD would have to be carefully considered in setting the operating point of the transponder, since several carriers will be transmitting at the same time. That is the case for VSAT networks using the various TDMA and FDMA modes, demanding that the operation of the transponder be carefully examined before full-scale operation is commenced.

5.5.3 Uplink and Downlink RF Interference

Electromagnetic compatibility (EMC) is often a complex and daunting topic in the design and operation of satellite communication systems. EMC problems may not even exist in some systems, while in others it is an overriding area of concern and may even prevent proper operation. Sources of EMC problems include various forms of self-interference that arise in the Earth station or the satellite on its own. Those types of problems can be prevented by proper design and testing of the applicable component, subsystem, or system before it is put into operation. What we are addressing in this section is RFI that comes from sources external to the satellite or Earth station, usually from other operating transmitters that use the same frequency band.

A variety of interference sources (called entries) can invade a satellite network. In addition, the operation of a satellite or Earth station can interfere either with another satellite network or with a terrestrial microwave network that shares the same band. We are not considering out-of-band RFI, which is produced by undesirable radiation from antennas not within the intended band of operation. We concern ourselves here with what can be called space system interference, which is RFI that is coupled between two different satellite networks. It involves both uplink and downlink interference, which are defined in Figures 5.36 and 5.37, respectively.

Uplink interference is the undesired radiation from an Earth station into an adjacent satellite that happens to cover the same portion of the Earth's surface on the same frequency. The undesired radiation is produced by the sidelobe pattern of the interfering Earth station's transmitting antenna. In FSS and BSS applications, the large uplink antenna typically radiates at a relatively low power level from its sidelobes; hence, uplink interference potentially is controllable through

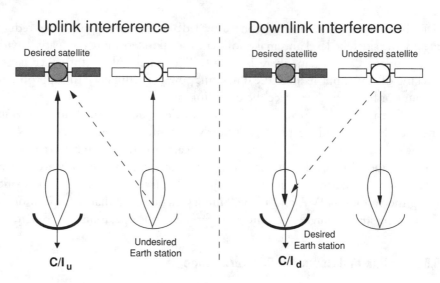

Figure 5.36 Simplified uplink and downlink interference geometry.

the process of frequency coordination. That might impose a limit on the transmit power applied to the antenna or on the sidelobe gain itself. In MSS systems, the small uplinking antenna may have nearly the same gain in both the desired and the undesired directions; hence, little can be done to reduce its effect on an

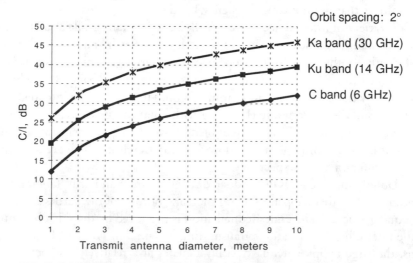

Figure 5.37 Uplink C/I from Earth stations of the same diameter, assuming standard CCIR radiation characteristics and 2-degree orbit spacing.

adjacent satellite. The only measure that can be taken here is to ensure that the interfered-with satellite either does not receive on that particular frequency or that its spacecraft antenna has reduced gain in the direction of the interfering Earth stations.

The formulation of the uplink interference problem is based on using the sidelobe pattern of the uplink Earth station along with the footprints of the desired and interfered-with satellites. To that is added the power level of the desired carrier coming from the desired uplink. The equation for the carrier-to-interference ratio (C/I) generally has the following form:

$$C/I_{up} = [G(0) - G'(\Theta)] + [P_t - P'_t] + [G_s - G'_s]$$

where

G(0) = the antenna gain of the desired Earth station in the direction of the desired satellite (e.g., the peak gain);

$G'(\Theta)$ = the antenna gain of the interfering Earth station in the direction of the desired satellite, which is at an offset angle, Θ, which corresponds approximately to the orbit separation between the desired and the interfering satellites; generally the form of the relationship between $G'(\Theta)$ and Θ is $G'(\Theta) = 29 - 25 \log (\Theta)$, in decibels, for Θ in degrees;

P_t = the transmit power applied to the desired antenna;

P'_t = the transmit power applied to the undesired antenna;

G_s = the receive gain of the desired satellite in the direction of the desired Earth station;

G'_s = the receive gain of the desired satellite in the direction of the interfering Earth station.

That relationship assumes that the path lengths from the satellite to both the undesired Earth station and the desired Earth station are the same. If they are different, an appropriate adjustment factor should be included.

The relationship looks complicated and confusing, but it really is nothing more than logic. Simply stated, the uplink C/I, in decibels, is the relative value of the three terms on the right side of the equation: the difference in gain of the transmitting antennas plus the difference in input power levels applied to the antennas plus the difference in gain of the satellite receive antenna. Figure 5.37 provides the uplink C/I as a function of uplink antenna size. Performance at three common microwave frequency bands is plotted, assuming a 2-degree fixed orbit spacing. If both P_t and P'_t are equal, the two networks are said to be homogeneous. That means the C/I performance is based only on the relative antenna gains. Otherwise, the power levels will have a decided impact on the compatibility of the two systems.

The case of downlink interference can be evaluated in much the same manner. This time, we formulate the problem in terms of EIRP differences:

$$C/Id_n = [P_t + G_s] - [P'_s + G'_s] + [G(0) - G(\Theta)]$$

where

> $[P_t + G_s]$ is the EIRP of the desired satellite in the direction of the desired Earth station;
>
> $[P'_s + G'_s]$ is the EIRP of the undesired satellite in the direction of the desired Earth station;
>
> $[G(0) - G(\Theta)]$ is isolation offered by the receiving antenna of the desired Earth station.

The downlink C/I generally is what we worry about in receive-only subscriber-based services like DTH TV. We have what should be a good signal from our satellite; however, a certain degree of interference is going to exist from other potentially interfering satellites that operate on the same frequency. An example of how the C/I is dependent on receive antenna diameter and orbit spacing is shown in Figure 5.38. We have assumed that the only protection from uplink and downlink RFI is that from ground antenna sidelobe suppression. In an actual case, additional isolation could be provided by offsetting carrier frequencies or by using cross-polarization.

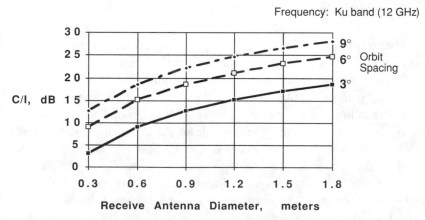

Figure 5.38 Downlink C/I versus dish diameter and orbit spacing for DTH service.

There is still the question of how that interference will affect the operation of our satellite network, particularly the resulting quality of reception by the user. For many signal and interference types, the effect is like an equivalent amount of white noise. We therefore only have to include the C/I term in the link budget, combining it through the general equation of the form:

$$C/N = (N_{up}/C + N_{dn}/C + I_{up}/C + I_{dn}/C + I_{xpu}/C + I_{xpd}/C)^{-1}$$

where C/N is the total value, considering all sources of noise and interference, expressed as a ratio (and not in decibels). In the case of C/I, the interference power is measured within the bandwidth of the desired carrier power, C. That may require careful analysis if the interference is wider in bandwidth than the information signal. On the other hand, narrowband interference from either an unmodulated carrier or one with analog modulation potentially could cause greater harm than its pure power would indicate. The formula for adding in the interference on a power basis (given above) would have to be replaced with a more rigorous approach, such as computer simulation or even direct measurement in the laboratory or over an actual satellite link.

References

[1] Sexton, MM., and A. Reid, *Broadband Networking: ATM, SDH, and SONET,* Norwood, MA: Artech House, 1997.

[2] Kondoz, A. M., *Digital Speech: Coding for Low Bit Rate Communications Systems,* West Sussex, UK: Wiley, 1994.

[3] Elbert, B. R., *The Satellite Communication Applications Handbook,* Norwood, MA: Artech House, 1997.

[4] Michelson, A. M., and A. H. Levesque, *Error-Control Techniques for Digital Communication,* New York: Wiley, 1985.

[5] Elbert, B. R., and B. Martyna, *Client/Server Computing: Architecture, Applications and Distributed Systems Management,* Norwood, MA: Artech House, 1994.

[6] Spilker, J. J., *Digital Communications by Satellite,* Englewood Cliffs, NJ, Prentice-Hall Electrical Engineering Series, 1977.

[7] Morgan, W. L., and G. D. Gordon, *Communications Satellite Handbook,* New York: Wiley, 1989.

[8] Schmidt, W. G., "The Application of TDMA to the Intelsat IV Satellite Series," *COMSAT Technical Review,* Vol. 3, No. 2, Fall 1973.

[9] Viterbi, A. J., *CDMA: Principles of Spread Spectrum Communication,* Reading, MA: Addison-Wesley, 1995.

[10] Glasic, S., and V. Branka, *Spread Spectrum CDMA Systems for Wireless Communications,* Norwood, MA: Artech House, 1997.

6

Communication Repeaters and Electronics

The purpose of this chapter and Chapters 7 and 8 is to explain how a satellite works and to review the critical factors in its design. These chapters should familiarize the reader with the key concepts and terminology that are common in the satellite industry. As a first definition, a *spacecraft* is the actual piece of hardware that is launched into orbit to become an artificial *satellite* for the purpose of providing a radio repeater station.

The hardware elements that a communications spacecraft comprises can be divided into two major sections: the *communications payload* (or just the payload), containing the actual radio communications equipment used for reception and transmission of radio signals, and the *spacecraft bus* (referred to as the bus), which provides the supporting vehicle to house and operate the payload. The emphasis in this chapter is on the requirements and specifications of the payload and how the general types of hardware designs can meet those requirements. Chapter 7 provides detailed information on spacecraft antennas because they probably are the most important single element of the payload regarding its communication mission. We will see that two bus configurations are popular: the *spinning* spacecraft (called the *spinner*) and the *three-axis* spacecraft (also called a *body-stabilized* spacecraft). Chapter 8 covers the various approaches to the design and operation of the bus in sufficient detail to give the reader an appreciation for the subsystems that ensure safe and continued operation of the payload. We begin with a general introduction to spacecraft design.

6.1 Overview of Communications Spacecraft

The design, manufacture, and operation of a communications satellite are no simple matter, complicated by the fact that it must survive the rigors of launch and deployment in orbit, followed by many years of satisfactory operation without physical intervention by human beings. In laypersons' terms, we assume that a certified repair person cannot travel to orbit to repair or reconfigure a satellite. That is certainly the case for GEO satellites, but it is not economically feasible even for large constellations of LEO satellites. It is critical that the payload and the bus work hand in hand to establish a highly efficient radio repeater on a stable space platform. The result is shown in Figure 6.1, which illustrates a typical geostationary communications satellite serving a country or region of a continent. At an altitude of approximately 36,000 km, a beam with dimensions 3 degrees by 8 degrees would cover an area the size of the United States. From the standpoint of a typical LEO satellite, the coverage is more limited in area simply because the satellite is much closer to Earth.

Figure 6.2 indicates the general arrangement of the spinner and three-axis bus designs. The spinner employs a drum-shaped solar panel that is able to produce continuous dc power as the spacecraft rotates at approximately 60 rpm. In contrast, the three-axis spacecraft uses flat solar panes that are separately pointed at the sun while the body of the spacecraft is pointed at Earth.

As will be discussed in Chapter 8, many other factors must properly be taken into account for us to have a stable and long-life communications satellite. Many of these principles are illustrated in Figure 6.3, which is the conceptual framework for overall satellite design. Obviously, the repeater will not operate on its own and needs the supporting functions illustrated around it. The electronics in the repeater need prime (electrical) power, which in the case of commer-

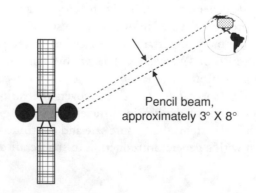

Pencil beam,
approximately 3° X 8°

Figure 6.1 Coverage of a land area from GEO requires the generation and control of a pencil beam onboard the satellite.

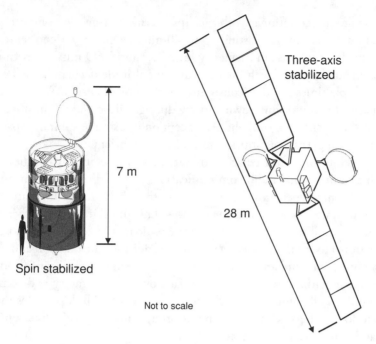

Figure 6.2 Basic satellite configurations: the spinner and the three-axis.

cial spacecraft is obtained by using a solar cell array to convert sunlight directly into dc electricity. An energy storage battery must be included to power the repeater when the sun is eclipsed by the Earth (or, less frequently, the Moon).

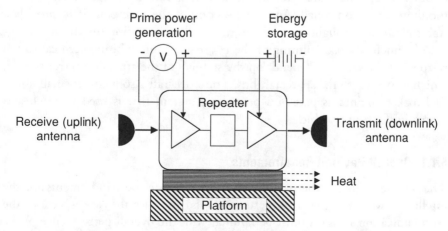

Figure 6.3 Functional diagram of an independent microwave repeater station for use in space.

For a satellite in geostationary orbit, eclipse occurs during a one-month period centered on the equinoxes, lasting up to 70 minutes per day. Requirements for eclipse operation increase in frequency for MEO and LEO missions because the angle subtended by the Earth increases as orbit altitude decreases. Such eclipses occur multiple times per day, consistent with the orbit period. Without batteries, a spacecraft would shut down entirely during eclipse and might not activate properly after reentering sunlight. An exception is a simple spinning spacecraft, which relies on gyroscopic stiffness to maintain its orientation.

A satellite or a constellation of satellites is operated from the ground through a bidirectional radio communications link dedicated to that purpose. Spacecraft control personnel can monitor the health and the status of all onboard subsystems using the *telemetry* downlink and can control active devices onboard the satellite (e.g., to configure the payload and to activate and disable bus functions) using the *command* uplink. In addition, the position and the orientation of the satellite are determined by measurements of distance and angle that are made with *tracking* signals that ride over the telemetry and command (T&C) links. Following the launch sequence, the T&C link provides the only means to locate the satellite in its initial orbit, transfer it to its final operating orbit, and configure it for service.

The repeater generates heat, much like an operating TV set or microwave oven. If not removed from the electronics, the heat will cause the temperature to rise sufficiently to affect or even damage the sensitive parts of the spacecraft, particularly the repeater electronics. The visible and invisible radiation from the sun also tends to raise the temperature of the spacecraft, while undesirable low temperatures can occur during eclipse and on darkened sides of the spacecraft at other times. Therefore, the flow of heat from and to the spacecraft must be monitored and controlled. Another function of the spacecraft is to provide a stable platform throughout the life of the satellite, including sustaining the rigors of launch. Precise pointing of the coverage beam or beams, critical to the purpose of the satellite, depends on the stability of the attitude control system, which is integral to the spacecraft bus. The spacecraft also includes small liquid fuel rocket engines as part of a propulsion system that is used to reach and maintain the proper orbit.

6.1.1 Overall Payload Requirements

The requirements for the payload effectively dictate the requirements for the satellite as well. For that reason, anyone who specifies the key aspects of the communication mission must be familiar with the overall performance of the satellite throughout its operating lifetime. We typically break the requirements into those for the repeater and those for the antenna system. Both depend heavily on the types of communication services that the satellite will provide. That is obvious because the payload is the central element in the overall communication network.

We typically start with a basic description of the overall system of satellites and Earth stations, along with the type of multiple access to be used. Then we move to the subject of the link budget, where the key performance parameters of all space and ground segments must be specified. It is the responsibility of the satellite system engineer to allocate those requirements to the elements, most notably to the payload. The initial discussion centers on the simplified payload block diagram shown in Figure 6.4, which is for a classical bent-pipe repeater of up to about 16 transponders per polarization or beam. This bent-pipe repeater, which relays the uplink carriers to the downlink with only a change in frequency and power, handles several broadband channels and includes extra equipment to restore onboard equipment failures. Several important steps are taken in its design because of the need to minimize weight and power and to operate the repeater for 15 years or more without hands-on maintenance. In Figure 6.4, all uplink signals are first amplified and then translated in frequency from the uplink band to the downlink band within a wideband receiver. Typical receivers have an operating bandwidth of 500 MHz, which is sufficient to handle the 12 channels of 36 MHz each in this example. Modern commercial satellites employ frequency reuse, which requires that there be a receiver for each polarization or beam.

Following the receiver (which provides the 500 MHz spectrum translated to the downlink band) is the input multiplexer (IMUX), individual power amplifier per transponder, and the output multiplexer (OMUX). The complete set of amplifier downlink transponder channels is then applied to the transmitting antenna of the payload. Other components and features are discussed later in this chapter.

There are many tradeoffs in satellite system design, and the requirements for the payload are a direct result. An incomplete or invalid tradeoff analysis

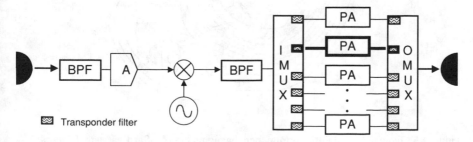

Figure 6.4 Simplified block diagram of a transponderized repeater.

will produce payload requirements that pose problems of one sort or another. For example, an imbalance in the allocation of antenna gain between the satellite and the Earth station might mean that an overly large antenna structure might have to be deployed in space or that the power to be transmitted could be unrealizable with practical equipment. This topic is always a source of controversy and is briefly discussed in later chapters. Resolving these dilemmas is well worth the time and effort of a thorough tradeoff study.

6.1.2 Transmit Effective Isotropic Radiated Power (EIRP)

At the highest level, the most important overall technical parameter of the payload is the transmit EIRP, which is the product of the antenna gain and the maximum RF power per transponder applied to it by the repeater. As illustrated in Figures 6.5 and 6.6 for typical GEO satellites, the EIRP specifies how the satellite serves the coverage area on the ground. (Alternatively, EIRP also indicates the radiated power for auxiliary services like T&C and intersatellite links used to connect satellites directly to each other in space.) It is industry practice to specify the minimum acceptable value over the coverage region, which is indicated by an outermost contour or set of definition points connected by straight lines.

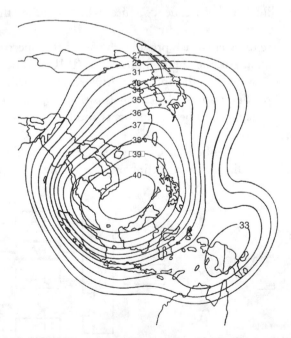

Figure 6.5 Typical C-band EIRP performance, based on the MeaSat 1 satellite. (*Source: Courtesy of Binariang Satellite Systems Sdn. Bhd.*)

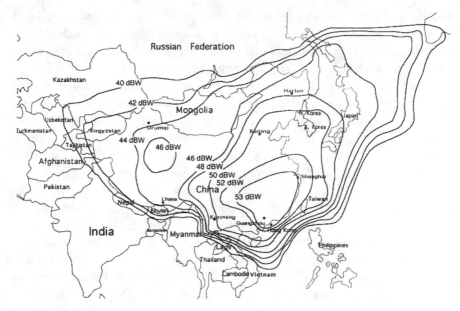

Figure 6.6 EIRP coverage patterns for AsraSat 3S. (*Source:* Courtesy of Hughes Space and Communications.)

Power amplification, illustrated to the right in Figure 6.4 for each transponder, is important to payload operation because of the long path length to the ground station. TWTAs are the mainstay, particularly at higher frequencies (above 10 GHz) and at higher power levels (50W and higher). Microwave power transistors are finding their way into commercial satellites for use in SSPAs at the lower power levels and for applications requiring improved linearity. (Typical power levels of TWTAs and SSPAs are reviewed in Chapter 3.) The particular power amplifier must be operable for essentially the entire life of the satellite; therefore, only the most reliable devices can be selected (hence the preference for TWTAs and proven SSPAs). Also, the amplifier design must be lightweight, typically weighing under 3 kg, and must be capable of converting dc power into RF with efficiencies of 50–70%.

From electromagnetic principles, we know that the antenna performance determines the shaping of the EIRP coverage. The majority of satellites that have been launched to date have fixed antenna beam patterns, which can be altered in space only by hardware switches. That has meant the buyer of the satellite must know and adequately specify the coverage requirements with precision long before services are offered. Some satellites have had provision for changing the transmit power, through switching and paralleling of amplifier stages, which causes the overall level of the EIRP to change accordingly. Changing of beam shaping or moving entire beams can be accomplished if the needed flexibility is included in the payload. Techniques such as movable reflectors, switched feeds, and phased arrays are required to allow beam reconfiguration and relocation. These topics are considered later in this chapter and further in Chapter 7.

Due to the availability of larger bus designs and increasing market demand, satellites have become more complicated and sophisticated in the way they provide channels and coverage. Capabilities such as multiple frequency bands (e.g., hybrid satellites), multiple antenna systems to service several countries or regions at the same time, and channel switching are available. In Figure 6.6, we see the various EIRP patterns of the AsiaSat 3S satellite, which has up to four beams available in its overall service areas. The antenna systems that provide that coverage are indicated in Figure 6.7, showing four reflectors that are used for transmit and receive in both C and Ku bands.

The actual power delivered to the transmit antenna will experience some RF loss along the waveguide path between the power amplifier and the antenna feed system. The loss comes directly out of the EIRP and therefore must be minimized. It makes no sense to design a satellite for very high transmit power and then to throw a significant fraction of the power away in the waveguide components that happen to be positioned after the final amplifier stage. For example, even 0.4 dB extra loss would require 10% increase in raw dc power to provide the necessary compensation (assuming the EIRP requirement is not

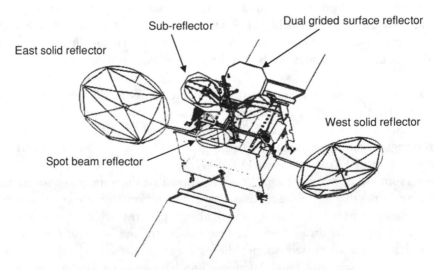

Figure 6.7 Physical antenna reflector arrangement of AsiaSat 3S. (*Source:* Courtesy of Hughes Space and Communications.)

reduced by a like amount). On a large three-axis satellite in the 10-kW class, that amounts to another 1 kW of power from the solar array and batteries. Readers should be aware, however, that reducing RF output loss by 0.4 dB may, in and of itself, be a considerable challenge, thereby making the design tradeoff more difficult on an overall basis.

As a general trend, peak EIRP values for C-band and Ku-band satellites generally are around 40 dBW and 55 dBW, respectively. That implies that Ku-band payloads are more demanding in terms of antenna gain and transmit power than those for C band. Higher EIRP also tends to increase the demand for prime dc power from the bus and makes the thermal design of the space-craft more complex. That may explain why Ku-band satellite users tend to pay higher prices for service as well. On the other hand, the higher EIRP of Ku band affords smaller receive antennas (which benefits the ground network) and also would be needed to overcome rain attenuation in tropical regions.

6.1.3 Receive Gain-to-Noise Temperature Ratio (G/T)

While EIRP defines downlink performance, it is the complementary parameter of G/T that is vital to the uplink operation of the satellite network. This param-eter is nothing more than the result of dividing the receive antenna gain as a true ratio by the noise temperature of the satellite receiving system, T_{sys}. The latter can be calculated as follows:

$$T_{sys} = \frac{T_a}{L} + \left(1 - \frac{1}{L}\right) \cdot 270 + T_{re}$$

where

T_a is the spacecraft antenna temperature, which typically corresponds to the Earth brightness temperature of 270K;

L is the input waveguide loss factor, for example, the ratio of input to out-put power (which, being greater than 1, is the inverse of the transmission line gain);

T_{re} is the equivalent noise temperature of the payload receiving system and is a specification for the payload wideband receiver itself.

The noise performance of the receiving system, like the transmit power, is under the control of the payload designer. Using the techniques of design analysis and optimization, the payload designer develops a gain and noise budget for the front end of the payload (e.g., those aspects of the payload that determine T_{sys}). From that point, subsystem and unit designers make hardware selections supporting that requirement, which in turn, produces the desired value of G/T across the coverage area. We will delve into some of those details in subsequent sections of this chapter and in following chapters.

The overall G/T characteristics of the payload after being converted to decibels per Kelvin are specified in the form of contours, as illustrated in Figure 6.6 for EIRP. In the simplest area coverage payload, the EIRP and G/T requirements have precisely the same shape, differing only in the value attached to the edge-of-coverage specification. The uplink Earth stations in these systems usually have large-diameter antennas and powerful transmitters, so G/T may not be critical to the application. However, satellite systems that support low-powered MSS hand-held user terminals and networks of low-cost VSATs are extremely challenging in terms of satellite G/T.

6.1.4 Transponder Filtering

The first GEO satellite to be launched, Syncom, had a single wideband receiver and offered one transponder. In that way, any signal could be uplinked and subsequently received on the ground, and the satellite did not constrain the RF bandwidth in any significant way. A substantial fraction of all communication payloads (GEO and non-GEO) use a transponderized design of one form or another. The basic principle is to subdivide the total RF bandwidth into RF channels that the payload can then route and amplify on a single carrier or multicarrier basis. That can be accomplished using analog filtering at microwave or IF frequencies or through digital processing onboard the satellite (to be discussed later in this chapter). Our simplified bent-pipe repeater in Figure 6.4 uses analog filtering directly at the downlink microwave frequency. Channels of this type are several tens of MHz in bandwidth each, to be consistent with wideband transmissions using FDMA, TDMA, and CDMA.

The term *multiplexer* in the context of the satellite repeater refers to any set of microwave filter hardware used to separate or to combine the RF channels, as distinct from the FDM and TDM multiplexer used to assemble a baseband at the transmitting Earth station. A detailed C-band repeater block diagram is provided in Figure 6.8 for reference. Draw your attention in Figure 6.8 to the series of circles (circulators) and attached boxes (input filters) composing the input multiplexer. The circulator uses ferromagnetism to create a microwave revolving door that causes any signal power entering a port to rotate in the direction of the arrow and exit at the next immediate port. Rotation is set up by a disc of ferromagnetic material sandwiched between a pair of magnets to create a magnetic field in a direction perpendicular to the paper. The polarity of the magnetic field determines the direction of rotation, which is fixed using permanent magnets or switchable using electromagnetism.

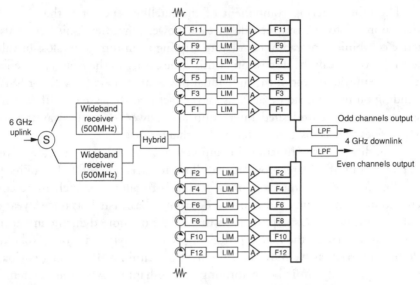

Figure 6.8 Block diagram of a typical 12-transponder C-band repeater.

Because all the channels pass through the circulator together, it is the function of the bandpass filter to select only one channel for amplification by that particular transponder. The input filter allows the specified transponder channel to pass through and reflects all others back to the same circulator. Thereafter, the five remaining channels are routed down the circulator chain with one more channel being dropped in each subsequent bandpass filter. By the time the last reaches its bandpass filter at the end of the chain, it has experienced the reflection process five times (all other channels experience the bouncing process proportionately fewer times). That reduces the power of the last channel by a significant amount due to losses in the circulators and filter input sections. The amount of loss, however, is significantly less than 1/6 (i.e., less than 10 log 6, or 8.8 dB), which is what would have been the result with a direct division of power.

At the final arm of the chain is a passive load (called a termination) to absorb any remaining channel energy and to prevent power from being reflected back up the chain. Circulator coupling is extremely effective for the purpose of separating channels, ensuring that only one channel enters each transponder chain. Leakage of adjacent channels produces an undesired effect called repeater multipath (which is different from propagation multipath discussed in Chapter 4). A step taken to reduce the problem is to narrow the IMUX filters as compared to those of the OMUX. (Repeater multipath is reviewed in Section 6.4.3.)

The other filtering component of the satellite repeater is the OMUX, which connects to the output side of each power amplifier. This device uses reactive combining instead of circulator coupling to minimize the loss of valuable RF power (recall the discussion of this issue earlier in the chapter). It effectively sums up the power of the individual power amplifiers (TWTA or SSPA, depending on the payload design), using the fact that they are on different frequencies to prevent undesired interaction and power loss. The way that the OMUX does that is discussed later in this chapter.

In addition to the functional components of the repeater, extra receivers and power amplifiers are provided to allow replacement of failed units by ground command. Another role of switching is to allow channels to be connected to different beams, either on an occasional basis (such as once a year or possibly monthly) versus requirements that might be more dynamic in nature. In regard to the latter, the digital processing type of payload is going to be the preferred approach as we move into the next millennium. Processing payloads also can support dynamic beam forming and reshaping, which are extremely useful in advanced MSS and multimedia applications.

6.1.5 Linearity

The EIRP, G/T, and transponder channelization requirements are among the most important in meeting the basic service requirements. The challenge from this point is to ensure that the secondary performance aspects of the payload are in keeping with the services that the satellite is supposed to provide. The first of those relates to the linearity of the transponder and its impact on distortion effects produced on the carriers that traverse the payload. In Chapter 5, we presented several of those aspects and explained the basic engineering methodologies used in satellite network design. The particular types of transponder linearity concerns were identified as well. Another important aspect is how susceptible the satellite is to uplink and downlink interference, which relates to the isolation properties (if any) from the spacecraft antenna system.

A growing area of interest is in how digital payloads influence the quality of service. That is because such repeaters incorporate A/D conversion as a key step in preparing the uplink channels for processing and D/A conversion to restore the analog format of the downlink. In Chapter 5, we explained that distortion introduced by the converter is directly affected by the sampling rate and the number of bits per sample. The onboard processor likewise is constrained regarding those parameters because they determine the amount of digital data bits that must be processed and ultimately the power requirements and the mass of the processor itself. That imposes critical tradeoffs between distortion on the one hand and digital payload complexity, cost, and spacecraft bus demands on the other.

6.1.6 Frequency Translation Effects

The last payload requirement to be considered here is one that can affect the frequency of all signals that pass through the repeater. That is the error introduced by the frequency translation process common to nearly every satellite payload design. The exception is the rare spacecraft that does not employ frequency translation but relies completely on time switching to separate transmit from receive (the Iridium spacecraft includes this feature, but nevertheless it still provides frequency translation to be able to route a high quantity of narrowband channels between beams, satellites, and gateway Earth stations).

The basic concept of frequency translation is similar to the operation of the upconverter and downconverter (discussed in Chapter 4 and further explained in Section 6.5). Most of the bent-pipe designs in use since the 1960s have a single translation from the uplink band to the downlink band, using a fixed frequency offset that should correspond exactly to the difference in frequency between the centers of the uplink and downlink bands, for example, $F_{up} - F_{dn}$. For that arrangement, there is no inversion of the band during translation; in other words, the lowest uplink frequency will reside at the lowest downlink frequency instead of being flipped to the higher end. That translation frequency, $F_{up} - F_{dn}$, is created by an oscillator in the payload. In any real design, there will be some error in that frequency, much like the time error of a quartz clock. However, unlike a wristwatch, the satellite oscillator cannot easily be adjusted. It is common for that error to amount to about one 10-millionth of the translation frequency, which is about 2 kHz at C and Ku bands. That amount of error can be ignored in the majority of applications.

In addition to the absolute error, the translation oscillator adds phase modulation noise, which could interfere with services. Phase noise is an undesired variation in oscillator frequency that produces spectral sidebands, as shown in Figure 6.9. During frequency translation, the sidebands are transferred to all carriers that pass through the mixer stage. In the receiving Earth station, phase noise is demodulated and contributes to the total baseband noise picked up along the satellite uplink and downlink. In a properly designed oscillator and mixer, phase noise is attenuated rapidly in the region of user baseband frequencies. That can be very important in analog FM and PM systems because they are particularly sensitive to the impairment.

Along with phase noise, an improperly designed downconverter stage on the satellite can introduce unwanted frequencies called spurious frequencies

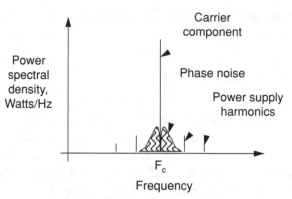

Figure 6.9   Illustration of phase noise sidebands on a carrier.

(spurs) that are radiated by the satellite along with the downlinked carriers. It is up to the receiver designer to hold spur levels to 20 dB or more below the level of the weakest signal to be transferred. There are situations in which spurs cannot be further reduced in power, and the system must find ways to accept or avoid them. It is possible that some spurs will reside at frequencies outside the specified downlink band and therefore will not interfere with the satellite network. While not a problem for our service, the spurs still may violate ITU rules for allowable radiation within the microwave spectrum. Alternatively, spurs actually may have to be coordinated just as if they were intended communication frequencies.

We have reviewed the primary specifications that are placed on the payload. Depending on the particular system design and user requirements, the specifications may be expanded to include other technical matters of importance. It is up to the system designer to work closely with the engineers, who will actually design and manufacture the satellite so that all the relevant factors have been identified and properly specified. In the real world, some of those effects cannot be controlled to everyone's satisfaction, so a compromise must be developed. To do that properly, trade studies can be performed to learn what parameters can be relaxed to allow the hardware to be built within cost and time constraints.

6.2 Analog Bent-Pipe Repeaters

The next step in payload design is to determine the general type of repeater that will meet the higher level system requirements. The basic bent-pipe design is intended for applications that involve wideband carriers for analog and digital transmission over relatively wide area beams, such as that illustrated for AsiaSat 3S. This approach is the simplest to implement and provides the most flexibility as to signal types, bandwidths, and multiple access. In fact, a bent-pipe repeater can accommodate all three multiple-access methods, that is, FDMA, TDMA, and CDMA.

The simple block diagram in Figure 6.8 provides a basic example of a bent-pipe repeater capable of 12 transponders in a single wide area coverage beam. Repeaters of this type are being built, launched, and used at the time of writing of this book and therefore will be available for service well into the next millennium. They represent the starting point for all repeaters, which must retain all or part of this structure.

While the bent pipe is the most versatile approach, it provides limited options for designers to optimize performance for a particular service. For example, a bent-pipe transponder has no intelligence to permit routing of channels between Earth stations. Nor can it help improve the throughput and error rate performance by restoring signal quality prior to retransmission to the ground. As the link budget clearly shows, satellite impairments and link noise are directly additive to the service during the bent-pipe turn-around process.

This brief introduction to the bent-pipe repeater gives some of the top-level guidelines for the payload designer. In Section 6.3, we discuss the new options for designing and employing digital processing repeaters. Many of the techniques can be combined with those of the bent pipe to realize greater overall service capability.

6.3 Digital Processing Repeaters

The service capabilities of the bent pipe have been well established during 30 years of evolution, culminating in large three-axis satellites with up to 100 active transponders. Owing to the limitations of the bent pipe coupled with advances in DSP and solid state devices and the availability of high-speed VLSI chips, digital processing repeaters are entering the satellite marketplace in greater numbers. Satellite communications engineers generally have applied those technologies on the ground to improve the cost and performance of Earth stations and user terminals. Examples include the digital modems found in data VSATs, DVB encoders and receivers used in DTH TV systems, and digital echo cancelers and transmultiplexers that support a variety of fixed telephone applications over satellites. We are on the verge of what can only be described as a revolution in the design and potential application of commercial satellite networks that integrate ground segments with these and newer DSP repeaters.

In years past, digital processing was limited to RF or IF switching for use in wideband TDMA systems such as INTELSAT's SS/TDMA. That type of service depends on large, expensive Earth stations that can develop the quantities of information traffic to justify the investment on the satellite. Current generations of MSS spacecraft have processing payloads that introduce narrowband channelization services for digital telephony and low-speed data. An important constraint in designs that exist at the time of this writing is the maximum processing speed and sampling rates allowed by existing DSP technology. We are awaiting the wideband versions that offer true multimedia applications to low-cost Ka-band VSATs, where the user can access the network directly over the satellite at megabit-per-second speeds. The fundamental approach to digital processing is to actively alter the signals in the satellite before they are retransmitted to the ground. That involves such functions as narrowband channel selection and routing, switching of time division channels or packets, decoding and recoding of data streams for onboard FEC, and complete reformatting of data for optimum downlink transmission. The technologies to accomplish that include high-speed A/D conversion, fast Fourier transform (FFT) computation, digital filtering, data speed buffering, and coding. It is difficult to try to describe all the possibilities in a short chapter; in fact, the technologies continue to evolve rapidly. As a result, what we present here is a snapshot at the time of this writing.

In this section, we review three digital processing payload architectures that are already in commercial service. They represent the early demonstrations of what such space-based digital switching and routing systems are capable of doing. They include the SS/TDMA repeater using wideband RF or IF switching, the narrowband digital processing repeater with channel routing and digital beam forming, and the demod-remod repeater, which allows complete reformatting of signals within the payload. In addition to the general overview, we provide a working example of each approach.

6.3.1 Satellite-Switched TDMA Repeaters

SS/TDMA is an offshoot of the wideband TDMA system discussed in Chapter 5. The first development in this area was performed at COMSAT Laboratories in the late 1960s as a way to expand the INTELSAT network without increasing the number of working satellites. Later, INTELSAT introduced SS/TDMA on the Intelsat 6 satellites launched in the early 1990s. The principle of the standard TDMA network is that Earth stations transmit their traffic at preassigned times according to a periodic frame format. That type of system is straightforward for an area-coverage beam in which every station can receive the burst transmissions from every other station. Also, it is convenient to identify one station to transmit the reference burst to be used for aligning and synchronizing the entire network. In doing that, the transponder power and bandwidth are nearly fully utilized.

There is a difficulty in transferring TDMA directly to a multibeam satellite that services different regions. An example of such a coverage is shown in Figure 6.10 for a system with 22 beams and frequency reuse in each beam (polarization isolation is used in overlapping and closely-spaced beams). The latter point means that the repeater will perform RF switching because the information is switched from beam to beam without additional frequency translation. Therefore, the frequency plan for the satellite contains only one transponder channel, with a bandwidth roughly equal to the total allocated spectrum. An ideal system would have 500 MHz of bandwidth in each beam, which gives a total bandwidth on the order of 11 GHz (e.g., 22 times 500 MHz). That can be converted into a data rate by assuming the use of QPSK at a bandwidth efficiency of 1.67 bps/Hz. The total capacity theoretically is 18 Gbps, assuming zero guardtimes. In actuality, guardtimes are needed for the TDMA frame and to allow for switch transitions by the onboard satellite switch.

The SS/TDMA repeater that goes with that beam pattern is given in Figure 6.11 (eight of 22 beams are shown). Uplink and downlink beams are paired to allow traffic to be properly coordinated between regions of coverage. The network allows Earth stations to connect through the repeater either to other beams or to stations in the same beam. Same-beam transfer is necessary

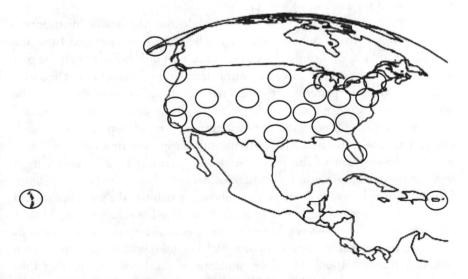

Figure 6.10 SS/TDMA satellite coverage with multiple spot beams.

Individual receive spot-beams, one per region

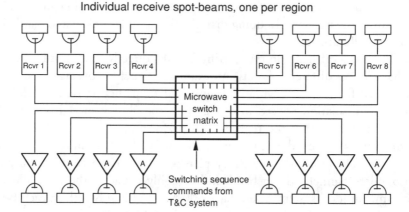

Individual transmit spot-beams, one per region

Figure 6.11 Configuration of an SS/TDMA repeater.

to permit each station to verify its transmissions and for traffic transfer among Earth stations in the same beam (if required). The uplink portion of the repeater consists of a dedicated wideband receiver for each beam, connecting the spectrum to the input of the TDMA switch. It is the role of the switch to set up connections between beams, during which time stations on both sides can transfer bursts of information. That is illustrated in the timeframe diagram in Figure 6.12 for a subset of four beams.

The actual switching pattern, which defines the traffic distribution matrix, is stored in an onboard memory. The pattern is uploaded from the ground control station using a data communication link (which might be part of the command system). Because traffic transfer requirements change over time, the switching pattern can be altered by uploading new data into the switch memory. One approach is to maintain several patterns in the onboard memory so a new configuration can be activated in rapid response to demand. The pattern also could be altered automatically at specific times of day.

Synchronization of the overall network is critical because bursts must arrive at the correct time to be transferred between beams without collisions. That, of course, is easy to do when the onboard switch ties the same uplink and downlink beams together. Under that condition, we have 22 separate TDMA networks operating. However, it gets more difficult when the beams are connected across so that a station's uplink will not be repeated back to its own downlink beam. Stations, therefore, transmit in the blind, except that they remain synchronized with the satellite switch. In addition, stations must be

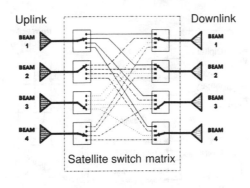

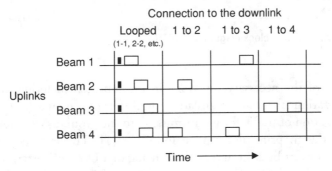

Figure 6.12 SS/TDMA using a time division satellite-switching matrix located in the satellite repeater.

given strict time slot assignments within the overall frame to ensure that bursts do not collide.

So far, the only working SS/TDMA networks were implemented on TDRS, Intelsat 6, and the Advanced Communications Technology Satellite (ACTS) of NASA. The IF switching payload for the ACTS is illustrated in Figure 6.13, and a block diagram of one of the experimental SS/TDMA Earth stations is shown in Figure 6.14. The network operates at an aggregate data rate of 250 Mbps and is capable of a wide range of services.

6.3.2 Flexible Routing and Digital Beam Forming

The SS/TDMA system discussed in Section 6.3.1 retains an analog type of design throughout the signal path. The digital aspect is simply that the path is switched in time to allow Earth station bursts to be routed from beam to beam. In contrast, digital processing for routing and beam forming means that the signal spectrum first must be digitized through an A/D converter before anything meaningful can be accomplished. That type of system utilizes mathematical

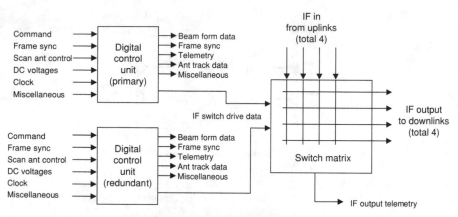

Figure 6.13 IF switch matrix block diagram for ACTS.

representations of the necessary functions in the repeater, including narrow-band channelization, frequency translation, multiplexing, digital beam forming, and aggregation of traffic for retransmission to the ground. More information on DSP design principles can be found in [1]. The last stage in the processor is to convert back from digital to analog in a D/A converter; the output is then filtered in the IF channel, upconverted to RF, and amplified for the downlink.

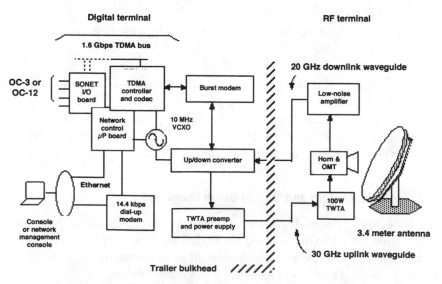

Figure 6.14 Block diagram of an SS/TDMA Earth station used in experiments with the ACTS satellite. (*Source:* Courtesy of NASA.)

A simplified digital processor architecture is shown in Figure 6.15. Like the SS/TDMA repeater design, each uplink beam is provided with a separate wideband receiver. In the case of L band, the bandwidth of the receiver is on the order of only 30 MHz, which is consistent with the available spectrum width and the overall capacity of the processor. The output of the receiver contains the spectrum of user channels at a low IF suitable for processing operations. In the processor, the bandwidth is digitized in the A/D converter to produce a combined bitstream. All functions from this point are performed with mathematical algorithms that represent analog functions like frequency translation, filtering, and routing. The frequency converters are used to move individual narrowband channels to different frequency slots in a time division stream. Channels can be routed using time division techniques similar to what is done in a conventional PBX; that is indicated in Figure 6.15 by vertical arrows. Next, the math is performed to convert the channels to the final frequency slot assignments and the resulting groups are transformed back to analog spectra with D/A converters. Final frequency conversion and amplification are then performed in the conventional manner.

The fact that the bandwidths involved are relatively narrow in comparison to bent-pipe satellites reflects the fact that the first generations of these systems are intended to support low-cost user terminals, particularly the handheld variety used for mobile telephone service. The actual transmissions can use FDMA or narrowband TDMA, as discussed in Chapter 5. Current technology probably can support CDMA as well but at a reduced capacity. That is because of the added complexity needed to despread each CDMA signal onboard the satellite.

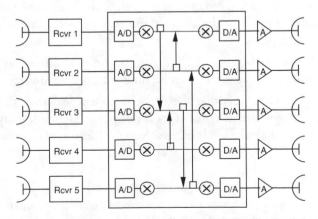

Figure 6.15 Example of a digital processing repeater providing narrowband channel routing.

Digital processor architectures in development at the time of this writing have been assembled with the following needs in mind:

- Narrowband channel routing, to allow users to transmit information within channels in the range of 50 to 200 kHz bandwidth. Within the constraints of this channel, any modulation or multiple access method can be supported.

- On-demand reconfiguration, to provide the network operator with the ability to change the channelization assignments in response to changing demand.

- Beam forming, multiple and flexible, to remove the restriction on the number of beams and their shape imposed by physical feedhorns. This requires that a given channel be able to drive multiple feed horns, after the replicas have been adjusted properly in amplitude and phase to produce the correct beam shape and gain.

Whenever analog signals undergo the A/D process, we can anticipate that there will be some distortion and information loss. These can be held to acceptable levels if the following constraints are addressed correctly (a detailed discussion of these characteristics is beyond the scope of this book):

- Satisfactory signal structure (multiplex, modulation, and multiple access). They must satisfy the requirements of the link as well.

- Bandwidth, which determines the size and complexity of the processor. Ultimately, it is a question about how the required processing functions can be performed with the current state of the art.

- Processor timing, speed, and complexity, which are constraints that are difficult to assess in the abstract and that require a careful analysis by relevant experts.

- Impairments, because the ideal processor that causes no significant impairment to signals cannot be built with real hardware.

- Dynamic range, because the processor must deal with the full spectrum of signals that enter at its input ports. Dynamic range refers to the range of power levels of these signals. It can cover as much as a 20-dB range or more in a typical application.

The needs on the one hand and the constraints on the other bound the design of an adequate processor that is also buildable through current technology. Systems engineers have to work closely with processor engineers so that all relevant factors are properly considered and that a practical design is developed. Any reduction in capability along the way should not pose a risk to the overall project. That can be handled only by proper management of the design.

As an example of a typical digital processing repeater design, consider the block diagram in Figure 6.16. The diagram represents a European developmental project for an MSS spacecraft that provides links between mobile users and a fixed gateway Earth station [2]. In the block diagram, the uplink at Ku band is fed through a wideband receiver to an A/D converter, followed by a demultiplexer in digital form. That, in turn, drives a memory switch that is used to separate out the individual frequency channels for further processing. The digital beam-forming element modifies the channels through complex multiplication by specific amplitude and phase values to produce the desired downlink beam. The values are uplinked from the network control station and can be changed as required over the satellite lifetime. The remaining functions are used to adjust the power level and translate the frequency to the correct beam channel. A D/A converter restores the analog format before RF frequency conversion, power amplification, and transmission from the antenna system.

Without going into details as to the design and supporting mathematical theory, the channel selection and beam forming are accomplished through the discrete Fourier transform (DFT) technique. That allows the processor to perform its functions in the frequency domain using mathematical representations of filters and mixers without using actual hardware. In that approach, the DSP designer must consider these factors:

- The number of bits used in A/D conversion;
- The required number of A/D and D/A operations;
- The size of the DFT used in processing, for example, the number of points, amplitude granularity, and sampling rate;

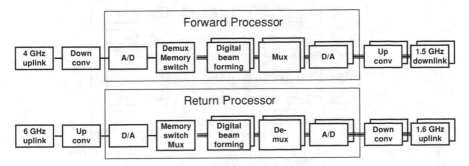

Figure 6.16 Example of a digital beam forming mobile payload.

- The window overlapping factor, to smooth the data for DFT processing and to minimize the amount of computation and memory; must be optimized for the types of channels on which the processor operates upon;

- The time window, which determines the amount of resulting ripple and inband interference;

- Techniques to ensure that all necessary operations are properly synchronized within the relevant stages of the processor.

Traditional payload engineers are accustomed to working with analog parameters, such as those discussed in previous sections. The factors listed in this section tend to be more difficult to assess and optimize. However, the situation may resolve itself as the tools used to simulate and synthesize the necessary circuitry are improved and enhanced. The day will come when digital processors can be designed using highly integrated software tools that accelerate the processes and perform the necessary validations in real time.

6.3.3 Demod-Remod Repeater

The last category of digital processing payload is one that places the demodulator function at the midpoint in the satellite link. This offers the opportunity to completely restore the digital format of the information and thereby improve signal quality before transmission to the ground. An example of a demod-remod transponderized payload is shown in Figure 6.17. Individual channels are filtered as is done in a bent-pipe design. From that point, the uplink is demodulated on an individual carrier basis. Its output in digital data form can

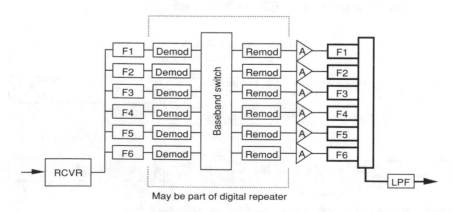

Figure 6.17 Demod-remod repeater configuration (transponderized design).

be switched and remultiplexed before remodulation and amplification. The last step is conventional output multiplexing to combine the powers of all the transponders prior to delivery to the transmit antenna.

This particular approach really provides only one basic benefit of the demod-remod payload, namely, the bits are restored onboard the satellite. That prevents the uplink noise and interference from being directly transferred to the downlink. Instead, those impairments simply introduce bit errors into the data. While such errors cannot be eliminated on the satellite (except through FEC decoding), the result on the final received data has fewer errors than if demodulation were done only on the ground. The maximum benefit of almost 3-dB improvement occurs if the uplink and downlink C/N values are equal. A substanitial difference in C/N values reduces the benefit essentially to zero. The demod-remod functions could be performed by discrete hardware or with a DSP repeater.

Another approach to applying demod-remod is to add complexity to the processing of information onboard the satellite. Once the uplinked channels are reduced to bits, the data can be decoded, demultiplexed, and switched as in a packet switching node or a PBX. The information can then be completely reformatted into a different multiplexed frame or packet stream and then modulated into a wideband digital carrier for the downlink. Additional FEC can be included as part of the process.

An effective use of demod-remod is illustrated in Figure 6.18. The Skyplex multiplexer was developed for the Eutelsat Hot Bird 4 and intended for use with the European DVB video transmission standard [3]. This pioneering payload design provides some unique benefits for digital video transmissions to low-cost DVB receivers. As shown in the figure, the payload can receive up to

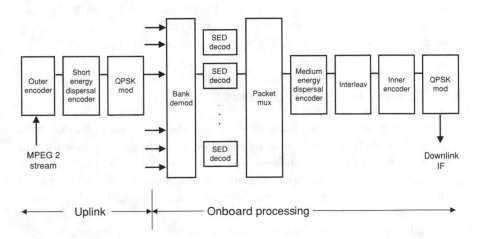

Figure 6.18 The Skyplex repartition of DVB channel.

six individual video carriers, each modulated by a DVB bit stream. The payload first downconverts the spectrum to IF, where it is applied to the digital processor for demodulation. The digital bitstream from each uplinked carrier is then rearranged to form a single TDM baseband that contains all six channels at one time. That, of course, is at an elevated data rate equal to six times that of an individual TV channel. The last step is to encode the data using FEC and to employ QPSK to modulate the single downlink carrier.

The Skyplex multiplexer payload allows the uplinks to come from six different Earth stations, thus providing additional flexibility for remote origination of programming. Of course, they are subject to being from a limited set of bandwidths and frequencies because of the internal IF architecture of the onboard unit. An alternative TDMA mode allows stations to transmit the channels in burst format. The payload has the capability to demodulate and reformat the channels on a single TDM carrier. In that way, standard DVB receivers can be used by subscribers, who cannot tell that the satellite performed any of those complex functions. In effect, the system behaves as if the satellite was a simple bent-pipe design. At the time of this writing, there is an open question on how the DVD program identification (PID) is to be created and maintained for the combined downlink.

The features of the previous digital processing payloads can be combined and expanded to provide new services not yet available to the public. In particular, an advanced processing payload can be used to deliver broadband multimedia services on a two-way basis. In conjunction with Ka-band spectrum, a digital processing payload might operate with high-speed packet switching using ATM, the international standard based on the 53-byte packet (called a cell). ATM already has been accepted for many local area and wide area networking systems and could well be the future standard for all high-speed telecommunications services on terrestrial networks. This type of satellite could receive, detect, and route ATM cells over a multibeam antenna system.

6.4 Standard Repeater Elements

Modern communication payloads have a wide variety of types of subsystems and elements, each of which must be relied on to perform an important and sometimes critical function. The previous sections of this chapter reviewed the requirements and the architecture of analog and digital payloads, relating their performance to the needs of the overall mission and applications. In this section, we go into detail on the building blocks of the repeater so that readers will have a full appreciation for the design and interaction of these important pieces of space hardware. We follow the functional sequence of major components in the typical repeater block diagram shown in Figure 6.4. Readers not needing this level of technical detail can proceed to Chapter 7.

6.4.1 Wideband Receiver

Satellite repeaters have as their first active task the job of amplifying the desired uplink frequency band, a concept that perhaps is unique to commercial satellite design. As shown in the block diagram in Figure 6.19, the range of channels in the uplink frequency band appears at the input; the same bandwidth appears at the output, but translated to the downlink frequency band. In addition, the power level of the signals has been increased greatly, by an order of 50 to 60 dB (i.e., amplified in power by a factor of 10^5 to 10^6), depending on the particular design. The difference, $F_u - F_d$, is a fixed value offset between uplink and downlink center frequencies. In C band, for example, the difference frequency is 2225 MHz, and the same difference applies for every transponder of the repeater. In the receiver, translation is produced through downconversion in the mixer. There are two inputs to the mixer: the uplink

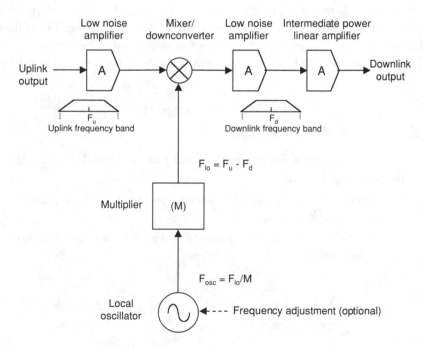

Figure 6.19 Block diagram of a typical wideband communication receiver that amplifies the 500 MHz uplink band and translates it to the downlink frequency range without modification.

frequency range and a pure single carrier from the LO. The frequency of the local oscillator, F_{lo}, clearly is the difference frequency. It is convenient to use a device called a frequency multiplier to generate F_{lo} as a harmonic from a crystal-controlled oscillator operating at a much lower frequency. For our C-band example, the oscillator typically runs at 139.0625 MHz, which means that the multiplication factor has to be 16.

Because multipliers and mixers are nonlinear devices, they generate a variety of harmonics, spurs, and intermodulation products in addition to the desired frequencies just described. Therefore, filters and frequency traps are inserted in the input and output legs of the mixer to preclude propagation of such products through the receiver or backward toward its input.

In addition to frequency translation, the receiver must provide a great deal of amplification to overcome the uplink path loss. Approximately 50 to 60 dB of gain is easily obtained with several transistor amplifier stages in tandem, as illustrated in Figure 6.20. An important consideration in satellite receiver design is the thermal noise, which is produced in all active and passive microwave devices. The term *low noise* (discussed in Chapter 4) is used to describe an amplifier with specially designed and selected transistors that have the property of minimizing the generation of internal noise. A tandem sequence of amplifiers, such as that in Figure 6.20, provides significant noise reduction. In the following equation, we see that the total noise temperature of the sequence is calculated based on the individual noise contributions, reduced by the gain of preceding stages:

$$T_{re} = T_1 + T_2 / G_1 + T_3 / G_1 G_2 + \dots$$

where

> T_1 and G_1 are the noise temperature and gain (as a ratio), respectively, of the first amplifier stage;
>
> T_2 and G_2 are the noise temperature and gain (as a ratio), respectively, of the second amplifier stage;
>
> $T3_3$ and G_3 are the noise temperature and gain (as a ratio), respectively, of the third amplifier stage, and so on.

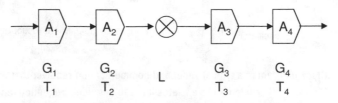

Figure 6.20 Use of amplifier stages in a typical wideband receiver.

This type of analysis shows that the total noise temperature is dominated by the noise contribution of the first amplifier stage. The contribution of all the stages that follow is reduced by a factor equal to the combined gain of the stages that precede it, expressed as a true ratio and not in decibels. If one of the stages provides frequency conversion in a mixer, we can anticipate a loss instead of a gain for that particular stage. That moves the factor from the denominator into the numerator of the equation. For example, inserting a downconverter after the second amplifier, as shown in Figure 6.20, results in the following modified equation for the receiver noise temperature:

$$T_{re} = T_1 + T_2/G_1 + LT_3/G_1G_2 + LT_4/G_1G_2G_3 + \ldots$$

where L is the conversion loss factor (L > 1).

The conversion loss has the unfortunate effect of amplifying the noise power contribution of the subsequent stages. There is a tradeoff in receiver front end design, in that there is a desire to increase the gain ahead of the mixer to reduce its impact on total noise, countered by the concern that the higher signal level produced by the gain can increase the spur level coming out of the mixer. For that reason, there are at most two stages in front of the mixer stage.

Following the mixer typically are two separate transistor amplifiers, the first still being designed for low noise. The final stage increases the power level of each of the channels sufficiently to drive the input to the TWTAs or SSPAs, as appropriate. While these stages do not add significantly to system noise, the relatively high power level does place considerable demand for linearity on the last stage. The intermodulation products produced in this wideband stage, which increases as the output power demand increases, must be kept at a minimum or the transponders will be subject to unwanted RF IMD interference. Because all carriers pass through this stage, it must have a peak power output capability considerably in excess of actual requirements (e.g., by 20 dB or more).

The elements of the receiver along with the associated power supplies are housed in a metal box to protect the components and facilitate placement on the repeater equipment shelf of the satellite. Because frequency translation is influenced by the stability of the internal LO, the typical receiver includes a thermostatically controlled heater. That minimizes the variation of temperature directly affecting the crystal reference, particularly as the satellite passes into and out of eclipse. A properly designed and compensated receiver produces a shift of less than 1 kHz during the eclipse cycle. More accurate frequency translation may be provided with an external frequency offset control.

6.4.2 Redundancy Switching

The active components of the repeater are the most likely parts to fail and thereby reduce or cease service performance. Included are the receivers just described, along with transponder amplifier and gain control devices. To counter that, all repeater designs include backup in the form of redundant components and subsystems. They usually are turned off and held in reserve until needed, at which time they are individually switched on by ground command. Manual operation of redundant equipment has been the practice and likely will continue until viable approaches for automatic restoration are introduced. (An exception is the altitude control system, described in Chapter 8.) In the following discussion, we review the RF switching devices that are used to replace a failed active unit by an appropriate spare unit.

Techniques for switching active and spare units consider two important attributes. First, the switches must pass microwave signals with the minimum amount of loss and distortion. Second, the switching scheme must permit the repeater to be reconfigured to allow access to spare units without disrupting the remaining operating equipment. Switches also must be reliable because a given switch often represents a potential single-point failure mode (i.e., failure of a switch could render a transponder or group of transponders useless). Switches that use coaxial connectors (similar to those used for interconnecting video equipment) are appropriate for RF power levels under 50W and where some measurable loss is acceptable. Waveguide switches, although heavier and bulkier than coaxial switches, are necessary for high power levels and low-loss applications, particularly at Ku and Ka bands.

The following paragraphs review the currently accepted switching devices and redundancy configurations, showing how they are used in the repeater. Because this subject is important but usually ignored in the literature, readers may find the discussion valuable. The illustration that is provided with each switch shows the symbol, the schematic diagram, and a typical application of the particular switch installed in a repeater.

6.4.2.1 S Switch

The S switch, shown in Figure 6.21, is the simplest and most basic form of a redundancy switch. The single coaxial input can be routed to either of two coaxial outputs; thus, it is a single pole–double throw switch. The application drawing with two amplifiers (or perhaps receivers) shows how a backup unit can be used to replace a failed or degraded operating unit. This scheme is often called two-for-one redundancy, meaning that two identical units are carried to provide one operating channel of service (not to be confused with the designation "one for one" used in terrestrial radio systems for the same configuration). Note how the second S switch is reversed to connect from two paths back to a single output. Two-for-one redundancy with S switches is required whenever the devices must cover the same frequency range or when full redundancy is needed to meet reliability requirements.

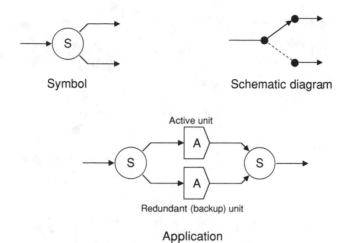

Symbol · Schematic diagram

Active unit

Redundant (backup) unit

Application

Figure 6.21 Definition and application of the S switch, which is used to switch in redundant equipment.

The S switch can be implemented with two different technologies. The most common approach is to use coaxial input and output connections and a mechanical wiper arm, similar to the A-B switch used to connect a TV receiver to either of two cables. Mechanical motion is produced by an electronically driven actuator controlled by ground command. A type of nonmechanical S switch uses ferromagnetism to direct microwave energy from the input to the selected output. It operates similarly to a circulator and is controlled by pulsing a magnetic coil to reverse the direction of the internal magnetic field. Ferrite switches often are used at the input to C-and Ku-band receivers because the ferrite cannot fail as an open circuit even if the control circuit stops in midposition. That failure mode introduces a 3-dB loss on both switch outputs, requiring only that one of the two units not be active.

6.4.2.2 2/3 Switch

As shown in Figure 6.22, two S switches can be packaged together to provide a versatile switch with two ports on one side and three on the other. If the two-port side is the input, it is called a 2/3 switch; with the three ports, it is a 3/2 switch (it depends only on how the switch is used in the repeater). The fundamental application, shown at the bottom of Figure 6.22, is to allow three-for-two redundancy, where one spare unit can back up two operating units. An extension of this method uses a pair of 2/3s and 3/2s along with a pair of S switches to implement a five-for-four redundancy scheme. Many C-band satellites use five-for-four redundancy for sparing of TWTAs. The spare amplifier in

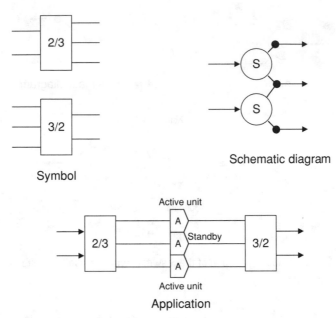

Figure 6.22 Definition and application of the 2/3 switch, which is used to share redundant equipment.

such cases must be capable of operating over the bandwidth of each of the transponders that it is sparing.

6.4.2.3   C Switch

One of the more innovative arrangements is the C switch, shown in Figure 6.23. The schematic diagram in the upper right corner shows how the two inputs and two outputs can be connected. In the straight-through position, the upper input goes to the upper output and likewise for the lower input. However, if the switch is commanded into the cross-over position, the upper input goes to the lower output and the lower input goes to the upper output, the only two positions in this particular design. The C switch uses coaxial input and output ports and therefore is suitable for low-level signals (i.e., under 50W). Alternatively, waveguide C switches are available that can handle hundreds of RF watts.

A typical application of the C switch, shown at the bottom of Figure 6.23, allows two operating units to be interchanged. If the upper amplifier were to fail, it would be possible to restore that particular channel of communication by commanding both C switches from straight-through into cross-over. Obviously, communications capacity is lost, but at least there is the option of

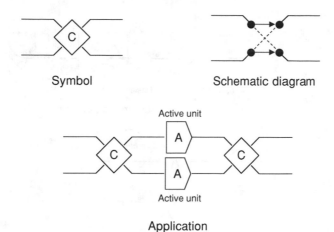

Symbol   Schematic diagram

Active unit

Active unit

Application

Figure 6.23 Definition and application of the C switch, which is used to interchange equipment or switch paths.

maintaining the more important of the two channels. The same switch also can be used as a routing device to rearrange communication paths within a repeater.

6.4.2.4 R Switch

The last switch to be described is designed to increase the ratio of active units to spare units (i.e., to anticipate a lesser need for spare units in relation to a larger quantity of operating units). The demand for the R switch came about at Ku band, where coaxial switches are not appropriate for high power transmission. That introduced the waveguide model, based on similar designs used in ground stations. The advantages of a waveguide R switch are low transfer loss, nearly perfect isolation, and high power capability. At Ku band, the dimensions and weight of an R switch are reasonable for a typical payload. The situation at C band is different because the corresponding waveguide R switch is three times as large and up to nine times heavier. The solution was to develop a coaxial R switch, which was first introduced in the late 1980s.

The device is called an R switch (R standing for redundancy) because it is used to restore a failed amplifier with an available spare by providing four separate waveguide ports, as shown in Figure 6.24. The schematic diagram of the switch is the same as its cross-sectional view, showing a metal disc having straight and curved channels carved into it. The channels provide waveguide paths inside the switch for microwave energy to pass between the four ports. It has been called a baseball switch, an obvious reference to the appearance of the cross-sectional view. The schematic diagram shows three possible positions: two

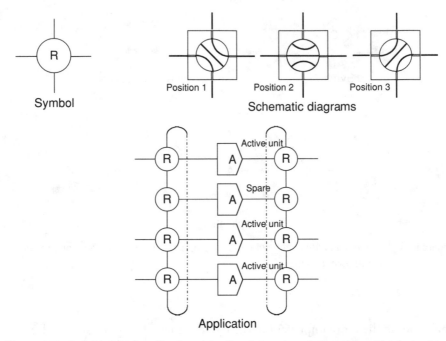

Figure 6.24 Definition and application of the R switch, or baseball switch, which is used to switch in redundant units or between waveguide connections.

cross-over positions and one straight-through position. In some applications, one port is terminated with a resistive load to absorb the full power of an amplifier if the switch is commanded to disconnect the active element.

The redundancy scheme shown in Figure 6.24 is called ring redundancy. A three-channel system is depicted with the R switches in the straight-through position (position 2 in the schematic diagram), placing the backup amplifiers completely out of the circuit. If the upper amplifier were to fail, then the spare amplifier could be used to restore service. To use the spare, the input and output R switches for upper and spare amplifiers are put in positions 1 and 3, respectively. That causes microwave signals to be routed from the upper channel away from the failed amplifier and into the upper spare. The ring is created by continuing the pattern of R switches and amplifiers upward and downward from the four sets shown in the diagram, until the chain connects back into itself.

The R switch designs and configurations are typical of what is used in the industry, but this discussion is not intended to be all inclusive. There are many variations to the schemes described here, and new switch technologies are appearing all the time. You can use this framework to understand the specifics of a given repeater and switch design.

6.4.3 Waveguide Filters and Multiplexers

Bent-pipe satellite repeaters can provide several wideband channels of communications (i.e., transponders) when the individual channels are separated from each other on the basis of frequency. As discussed in Section 6.1.4, the input multiplexer separates the channels for amplification, and the output multiplexer combines them again for transmission through the downlink antenna. The typical multiplexer contains a microwave filter for each individual channel, the filter being tuned to pass the full bandwidth of that channel and to reject (or reflect) all others. In addition to microwave filtering, repeaters that operate on narrowband channels such as applied to MSS can use alternative technologies that filter at much lower frequencies. An example is surface acoustic wave (SAW) filters, which can provide excellent channel filtering for bandwidths down to a few kilohertz. Our focus in this section, however, is on the microwave variety.

The input multiplexer is the most critical repeater element with regard to creating the transponder channel. Figure 6.25 is an example of an input multiplexer taken from a typical C-band satellite design. There is a total of 12

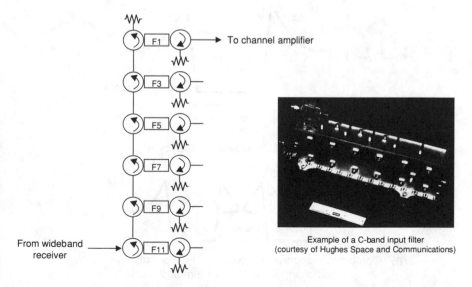

Example of a C-band input filter
(courtesy of Hughes Space and Communications)

Figure 6.25 A six-channel input multiplexer composed of individual microwave filters.

channels, each with 36 MHz of usable bandwidth, which are separated into even- and odd-numbered groups. We will explain why the channels are split that way later in this section. The hybrid, at the bottom left is used to split equally the total power of the downlink spectrum, transferring all 12 channels to each output leg. An input multiplexer typically uses circulators to route the entire spectrum of channels to the filters in a sequential manner, each filter being used to drop off one of the channels. One nice feature of using circulators is that they essentially isolate the filters from each other, which from a technical standpoint tends to simplify multiplexer design, manufacture, and final tuning of the complete assembly.

We have been describing the action of a microwave filter without being specific about how it affects the signals that pass through it. Figure 6.26 gives a qualitative view of the frequency response of the filter by itself and of the combined action of filters in a multiplexer. The characteristic of a single filter (shown at the top of the figure) includes a passband of sufficient bandwidth to allow all the signals within the particular transponder to be transmitted with minimal distortion. A perfect filter would have a flat passband and infinite loss for all frequencies that are outside the passband (referred to as the out-of-band response). Real filters are not perfect, so there are variations from the ideal. For example, there is some residual loss, called insertion loss, within the passband. The perfectly flat insertion loss characteristic shown is impossible to achieve,

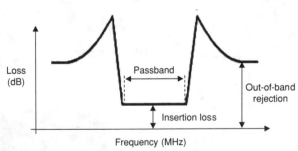

(A) Single input bandpass filter frequency response

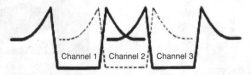

(B) Combined response of a two-channel input multiplexer (odd channels on the same multiplexer)

Figure 6.26   Requirements for input multiplexer.

and we must be satisfied with variations in loss of typically ±0.5 to ±1 dB across the passband. That variation appears as a rounding of the edges of the passband, a ripple within the passband, or a combination of both. The out-of-band portion of the characteristic shown in Figure 6.26 is more representative of real filters, where the loss increases rapidly starting at the edge of the passband until a point of maximum loss is reached. Beyond that point, the out-of-band loss can decrease but generally stays above some minimum level, which is called the out-of-band rejection.

A well-designed filter has low insertion loss, a reasonably flat passband, rapidly increasing attenuation at the edge of the passband, and high out-of-band rejection. Another important property is delay distortion (also called group delay), which was mentioned as an impairment in Chapter 5 and earlier in this chapter. Taken together, those characteristics represent tradeoffs in filter design and construction. For example, rapidly increasing attenuation at the edge of the passband requires that many filter sections (which are resonant cavities) be used, which increases both the insertion loss and the delay distortion. (See Figure 5.31 for examples of both characteristics for a real input filter.)

Combining several filters into a multiplexer implies the overlaying of their individual frequency responses, as shown at the bottom of Figure 6.26. The two filters in that ideal multiplexer are each tuned to different center frequencies corresponding to two transponders. The slight gap between the passbands (the guardband) is necessary to prevent repeater multipath, which occurs when a signal on the edge of one channel finds a second path through the adjacent channel filter. If not adequately attenuated, repeater multipath can cause problems in the operation of the communications payload, such as tilted transponder passbands and frequency-selective valleys in the antenna coverage pattern.

Another problem is multiplexer spurious response, which is the leakage of power through a multiplexer on undesired frequencies. Again referring to Figure 6.26, the amount of loss at the point where the filter skirts cross over is important to the design of the multiplexer. The greater the loss at that point, the easier it is to match the actual performance of the multiplexer to theory. The loss at the cross-over can be significantly increased by splitting the channels into two groups, with each group on one multiplexer having every other channel. That is why there are odd-numbered frequency channels in the actual multiplexer design shown in Figure 6.25. One group would consist of the odd-numbered frequency channels (the darkened curves at the bottom of Figure 6.26), while the other would consist of the even-numbered channels (the dotted curve). That greatly improves the cross-over performance of the multiplexer because an entire empty channel now lies in the gap.

The actual filters used in the input multiplexer have evolved over the past 20 years. In the first generation, waveguide cavity filters were made of rectangular waveguide sections, which tended to be heavy and somewhat less stable over a typical spacecraft temperature range. Second-generation input filters were constructed from coaxial structures, which are much more compact and can include temperature compensation. An example of such a filter is shown in Figure 6.25. Materials used in both cases evolved from invar, a very stable metal alloy, to aluminum with temperature compensation, to even graphite-plastic composite in some cases. Another improvement is to reduce the group delay through delay equalization built into the filter itself.

The output multiplexer is constructed a little differently and without the use of a lossy circulator chain. Each filter is attached to a section of waveguide called a manifold. Without circulators to provide isolation, the filters must be positioned in accordance with wavelength along the manifold and manually adjusted (tuned) as an integrated system. A schematic drawing of a circular waveguide cavity filter is shown in Figure 6.27. The corresponding block diagram of the multiplexer is shown in Figure 6.28. This approach is less effective against multipath and provides very little out-of-band filtering of intermodulation products or other undesired radiation. This situation is driven by the need for low insertion loss, which demands fewer filter sections and a more gentle slope of the out-of-band rejection.

Improvements in filter design have produced the contiguous multiplexer, which contains filters on adjacent channels (e.g., without the odd-even split). It has the advantage that there is a single output port to connect to the antenna, as opposed to the two output schemes found with the odd-even split. On the other hand, the contiguous multiplexer has the disadvantage of limited control

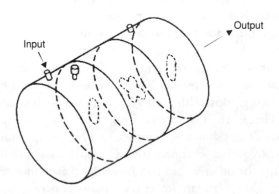

Figure 6.27 Schematic diagram of a circular-waveguide output filter.

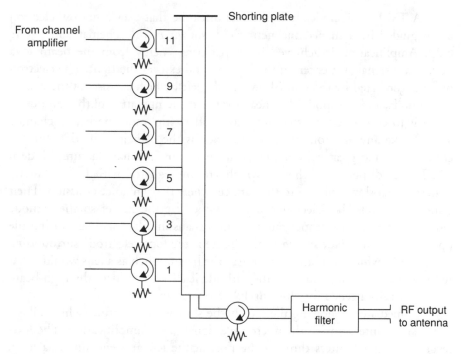

Figure 6.28 A six-channel output multiplexer configuration.

of the cross-over level and frequency response and so has a somewhat poorer overall response than the noncontiguous multiplexer. The performance advantages of the single connection to the antenna are covered in Chapter 7.

6.4.4 Traveling Wave Tube Amplifiers

While the conversion to solid state electronics is nearly complete in the telecommunications industry, there remains one application for vacuum tubes in space. The TWT continues to demonstrate its qualifications as a reliable, efficient, high-power device for use as a transmitter at microwave frequencies. The TWT is the tube itself, while the TWTA is the TWT and complete amplifier with its associated power supply. The two are provided as a set because the necessary dc high voltages must be applied on an individual select basis. In some repeaters, two TWTs are paralleled and only one power supply is provided for the matched set. An alternative term for the power supply is the electrical power conditioner (EPC). As discussed later, the EPC must be built for long-life operation and efficiency of conversion from low-voltage to high-voltage dc.

A TWT is a physically linear vacuum tube that uses a narrow electron beam guided by a shaped magnetic field, shown in the schematic in Figure 6.29. Amplification is achieved by transferring energy from the beam to a microwave signal as it enters at the left end and exits at the right. The electron beam is generated by the gun (shown at the left), which consists of the heater, the cathode, and the anode. The heater raises the temperature of the cathode to cause it to emit electrons that are drawn off by the more positively charged anode. Like any electron tube, the cathode is the principle life-determining component, and great pains are taken in its design and manufacture. Modern TWTs use "dispenser" cathodes, which are impregnated with low-emissivity materials that slowly diffuse to the surface, where they enhance emission. Their temperature must be selected and maintained by the heater so the cathode emits electrons at the proper rate without exhausting the materials during the operating lifetime. Not shown in the figure is the focus electrode surrounding the cathode, which aims and compresses the beam, much as a lens would. That reduces the stress and strain on the cathode itself yet produces the high beam current density needed for efficient TWT operation.

Immediately following the gun is the slow wave structure, which allows the beam to interact with the microwave signal under amplification. The low-level input signal enters through the port at the left and encounters a long, tightly wound coil called the helix. Microwave energy in free space travels at the speed of light, which is considerably in excess of the speed of the electrons in the beam. The microwaves, which are going around the spiral helix, have their axial velocity slowed down to a little less than that of the electron beam. The beam is kept tightly formed within the helix by a string of permanent ring magnets. Electromagnetic interaction of the microwave signal, which is a wave phenomenon, with the beam causes bunching of the electrons as they move along. Ultimately, that bunching transfers energy back to the microwave signal, adding considerable power to it. When the signal reaches the end of the tube,

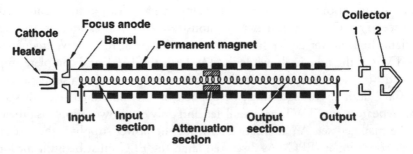

Figure 6.29 Schematic diagram of a TWT.

the electron beam has lost the majority of its energy of motion, while the microwave signal has become greatly amplified in power. The attenuation section at the center of the helix is an added feature to prevent self-oscillation and thereby stabilize TWT operation. The amplification (gain) of the TWT is roughly proportional to its length, which is easy to see because the interaction and transfer of energy increase as the beam and signal travel along the slow wave structure. Efficiency is enhanced by using a tapered helix, which means that the pitch of the helix coil is altered along its length. An output port at the end of the helix couples the microwave power through an isolator to the output filter.

The collector, located at the right end of the TWT in Figure 6.29, is critical to efficient operation. It is the objective of tube operation that energy be transferred from beam to signal, which implies that the velocity of the beam gradually reduces as it moves toward the output end. By the time the electrons reach the collector, they still posess considerable energy and are moving at less velocity than at the gun. Efficient operation is achieved if the electrons are collected in such a way that their velocity is reduced to near zero before they actually hit the collector. That is accomplished by placing a negative voltage on certain collector stages to repel the electrons and thereby slow them down. The use of negative voltage for this purpose is called collector depression and is done with a multiple staged collector, as shown in Figure 6.29.

A typical high-power TWT can deliver over 60% of the dc input to RF output, which means nearly 40% of the dc power must be given up as heat. That can pose a challenge to the TWT and spacecraft designer, so proper measures must be taken. An option is to use a direct-radiating collector that is exposed directly to cold space. It turns out that about half the total heat dissipation of a TWT is at the collector end, so that is an excellent way of dealing with the problem. The rest of the heat must be conducted away through the base of the TWT, either through the equipment shelf or by radiating it away using a thermal doubler. Those aspects of thermal design are discussed in Chapter 8.

The TWT has evolved over the years and can be depended on to operate for 15 years or more. As discussed, the principal life-limiting element of the TWT is the cathode, which is operated at a high temperature to provide an adequate electron beam. The materials and processes used in cathode manufacture are critical to that aim; currently, only three manufacturers produce commercial space TWTs: Hughes Electron Dynamics, Thomson CSF, and NEC. When properly designed, tested, installed, and operated, a space TWT can be depended on to meet mission needs. However, any deviation from proper practice is serious cause for concern.

This discussion is by no means all inclusive, and an examination of the detailed design and operation of space TWTs is beyond the scope of this book.

However, it should give the reader some idea of how a TWT works and generate an appreciation for the sophistication of the technology. Areas of critical importance include the materials and the manufacturing processing techniques used. Those aspects determine the physical properties of the TWT (weight and ruggedness) as well as the lifetime and reliability of the amplifier when operating in orbit.

It is important to say a few words about the power supply that must accompany the TWT for the purpose of providing the various voltages and currents. Examine the typical TWT voltages in Figure 6.30 to get an appreciation for the complexity of the design requirements. The circuitry to obtain the required voltages usually consists of an ac oscillator, high-voltage transformers, rectifiers and capacitor-type voltage multipliers. To provide stable operation of the TWTA, it is vital that the voltages be held constant under all conditions of TWT signal loading and unit temperature and throughout the lifetime of the repeater. Also, transient conditions on the power line coming into the power supply should not be transmitted to the TWT. That is an important consideration if the TWTA will be used in wideband TDMA service. A well-designed TWT power supply will meet those stringent requirements and also have an efficiency (i.e., the ratio of power actually applied to the TWT to the dc input power from the spacecraft power system) in the range of 95–98%. For C- and Ku-band TWTAs, the weight of such a power supply should equal approximately the weight of the TWT by itself.

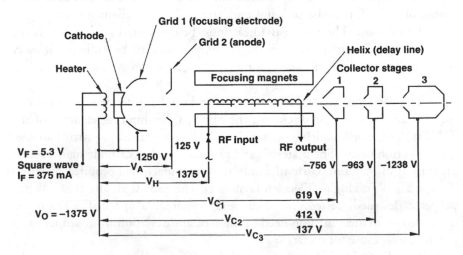

Figure 6.30 The dc voltage requirements for a C-band with a three-stage collector TWT.

6.4.5 Solid-State Power Amplifiers

Historically, the TWT was the only active device on commercial satellites that was not solid state. That was because while transistors have always been found in the digital circuitry used to control the various bus and payload elements, the TWT was the only practical device that could efficiently amplify microwave signals to power levels in excess of 1W. Providing a lightweight power amplifier for the original Syncom with up to 30% efficiency was, in 1960, quite an accomplishment. Keep in mind that the context here is the operation of a microwave repeater in the space environment for many years.

The quest for an all-solid state amplifier led to the introduction of the first devices on the Telstar 3 and Satcom 4 series of satellites. Both of those satellites were intended to break through the nine-year lifetime barrier that seemed to exist due to the wearout properties of the TWT. (Of course, innovations in TWT cathodes and other aspects of tube design have overcome that impression.) SSPAs were introduced in the early 1980s at a relatively low power of 5W to 10W. Linearity of the amplifiers proved to be significantly better than that of a comparable TWT, so SSPAs were targeted to FDMA applications, where they could provide somewhat greater channel capacity. To make the SSPA practical, the key was the GaAsFET, which quickly established itself as a reliable and efficient power amplifier device and which is in use in a number of C-band satellites. SSPAs are also found on L- and S-band satellites used in MSS and DAB services. Ku-band SSPAs are under development for medium-power applications and have been in use for several years in VSAT transmitters.

An SSPA is not a single amplifying device like a TWTA; rather, it is built up from individual amplifier modules. The basic building block, shown at the top of Figure 6.31, is the single-stage microwave transistor power amplifier. The triangle symbol represents a GaAsFET and its associated bias circuitry. Circulators at the input and the output provide considerable isolation (hence, the term isolator) from the rest of the SSPA and repeater. Note how the third port of the circulator is connected to a dummy load, which catches any microwave energy reflected back into the circulator by the amplifier output. Without the isolators, the GaAsFET would tend to produce unpredictable results and may even damage itself.

The basic module can have from 6 to 10 dB of gain and a maximum power output of from 3–10W. Higher levels of gain or power out are obtained by combining stages, as illustrated in the rest of Figure 6.31. Gain is increased by adding stages in tandem, where the individual gains (in decibels) are simply added algebraically. That does not increase the maximum output power capability because all the signal still must come from the last stage; however, the input level can be

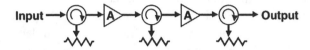

(A) Single stage microwave transistor power amplifier

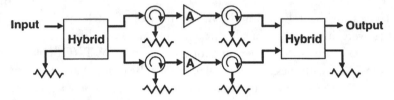

(B) Two transistor stages in tandem

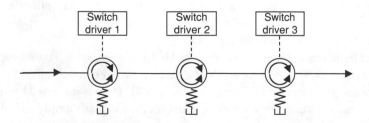

(C) Two hybrid-coupled stages in parallel

Figure 6.31 Three versions of microwave transistor solid state power amplifiers.

reduced to a much lower level, corresponding to that delivered by the wideband receiver or digital processor. Achieving a total gain comparable to that of a TWT (roughly 50 dB) takes up to 10 stages in tandem. They usually are split into two different units for ease of integration into the overall repeater.

The method of paralleling stages, shown at the bottom of Figure 6.31, increases the maximum output capability in direct proportion to the number of stages in parallel. At the input, the signal power is split into two equal parts with a four-port hybrid. Note that the fourth port of the hybrid is not required and is therefore terminated, that is, connected to a passive load. Amplification

Figure 6.32 Three-stage step attenuator using circulator switches and fixed attenuators.

is accomplished by a matched set of single-stage microwave transistor power amplifiers, whose outputs are combined by a second hybrid. The unique structure allows the power to be split and recombined efficiently, without more than about 0.25 dB of total loss at the output. The gain of the parallel combination is equal to that of a single stage, but it can output approximately twice the power since two amplifiers are working together. The key to successful operation is that the hybrids split and combine the signals with the proper phasing and that the individual GaAsFET stages be nearly identical in their characteristics. Not shown in the figure are the circuit components that allow a technician to adjust the balance and phasing of the signals to achieve the desired performance. With everything properly aligned, virtually all the output power will appear at the upper port of the hybrid on the right and none will appear at the terminated port.

The approach to further increase power is to treat the parallel stage as yet another type of stage that can be further paralleled. For example, paralleling two of the parallel stages gives nearly four times the power of the single transistor. However, it is easy to see that the achievement of high power and high gain comes at a price in SSPAs.

There are some unique benefits in using SSPAs instead of TWTAs. GaAsFET devices, being solid state in nature, do not contain a cathode and do not require high voltages to produce the electron beam. Unlike a TWT, which operates in a hot condition for its entire lifetime, the delicate internal structure of the GaAsFET is sensitive to temperature and should be kept rather cool. The high field and power densities inside a GaAsFET can limit life. Another consideration is that an SSPA is somewhat complex in terms of the number of internal microwave connections, which can increase the chance of failure of a random nature. For example, the solder joints or welds where the GaAsFETs are connected into their associated circuits can deteriorate and open up. On the performance side, the linearity of an SSPA generally tends to be superior to a TWT by itself. SSPAs can deliver 20–50% more FDMA channel capacity for the same output power level. TWTAs can be improved in terms of linearity at the expense of using a device called a linearizer (discussed in Section 6.4.6).

The areas where the TWTA continues to shine are in power output and efficiency. The current maximum practical output for a C-band SSPA is approximately 20W, while TWTs with powers in excess of 175W are being built. The efficiency of an SSPA tends to decrease as power is increased (because of the need to parallel stages and suffer losses through the hybrids that are required), while just the opposite is true for a TWTA. In some applications, TWTAs also have been paralleled because that provides greater output power for DTH services to a large footprint.

6.4.6 Transponder Gain Control and Linearization

While earlier communications satellites generally employed fixed gain repeaters, the trend has been toward providing each transponder with commandable gain control. That is illustrated in Figure 6.32, where a typical transponder chain has an attenuator installed at the input to the power amplifier stage. The satellite operator can preset the gain of the transponder to adjust the uplink segment for the desired overall effect on service quality. For example, if the uplink is coming from a large TV broadcast facility, the gain is set to minimum (e.g., maximum attenuation) to ensure both a high uplink C/N and a margin against uplink fade. In some instances, interference can be rejected by reducing transponder gain and raising uplink EIRP accordingly.

Alternatively, if the uplinks are coming from low-power VSATs, it is more desirable to increase gain (e.g., remove the attenuation) to allow the lowest possible transmit power from the ground. Networks of small VSATs need to allow ground SSPA power of 1W or less, which can place a considerable burden on the overall link. That is because there is a penalty in uplink C/N equal to the amount of attenuation provided at the input to the repeater power amplifier. There are three possible ways to compensate for that reduction: (1) directly compensate by increasing downlink C/N through the use of a large hub antenna to receive the signals; (2) provide more uplink EIRP through a larger Earth station antenna and/or HPA; or (3) employ a demod-remod style of repeater.

First-generation commandable attenuators consisted of one or more fixed steps of insertion loss, each providing a few decibels of attenuation. Those are placed on the input side of the TWTA or SSPA. An example of such a device is a ferrite switch with the third port connected to a passive attenuator with half the desired amount of loss. The opposite end of the attenuator is connected to a metal reflector element that bounces the microwave energy back through the loss element. A typical arrangement, such as that shown in schematic form in Figure 6.32, has three step-attenuator stages that allow gain adjustments in steps. The advantage of that approach is that the signal path does not go through any active devices. In the event of a switch failure, the only result is that the amount of fixed attenuation cannot be changed. More recently, the attenuation function has been incorporated into a channel driver amplifier (Figure 6.33) that provides adjustable positive gain to the input to the power amplifier. The device can be switched to an automatic gain control (AGC) mode, which provides a constant drive as the uplink varies over a range of 10 dB or so. Gain control also can be provided with a limiter that effectively clips the input signal and holds it constant. The only consideration with a limiter is that the input must not contain multiple carriers or have significant amplitude

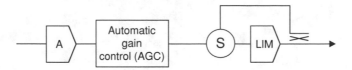

Figure 6.33 Driver-limiter-amplifier block diagram.

variations. That is because the limiter is a highly nonlinear device that would produce undesirable intermodulation products or sidebands.

The service quality of a bent-pipe transponder can be improved through the introduction of a linearizer. By designing the circuit of the driver amplifier to compensate for the property of the SSPA or TWTA, it is possible to provide a more linear overall service for applications involving digital transmission or FDMA. The linearizer must be properly adjusted to cancel a portion of the nonlinearity and, at the same time, to provide good overall performance for any type of signal. Because that may not be feasible under all situations, the operator would have the option of switching the linearizer out of the path.

We have described the elements of the repeater section of typical communication payloads. These devices support analog and digital services and can be used with or without a digital processor. In Chapter 7, we review the design and performance of the remaining critical element, the spacecraft antenna system.

References

[1] Lyons, R. G., *Understanding Digital Signal Processing,* Reading, MA: Addison Wesley Longman, 1997.

[2] Craig, A. D., and F. A. Petz, "Payload Digital Processor Hardware Demonstration for Future Mobile and Personal Communication Systems," in E. Biglieri and M. Luise, eds., *Signal Processing in Telecommunications,* London: Springer-Verlag, 1996.

[3] Carducci, F., and R. Novello, "On-board Multiplexing: The Skyplex Transponder and the Multimedia Highways," *17th AIAA Internat. Communications Satellite Systems Conf. and Exhibit,* Yokohama, Japan, February 23–27, 1998, Collection of Technical Papers, American Institute of Aeronautics and Astronautics, New York, 1998, p. 461.

7

Spacecraft Antennas

The spacecraft antenna system, which is often viewed as part of the communications payload, has many special and important characteristics and therefore is deserving of its own chapter. Modern antenna systems provide a variety of communication coverage patterns such as domestic and spot beams, allowing the repeater to be utilized in an efficient and flexible manner. In the classical bent-pipe repeater design, the antenna coverage typically is fixed prior to launch. Changes on orbit are limited to what can be physically switched within the repeater. Onboard digital processing provides more flexibility because beams can be reshaped and even created dynamically in accordance with demand. The use of such processing is revolutionizing antenna system design because of the much wider range of performance options that can be obtained. Still, microwave antennas have not changed fundamentally since the bulk of the initial R&D was performed nearly 60 years ago.

This chapter explains how typical antennas are physically constructed and how they create the appropriate coverage patterns. An exact description of antenna operation would require use of electromagnetic theory, which (while mentioned in Chapter 4) is beyond the scope of this book. However, the somewhat simplified explanations that follow should provide a practical understanding of spacecraft antennas. Systems engineers need to develop requirements for payload antennas in cooperation with antenna experts who understand the details of the theory, the appropriate software tools, and the means to implement the physical hardware. Those are the stock and trade of the antenna system designer and are beyond the scope of this book. After the antenna electrical design is complete, there are still the critical matters of the choice of lightweight materials that provide sufficient strength, conductivity, and thermal control properties. Finally, the antenna must be physically attached to the spacecraft and adequate means provided for stowage and deployment in orbit. Pointing of the antenna beams, discussed briefly in Chapter 6, is another important aspect of the overall RF performance of the payload.

7.1 Horn Antennas

The basic microwave properties of waveguide were introduced in Chapter 6. Waveguides function in satellite communications as either a component of an antenna or as an antenna in its own right. If we were to take a fixed-length piece of waveguide and drive microwave energy at the proper frequency into one end, the opposite end would radiate some percentage of that power into space. The efficiency of radiation is simply the ratio of the power actually radiated into space divided by the total input power. Power not radiated is reflected back to the transmitter or dissipated as heat at the outer edge of the waveguide.

What this open-ended piece of waveguide has become is the simplest type of nonreflector antenna horn. Figure 7.1 shows a pyramidal horn in which the end of the waveguide is flared outward in the shape of a pyramid or a cone. The dimensions of the opening are in direct proportion to the wavelength and are dictated by the shaping of the far field antenna pattern. Also, the ratio of opening to waveguide dimensions directly affects the match between the impedance of the waveguide to that of free space, which is approximately 377Ω. That match is needed to reduce the power reflected back toward the transmitter. For a receiving antenna, a good match ensures that the received signal energy is delivered to the repeater input port. It is that match, measured in terms of the voltage standing wave ratio (VSWR), that defines the fundamental electrical property of the antenna when connected to either the input or the output of

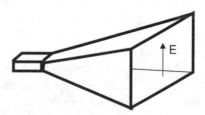

Figure 7.1 A common type of horn antenna called the pyramidal horn. The opening is wider than the waveguide to increase gain and to match the impedance of the wave-guide to free space.

the repeater. The ideal condition of zero reflection occurs at VSWR = 1, which cannot be achieved in practice across the entire operating band. However, it is not uncommon to obtain a maximum-value VSWR of about 1.25 to 1.50, with 1.00 being measurable at certain frequencies. Of course, that does not define the actual radiation pattern, which is discussed below in more detail.

Examples of different types of horns are provided in Figure 7.2. Selection of the pyramidal shape over the conical one is based on the particular application. The pyramidal horn is well suited to linear polarized systems because of the natural straightness of the extending rectangular waveguide. The standard waveguide mode of propagation is maintained, resulting in a symmetrical main-lobe pattern. In contrast, the conical horn flares from either circular or square waveguide (in the latter case, there must be a square-to-circular transition) into a circular aperture. This type of horn is best suited to CP, while the LP performance of the rectangular horn is superior.

Standard pyramidal and conical horns have smooth interior walls, as in the case with waveguides. That minimizes losses and reduces the generation of higher order modes that affect both efficiency and polarization performance. (We will see that another type of horn, called the corrugated horn, intentionally generates higher order modes to improve pattern and isolation performance.) Either type of horn can be machined ("hogged") out of a solid block of copper or aluminum or assembled from flat stock using welding techniques. The latter method is somewhat prone to higher loss and the possible generation

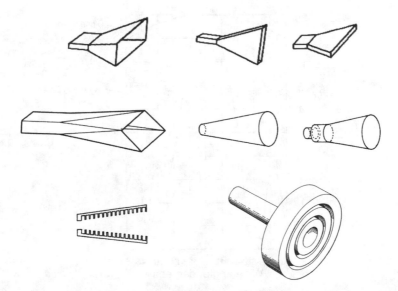

Figure 7.2 A selection of various microwave horn geometries.

of passive intermodulation (PIM) products, discussed later in this chapter. Another concern relates to high-power transmit service, as on a DBS satellite. Imperfections and sharp edges inside a high-power OMUX or transmit horn can generate multipaction, a form of high-temperature arcing of waveguide internal structure.

More precise horn geometry and quality can be obtained through the process of electroforming, which, by its nature, is considerably more costly and time-consuming than the previously described methods. In electroforming, the horn is electroplated onto a reverse mold, called a mandrel. The mandrel metal material has a lower melting point than the electroplating. After the horn is electroplated and the exterior machined to the desired shape, the interior mandrel simply is melted out. The resulting horn is as physically perfect as one can produce, based on the tolerances used in machining the mandrel itself.

Horn antennas are used extensively on commercial satellites. For example, full Earth (global) coverage from GEO, requiring a beamwidth of approximately 17 degrees, is efficiently achieved with a single horn pointed toward the Earth. The horn produces maximum power toward the subsatellite point and tapers downward to become about 3 dB below the peak at the edge of the Earth. An example of such a coverage pattern is shown in Figure 7.3, in which the upper (wide) curve applies for ±10 degrees (the upper degree scale on the

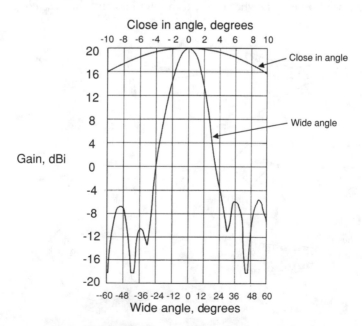

Figure 7.3 Antenna patterns of a global coverage horn on a GEO satellite.

X-axis) and the lower (sharp) curve applies for ±60 degrees (the lower degree scale). Antennas of this type are not used extensively because the peak gain is only about 20 dB, resulting in relatively low G/T and EIRP performance. As one would expect from basic antenna properties, the peak gain of a horn is inversely proportional to the square of the horn opening dimension times the square of the frequency. Therefore, to increase gain at a certain frequency, the physical size of the horn must increase. The additional constraint is that the waveguide input must be maintained at proper dimensions to permit efficient coupling of microwave energy from the transmit end to the radiating end. The same can be said of the receiving direction.

The most common use of the horn is as a radiating element for a reflector type of antenna. (Section 7.2 reviews some of the design and performance aspects of current reflector antenna systems.) The horn is placed at or near the focus of a parabolic reflector to illuminate the surface, in either a transmit or receive sense, depending on the application. As illustrated in Figure 7.4, the electromagnetic radiation on the reflector surface produces electrical currents on the surface. From those currents, another electromagnetic field is produced that eventually becomes the far field radiation pattern of the entire antenna system. The process is complex and only recently has been visualized through the facilities of high-speed computer graphics animation. In spite of that, the design principles are well understood in an engineering sense, allowing antenna designers to rapidly synthesize reflector antennas for virtually any satellite communication mission.

The other way to view reflector operation is in terms of an optical model, in which the horn produces rays of energy that are reflected off the reflector surface to form a collimated beam. As discussed in Chapter 4, the beam does

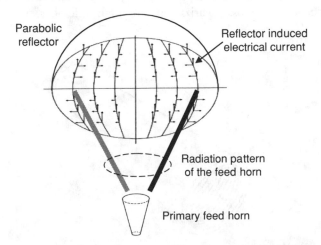

Figure 7.4 Principles of operation of a parabolic reflector microwave antenna.

not actually materialize until the far field of the antenna is reached. The horn becomes the primary radiator of the reflector and therefore must be designed to introduce the proper radiation of microwave energy across the reflector. Also, the impedance match between horn and waveguide, which controls the amount of reflected power, is corrected by adjusting horn dimensions and spacing from the reflector. For those reasons, open-ended waveguide seldom, if ever, is used as a primary radiating element. More popular are pyramidal horns and conical horns, which are applied to LP and CP antennas, respectively. (There are exceptions when a square cross-section can support dual polarized CP and, conversely, a conical horn is used in dual LP systems.)

Returning to the horn itself, its ability to produce the proper primary radiation pattern is determined by its size, shape, and internal structure (which interacts directly with the microwave energy in the waveguide). Other important factors in horn design are cross-polarization purity (and hence isolation) and sidelobe performance. To raise performance in those categories, the corrugated horn (Figure 7.5) was introduced. The desirable radiation properties of the corrugated horn include good pattern symmetry (along the vertical and horizontal dimensions, across the pattern itself), low sidelobe levels, and high beam efficiency (approaching 95%). However, the design and manufacture of the corrugated horn are more complicated than for smooth horns, which is expected due to the intricacies of the internal structure. Electroforming is the norm, rather than the exception, in this case.

In basic terms, the indentations allow two simultaneous modes (transverse electronic, or TE and transverse magnetic, or TM) to propagate up the horn to its opening. At the same time, the depth and the spacing of the corrugations cancel energy in the other propagation modes that could produce undesired properties like cross-polarization and high sidelobes. Furthermore, the

Figure 7.5 A C- or Ku-band corrugated horn.

pattern that the horn produces is symmetrical, which means that it is as nearly circular as can be achieved in practice. That lends itself to applications using CP with low sidelobes.

Horns often are combined into a feed array and used with a reflector. One popular application is the shaped beam antenna like that provided by AsiaSat 3S, shown in Chapter 6 and discussed in Section 7.2. Figure 7.6 provides an example of a multiple-feedhorn array for frequency reuse through Ka-band spot beams. Antenna arrays of this type are used for narrowband L-band MSS cells that reuse spectrum and provide increased receive and transmit gain. Another application is for Ku-band wideband spot beams to transmit high data rate digital carriers. There, the feedhorns are not connected to a power divider but are used separately. Important considerations in both of those horn arrays are the physical location and the proximity of the individual horns. The radiation pattern of one of the horns is determined by the size of its opening. However, that size still must allow for the required quantity of feeds to be placed within the prime focus of the reflector antenna. If that is not physically possible, then smaller, less effective horns may have to be used. Another solution that is available only in multiple spot beam antennas is to split the horns into separate groups that illuminate different reflectors. That adds to the physical complexity of the satellite and might even require that multiple satellites be operated at the same orbit position.

The last application of horns is a part of a direct radiating array antenna system, such as that shown in Figure 7.7. As in the feedhorn array for a reflector antenna, the direct radiating array uses multiple radiating elements, such as horns or crossed dipoles to produce a combined pattern or set of patterns. A group of elements are fed the same signal but are individually adjusted in

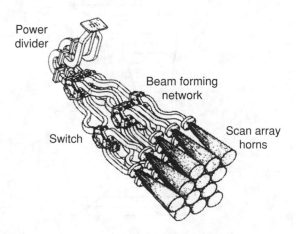

Figure 7.6 Multibeam feed array for NASA ACTS.

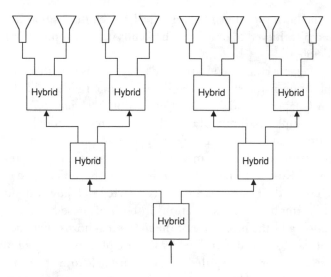

Figure 7.7 Schematic diagram of a direct-radiating horn array.

amplitude and phase. The result can be a far field pattern, with remarkably high efficiency and gain, and low sidelobe levels. The elements also can be driven with different signals to produce multiple beams for frequency reuse applications, which is the approach taken in MSS satellites like Globalstar.

A final application is as a microwave lens antenna, in which the input end of each waveguide literally is left in the air to collect energy from a single primary radiating horn. As illustrated in Figure 7.8, the horn array refocuses the

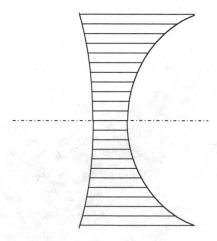

Figure 7.8 The basic configuration of a waveguide lens antenna.

energy into a precisely defined beam, aligned to the direction of propagation. The lens also can support multiple beams at the same time, provided there is a proper primary radiating antenna system.

7.2 Reflector Antennas

The simplest and most effective antenna system for a conventional FSS and BSS satellite employs reflectors and feed systems. That is evident from looking at the typical GEO satellite, such as that shown in Figure 7.9, with its reflectors being among the most obvious physical appendages. The feed system normally consists of one or more horn-type radiators. In the transmit mode, microwave energy enters the feedhorn from a piece of waveguide that carries the output of the repeater. The feedhorn radiates microwave energy from the focus of the parabolic reflector into space in the direction of the reflector. Those waves, on reaching the conducting surface of the reflector, induce electrical currents on its surface in direct proportion to the local level of the energy. The combination of all those currents reradiates the microwave energy in the direction of the Earth and over an angular range corresponding to the desired coverage pattern. Viewed another way, the reflector intercepts the feedhorn's radiated beam and forms a new collimated beam aimed toward the Earth. That is

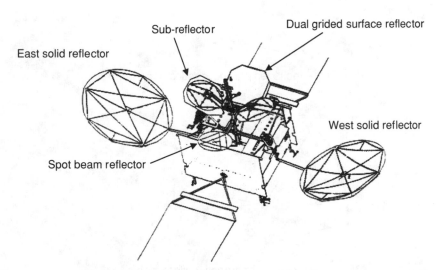

Figure 7.9 The general arrangement of a three-axis spacecraft with two offset parabolic reflector antennas. A subreflector attached to the front of the body extends the focal length (referred to as Gregorian optics).

why a light beam analogy is often used to describe the operation of the far field of the reflector antenna, even though that is not strictly the case.

Reception works in the opposite direction, although with a much weaker signal. The energy arriving at the satellite produces very weak electrical currents on the surface of the reflector, resulting in reradiation toward the feed. In this case, the reflector acts as a gatherer of signal energy, focusing it on the feed. A nice feature of such passive antenna designs is that a single reflector with feed can function for transmitting, receiving, or both simultaneously. This property, called reciprocity, was mentioned in Chapter 4. It is an interesting curiosity of history that this property was discovered by Hertz in 1888, well ahead of any knowledge of microwave propagation and only when the famous genius was experimenting with spark-gap transmitters and diode detectors. Much of the real work on microwave antennas, particularly that using reflectors, was carried out in the west during World War II. The results of that vital research are available in a classic text [1], which remains a good reference on microwave antennas.

Two typical antenna geometries using parabolic reflectors are shown in Figure 7.10. The center-fed parabola is circularly symmetric with the feed

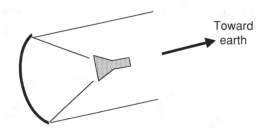

(A) Prime focus fed parabolic reflector antenna

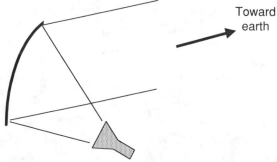

(B) Offset focus fed parabolic reflector antenna

Figure 7.10 The prime focus-fed parabolic reflector antenna is subject to feed blockage, while the offset feed design provides an unobstructed beam.

located at the focus, while the offset reflector parabola allows the feed to be placed below the line of transmission. The reflector surface is formed by taking a parabola and rotating it about a line drawn from the focus to the center of the parabola (forming a segment of a paraboloid). For the offset case, the required surface is cut from one side of the paraboloid (see Figure 7.17 later in this chapter for an example.) This type of antenna continues to be popular for spacecraft and ground applications because it is straightforward to design and build. Also, extremely cheap versions, which are still effective, have been produced for DTH reception around the world. The spacecraft version is not cheap because of the need to use lightweight materials for the reflector and support structure. They must be rigid to hold the beam properly and must not undergo significant changes in dimensions under the varying temperatures of space. In many cases, the reflectors must be stowed for launch and then reliably deployed in space prior to start of operation.

The size of the reflector is dictated by the required gain, as discussed in Chapter 4. Beamwidth, which is equally important in payload performance, varies inversely with diameter. That provides the primary basis for determining the physical size of the reflector. Another factor is the actual frequency of operation, which must be sufficiently high to permit the reflector to perform its function and to allow the use of a reasonably sized feedhorn. The horn opening, which determines the radiation pattern across the reflector, also is dictated by the frequency of operation. Low frequencies mean a large opening. If the horn opening is too large, it could block the energy after it leaves the reflector. That is why there is a preference for the offset geometry, since blockage is avoided.

Reflector antennas may be designed with two reflecting surfaces, as illustrated in Figure 7.11. There are technical and performance advantages to using two reflectors, the main one being that it allows the feed to be located behind

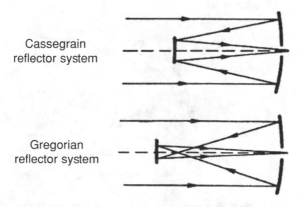

Figure 7.11 Dual reflector geometries: the Cassegrain and the Gregorian.

the main reflector instead of ahead of it (and potentially in the line of propagation). In ground antennas, the Cassegrain approach, popular in optical telescopes, allows the feed, LNA, and potentially the RF electronics as well to be placed behind the reflector in a protective housing or shelter. A spacecraft version of the Gregorian type has been used on FSS and BSS spacecraft (Figure 7.9) to increase the focal length and permit more convenient placement of the feed in close proximity to the output multiplexer.

Another technical approach to reflector antenna design can provide a shaped beam with a single feedhorn. It is the shaped reflector, a technology from Japan that became an industry standard in the mid-1990s for Ku-band BSS satellites. As show in Figure 7.12, the reflector is contoured in such a way as to induce the appropriate current density and phasing on the reflector surface. That creates a shaped beam in the secondary pattern. To build the shaped reflector, precise dimensions are machined in a mandrel, which is the metal mold onto which softened carbon fiber is laid out. After curing, the carbon fiber surface is lifted off the mandrel and fabricated into the reflector assembly.

We begin our discussion of the two configurations with a review of the center-fed parabola, which is the basis for all classes of reflector antennas and the simplest to design and construct. Following this, we consider the offset-reflector antenna, popularized on the first generation of domestic satellites like Anik A and Satcom 1.

7.2.1 Center-Fed Parabolic Reflectors

Shown in Figure 7.13 is the prime focus-fed parabola, which operates in principle like a reflecting mirror. It produces a collimated beam of parallel rays from a source located at the focus. The prime focus approach is efficient and effective

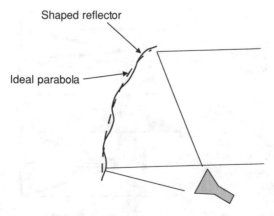

Figure 7.12 Conceptual drawing, in cross-section, of a shaped offset reflector.

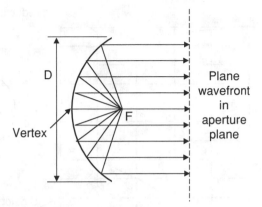

Figure 7.13 Geometry of paraboloidal reflector.

because of the ideal geometry that it possesses; however, it has the drawback that the feedhorn (or system of feedhorns) can block and deflect a portion of the collimated beam. That is particularly troublesome in frequency reuse spacecraft antennas in which feedhorn blockages cause high sidelobes, scatter energy outside the desired coverage, and degrade cross-polarization isolation. The effect of blockage on sidelobe level for a hypothetical center-fed parabola is graphed in Figure 7.14 (it is approximate and is shown for illustrative purposes only).

The typical parabolic reflector is round to provide maximum symmetry for a curricular beam. Other shapes are possible and permit the beam to have a noncircular cross-section as well. For example, an oval-shaped dish produces an oval-shaped beam, except that the major (longer) axes of the two are at right angles. The reason is that the wider dimension produces the narrower beamwidth (due to better concentration of the beam) and vice versa.

The reflector surface is smooth to provide the cleanest output wavefront. There is a relationship between the surface accuracy, called the surface tolerance, and the resulting gain, as provided by the following [2]:

$$G / G_0 = \exp[-(4\pi\varepsilon/\lambda)^2]$$

where

G = antenna gain with error;

G_0 = antenna gain without error (determined by the diameter, frequency, etc.);

ε = surface (root-mean-square, rms) tolerance in the same units as λ;

λ = the wavelength.

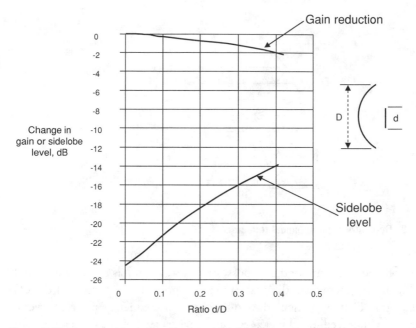

Figure 7.14 Aperture blockage effect on sidelobe level and gain in a center-fed parabola.

The equation is an approximation that represents the average gain loss over a large number of antennas having the same physical properties. An individual antenna would have its own unique error profile and a corresponding gain loss, which would be larger or smaller than the formula would predict. For example, at an average gain loss of 1 dB, the rms surface tolerance would have to be only 1/30th the wavelength. That allows large L-band reflectors to be somewhat imperfect in absolute terms, since the required surface accuracy is relatively easy to achieve. At Ka band, the problem is worsened by the ratio of wavelengths of more than 12 to 1.

One interesting approach for a center-fed parabolic antenna, shown in Figure 7.15, is to package a very large deployable center-fed reflector constructed like an umbrella. That allows stowage as a long post and deployment by extending the umbrella substructure. A subreflector is shown in this configuration to create a Cassegrain system with the feeds attached to the spacecraft body. The main surface normally would be made of a flexible metallic mesh or metal-covered fabric. Issues still remain regarding that design because of the potential for PIM products that are generated from high-power transmission. PIM, in turn, can radiate and conduct back into the receiving side of the spacecraft, interfering with proper operation of the payload. Control of PIM requires careful attention to the design of the mesh or fabric and good consideration of

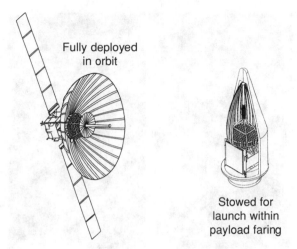

Fully deployed
in orbit

Stowed for
launch within
payload faring

Figure 7.15 An example of a large deployable center-fed parabola for use at L band.

the overall payload and system design to avoid PIM from causing interference in the first place.

7.2.2 Offset-Fed Parabolic Reflectors

The offset-fed parabola, illustrated at the bottom of Figure 7.10, eliminates feed blockage and substantially reduces reflection of energy from the reflector back into the feed. While the physical characteristics have less symmetry and require more careful design and manufacture, experience has produced very efficient and high-performing spacecraft antennas of this class. The offset reflector permits placement of feeds much closer to the repeater electronics and suppresses sidelobe levels that would result from feed blockage. Subsequently, nearly all bent-pipe type satellites adopted the offset reflector type of design, taking advantage of the fact that the reflector can be stowed for launch by simply folding it directly on top of the spacecraft. As illustrated in Figure 7.16 for a spinning satellite, a single hinge and drive mechanism can be used to raise the reflector for service.

The specific dimensional characteristics of the offset reflector are shown in Figure 7.17. There are a few potential technical drawbacks of the offset approach, the majority of which can be overcome through certain design optimizations and device selections. The first is that the offset parabola experiences some efficiency loss because the beam is tilted slightly upward from optimum. An illustrative example of how gain performance would degrade as the beam is offset from boresite is plotted in Figure 7.18. Gain reduction, which increases

Figure 7.16   Offset reflector on a spinning spacecraft.

more rapidly as one moves away from boresite, is the result of beam defocusing. Any such gain loss usually is countered, to some degree, by the complete elimination of feed blockage. The second effect is on sidelobe levels and cross-polarization, which are different depending on which direction one looks from boresite. Some of that can be controlled through proper selection of feed type and dimensions, including the use of more polarization-pure designs like the corrugated horn. Another concern is cross polarization isolation, which could

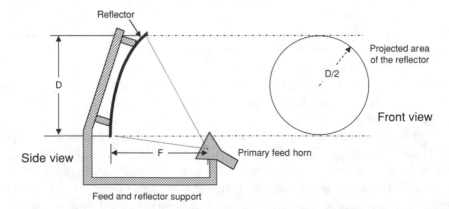

Figure 7.17   Offset-fed paraboloidal reflector with the feed and feed support located so as not to produce blockage.

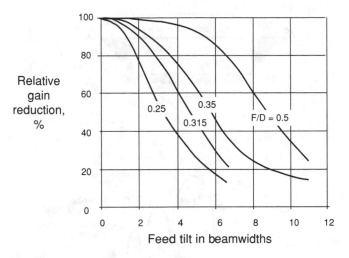

Figure 7.18 Gain degradation due to beam offset.

be degraded by the lack of circular symmetry of the antenna system. That can be controlled by proper selection of the feedhorn and its placement relative to the reflector.

Finally, in CP offset reflector antennas, there is a slight squint of the beam to one side or the other. The direction depends on which sense of CP is used: left-hand or right-hand. That has an impact on the deployment of transponders on particular satellites and could result in using twice the number of reflectors.

Another important consideration is the deformation of the reflector surface and feed support from differential solar heating and the resulting thermal expansion. That misshapes the reflector, causing the pattern to distort. Deformation of the feed support relative to the reflector causes the beam to mispoint. Two solutions are available: make the antenna less sensitive to thermal effects (e.g., prevent temperature variation) and provide a mechanism to realign the reflector on orbit.

7.3 Antenna Patterns

For most spacecraft antenna systems, the feedhorn works with a reflector of some sort to produce the desired coverage pattern. At the top of Figure 7.19, a reflector is illuminated with microwave energy by a single feedhorn located at the focus. The feedhorn has a radiation pattern (indicated as the shaded area in front of the reflector), which has a maximum along the centerline and decreases

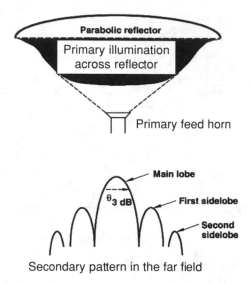

Figure 7.19 Illumination of a parabolic reflector by a waveguide horn producing a secondary pattern, which contains a main beam and sidelobes.

in intensity toward the edge of the reflector. The specific manner in which the
intensity is adjusted is called the taper. As you will recall, the primary illumination produces a corresponding distribution of electrical currents across the
reflector surface. The currents reradiate their energy toward the front of the
reflector, resulting in the desired secondary pattern in the far field of the
antenna. Energy not radiated is dissipated through the electrical resistance of
the reflector metal (or metallic) surface.

The secondary pattern of a reflector antenna system is a three-dimensional
Fourier transform of the horn illumination. That statement has meaning to electrical engineers and points out that reflector antennas share a mathematical
property with pulse spectra. An example of a secondary pattern is shown at the
bottom of Figure 7.19. The main lobe in the center provides the desired coverage, a concept introduced in Chapter 1. Visualize that the peak of the lobe is
pointed at the center of a country to be covered. The main lobe extends across
an angular dimension, Θ_{3dB}, which roughly corresponds to the width of that
country. The parameter Θ_{3dB} is the 3-dB beamwidth, defining the point where
the gain of the antenna is 3 dB down (i.e., one-half the power) from the peak.
That natural characteristic of an antenna beam explains why reception from
domestic satellites generally is better in the center of the country than at the borders. Real antenna patterns are shaped to match the desired service area.

The secondary pattern also shows sidelobes, located on either side of the main lobe. The sidelobes represent signal energy taken from the main lobe that radiates into adjoining regions on the Earth. An actual example of coverage sidelobes is provided in Figure 7.20. Sidelobes, of course, are potential sources of radio interference to and from Earth stations in the geographic regions where they fall. In satellites that use multiple beams to reuse frequency, it becomes particularly important to control sidelobes. That can be accomplished through the selection and location of feedhorns as well as the use of the proper taper of the primary pattern.

On digital processor repeater systems, it also is possible to electronically alter the sidelobe levels through digital beam forming. That is accomplished inside the processor, where the carriers to be downlinked are altered in amplitude and phase so they combine in the far field in an optimum manner.

In rough terms, the 3-dB beamwidth (in degrees) of the far field pattern is related to the diameter of the reflector through the following approximation:

$$D = 70\lambda / \Theta_{3dB}$$

where D is the antenna diameter measured in the same dimensions as λ. The formula is plotted in Figure 7.21 for the microwave bands of interest. The data show that the beamwidth decreases as either the diameter or the frequency

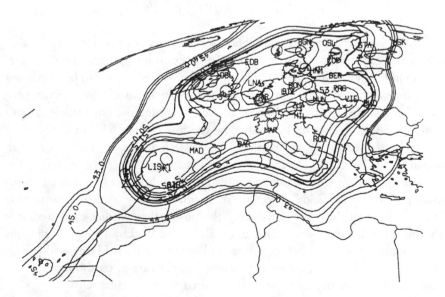

Figure 7.20 Spacecraft coverage pattern, including main lobe and sidelobes.

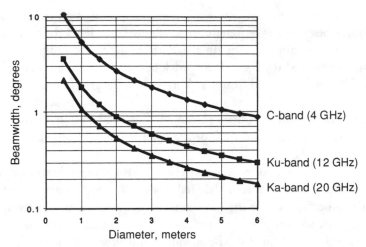

Figure 7.21 Relationship between diameter and beamwidth at C, Ku, and Ka bands (down-link frequency assumed).

increases. The gain of the antenna, on the other hand, increases as the square of either the diameter or the frequency. Also, the relationship between Θ_{3dB} (i.e., the coverage area) and the gain of the antenna is independent of diameter and frequency, for example,

$$G_0 \cong 27{,}000/(\Theta_{3dB})^2$$

At C-band frequencies and higher, it is practical to generate a beam that follows the contour of a country or region by combining the action of several individual feedhorns. Figure 7.22 shows how three feedhorns can be combined to produce a shaped beam. Horn B, located at the focus of the reflector, produces the expected main lobe labeled B. Now, examine the horn labeled A and its associated far field main lobe. Shifting the horn to the left causes the main lobe to be shifted to the right, much like what happens with a glass mirror. Similarly, shifting horn C to the right causes the far field pattern to shift to the left. The sum of the three patterns is illustrated by the wavy line labeled "combined pattern in far field." The pattern is produced by a feed network that divides the signal three ways, reducing the power that reaches each feed. For example, an equal three-way split reduces the signal level by 10 log(3), which is approximately 5 dB. The power will recombine in the far field due to the overlapping of the individual secondary patterns. In the case of reception, the same signal coming from the three horns is summed in the feed network.

Beam shaping is a powerful technique for increasing the effectiveness of a communications satellite. In Figure 7.23, the coverage of Mexico provided by

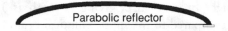

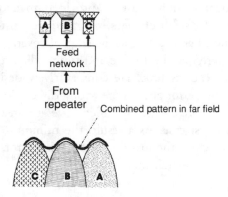

Figure 7.22 Generation of a shaped beam antenna pattern using multiple feedhorns and an associated feed network.

a single elliptical beam is compared with a shaped beam coverage from combining four nearly circular beams (indicated as the shaded area). Both the single and the shaped beams cover the entire Mexican land mass. However, the single beam radiates a substantial fraction outside Mexico, which represents a loss of about half the available signal power. In comparison, the shaped beam

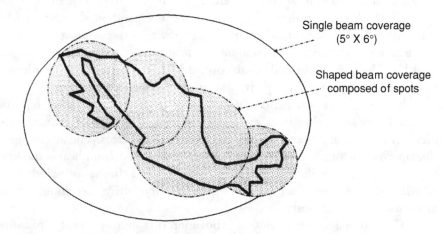

Figure 7.23 Spacecraft antenna coverage of the Mexican land mass demonstrating that a shaped beam is more efficient than an elliptical beam.

provides a relatively tight fit of Mexico, giving as much as 3 dB more gain at the edge of coverage, compared to the single elliptical beam.

The overall communications performance in terms of EIRP and G/T depends on the spacecraft bus as well as the repeater and the antenna system. Some additional margin must be provided to allow for movement of the antenna caused by small changes in spacecraft attitude. As a result, there are periods of time when the gain at the edge of coverage is reduced relative to the nominal (properly pointed) value and times when it exceeds nominal as well. Most communications links are conservatively designed for the worst case, when the pointing error goes in the wrong direction. To take that into account, the points that represent the specified coverage area are extended outward to represent the worst case. As a result, the minimum required EIRP and G/T automatically include the loss due to pointing error. Alternatively, the pattern could be shrunk to match the worst-case error.

7.4 Innovative Spacecraft Antenna Designs

Spacecraft antenna systems have always been a challenge due to the difficulty of the mission and the demand for tight performance requirements. More recently, systems engineers and antenna designers have gone to approaches that previously were used in government and military applications. For example, the phased array antenna found on the Iridium antenna was developed by Raytheon, a company with decades of experience in advanced microwave antennas. Up to that point, no commercial spacecraft had yet flown that type of system. The design of an array antenna is more akin to that of a microwave filter, which also uses multiple elements to produce the desired response. One needs individual elements that are driven with the appropriate power and phase of the desired signal. An illustration of a basic array antenna is shown in Figure 7.24 [3]. Through power and phase control, the final pattern can be adjusted to move the beam without any physical motion of the elements.

Individual array elements can consist of feedhorns, miniature reflector antennas, dipoles, or slots in waveguide walls. Some systems even use active elements consisting of power amplifiers with integral radiators. The advantage of that approach is that it reduces feedline loss, although it complicates the matter of providing amplifier redundancy. A way around that is to include extra amplifier-feed elements and then to adjust the input drive accordingly in the processor or input circuitry.

The waveguide lens is another innovation that soon will find application in the coming generation of Ka-band systems. An example of a waveguide lens is shown in Figure 7.8. In the figure, the signal is transmitted from a primary

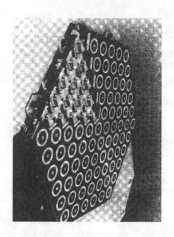 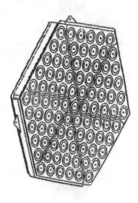

Figure 7.24 Globalstar active MMIC phased array antenna.

feedhorn or horn array toward the lens. The individual waveguides in the lens receive, transfer, and reradiate the energy along a modified wavefront that produces the desired far field pattern. In technical terms, the lens transforms the nonuniform wavefront from the input side into a collimated beam by providing the correct line lengths for phase adjustment. There may be multiple focal points to allow the beams to traverse the region of coverage. Because it is constructed of waveguide segments, a lens antenna can introduce significant losses.

Literally thousands of possible antenna structures have undergone research and development over the years since World War II. It is doubtful that all useful configurations have been discovered, so we can expect to see innovations as satellites are delivered to meet greater performance demands. We recommend that anyone considering a special requirement first review the past research so as not to reinvent the wheel.

References

[1] Silver, S., ed., *Microwave Antenna Theory and Design,* London: Peregrinus, 1984.

[2] Ruze, J., "Antenna Tolerance Theory—A Review, *IEEE Proc.,* Vol. 54, pp. 633–640, April 1966.

[3] Gibson, J. D., ed., *The Communications Handbook,* IEEE Press, 1997, p. 1050.

8

Spacecraft Bus Configuration and Subsystems

The spacecraft bus is the platform that supports the communications payload during all phases of the mission, from launch, through injection into orbit, and throughout its operating lifetime. That is as challenging a task as processing and relaying the communications signals themselves and demands a thorough understanding of both the internal workings of the vehicle as well as its interaction with the various environments that are encountered both inside and outside the atmosphere. Bringing payloads up to GEO was long considered the most difficult type of mission, yet now we see new constellations of smaller non-GEO spacecraft entering the marketplace for telecommunication services. Across that spectrum of orbits and applications, the basic principles of bus design have not changed appreciably. It is now a matter of adjusting the bus requirements to fit the mission.

8.1 Mission Summary

The steps necessary to provide a working communications satellite or constellation are generally termed the mission. Just as in the conventional military counterpart, each step requires careful planning and proper execution. System engineers and mission planners spend considerable time and effort to anticipate all possible situations and contingencies that the spacecraft will or might experience and to provide the hardware, software, and procedures necessary for a satisfactory outcome. The problems associated with this type of communications technology are exacerbated by the fact that launching spacecraft is inherently both expensive and risky, while on-orbit operations are possible only through the facilities of telecommand and telemetry. Communications satellite systems engineering, then, is both a science and an art. A detained discussion of the processes and technologies is beyond the scope of this book; however, many of the more important aspects are reviewed in this chapter. Some additional methodologies can be found in [1].

8.1.1 GEO Mission Profile

As an aid in understanding the functions of the bus, we use the standard get-to-orbit mission as a prime example. The bus design must deliver the payload to orbit and support it throughout the operating lifetime. The typical GEO mission, illustrated in Figure 8.1, goes through literally every phase that an orbiting vehicle could experience. Table 8.1 identifies the primary steps in the process, indicating that several, while brief, are critical to mission success. This is a somewhat lengthy mission in terms of time, due to the need to test and verify every aspect of performance along the way. It may be possible to abbreviate several of the steps, something that has been done in the past when the satellite was urgently needed. The design of the bus must be consistent with the plan and vice versa.

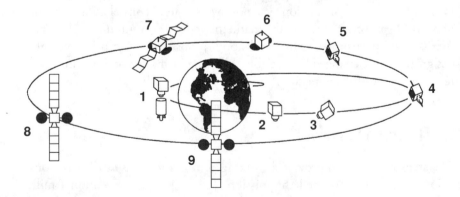

1	Separation from Launch Vehicle	6	Deploy reflectors and spin down
2	Reorient to sun-normal	7	Deploy solar panels
3	Reorient for apogee injection attitude	8	Sun acquisition and bus testing
4	Apogee injection boost	9	Earth acquisition and start of operations
5	Reorient for deployments		

Figure 8.1 A typical GEO mission sequence for a three-axis spacecraft.

Table 8.1
Steps in a Typical GEO Mission

Step	Time from liftoff	Event
1	0–30 min	Main boost phase
2	30 min	LEO achieved (GEO or non-GEO mission)
3	45 min	Transfer orbit injection (GEO mission)
4	45 min	Separation from launch vehicle
5	1 hr	TT&C link established with ground tracking station
6	Day 2	Geosynchronous orbit injection
7	Day 2 to 5	Orbital adjustments
8	Day 6 to 8	Initial bus configuration by ground command
9	Day 9 to 11	Drift to assigned orbit position
10	Day 12 to 13	Orbit and attitude adjustments
11	Day 14 to 18	Reconfiguration of spacecraft for operation
12	Day 19 to 23	Bus testing
13	Day 24 to 30	Payload testing
14	Day 31	Start of operation

8.1.2 On-Station Operation Requirements

The technical requirements on the bus after the spacecraft is ready for service are somewhat less severe that those that preceded. The space environment is harsh and places numerous demands on the bus. To properly take that into account, the bus is divided into functional subsystems, each of which is the subject of critical scrutiny by the relevant engineering and scientific experts. Designing and manufacturing the bus is the domain of rocket scientists, as well as some of the best minds in space physics and astronomy, thermodynamics, electronics, structural engineering, aeronautics, and computer science. It is a multidisciplinary field in which the experts must cooperate and communicate effectively. The final critical element is the qualification and testing of the hardware to guarantee that all mission requirements can be met even before the spacecraft is launched.

8.1.3 Non-GEO Requirements

The mission concepts for non-GEO satellite constellations are evolving even at the time of this writing. Non-GEO systems intended for Earth observation, positioning, and science have been with us for many decades. However, they generally do not have to provide continuous service to all or a substantial part of the globe. Therefore, we anticipate many innovations in that area to ensure that commercially viable telecommunications services can be rendered to consumers. An interesting observation is that, while current-generation GEO satellites are maintained tightly within their orbital constraints, non-GEO constellations may not require stationkeeping for satisfactory performance [2].

The first non-GEO system to initiate service is ORBCOMM, which launched its initial pair of satellites in April 1995 [3]. When the network of 36 satellites is fully deployed, ORBCOMM will provide two-way data messaging anywhere in the world. In contrast to Globalstar, ORBCOMM offers nearly continuous coverage for most of the inhabited areas of the planet. The following discussion of the ORBCOMM mission is illustrative of the steps involved. This example demonstrates that each non-GEO constellation has its own unique features and constraints. Spacecraft are launched by Orbital Science's Pegasus XL air-launched vehicle. It is a three-stage winged rocket booster, lifting off from a conventional jet aircraft flying at 11,500m (38,000 ft) and Mach 0.79. Carrying a stack of eight compact ORBCOMM spacecraft, the Pegasus XL will first eject four in the forward direction and the remaining four after reversing direction. Passive springs cause the spacecraft to be deployed in precise sequence, as illustrated in Figure 8.2. The separation system, first demonstrated for the initial spacecraft pair, performs the necessary ejection of individual spacecraft to ensure acceptable shock, low tumble rates, and enough energy to avoid collisions.

Once the ORBCOMM spacecraft have achieved initial orbit, ground-commanded burns of the onboard nitrogen propulsion system will be used to increase the separation rate of the spacecraft within the same plane. Several short bursts are used to accelerate and, subsequently, decelerate once the spacecraft near their proper positions in the constellation. Subsequent orbit maintenance corrections obtain the necessary force from aerodrag, using either "feathering" or "flaring" of the solar panel. That will maintain proper spacing within a plane, a separation between planes of 45 ±5 degrees and 180 degrees apart for polar planes. T&C operations depend on data relay from the remote gateway Earth stations to and from the SCC, located at ORBCOMM headquarters in northern Virginia. As each satellite makes contact with a gateway, its stored position and velocity data previously gathered from the global positioning satellite system (GPSS) are transmitted to the network control center for processing and storage. The data are used for subsequent corrections to maintain the proper orbit spacing and geometry. From an overall standpoint, the ORBCOMM mission is based on a simple, low-cost bus design.

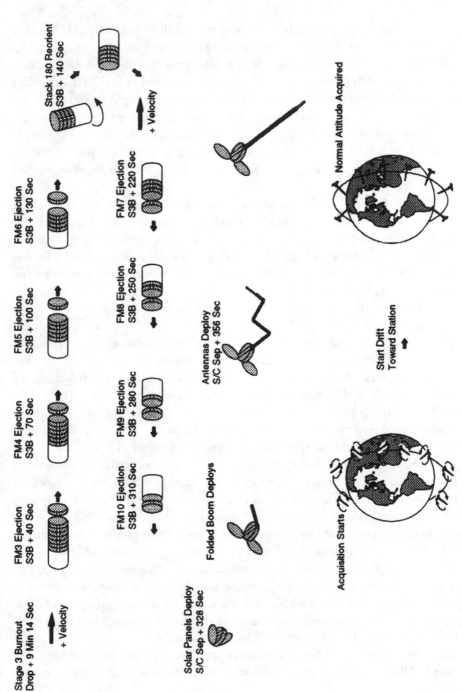

Figure 8.2 ORBCOMM constellation plane deployment scenario [3].

Mission design for Globalstar has many parallels to ORBCOMM, with the number of operating and spare satellites being 48 and 8, respectively. The constellation is designed to provide the best service to the middle latitudes, which corresponds to where the bulk of the population live [4]. Each operating orbit is at an altitude of 1,414 km and is inclined by 52 degrees. Spare satellites are kept in a lower orbit at 900 km, where they are available to be raised to the operating orbit when needed to replace a failed satellite. The lower altitude causes those satellites to move relative to the higher orbit operating satellites, allowing more convenient replacement when a replacement satellite reaches proper alignment.

Deployment of Globalstar satellites will be in batches of four using Delta or Long March (a Zenit rocket failed to deliver 12 satellites to orbit). They will be released at the lower orbit of 900 km and a stored command sequence will be used to deploy the solar panels, align them with the sun, and turn on the electronics. Once under ground command, a series of burns is used to move the satellites to the higher orbit.

All the Globalstar satellites will be controlled from a single operation center, where data from the global network are available for analysis and storage. Useful data per satellite can be obtained for about 10 minutes per 114-minute orbit. Data from noncovered periods of time are stored onboard the satellite and downloaded during a monitored pass. The operational philosophy of Globalstar is to track differences from the norm and not to rely on 100% data availability and analysis, an approach common in GEO satellite operations. That is because of the large number of satellites to be monitored, the limited observation time, and the similarities among the satellites in operation.

The requirements for non-GEO satellites and constellations, as well as the operational procedures needed to ensure reliable service, can be expected to evolve over time. There clearly is a strong motivation to cause the systems to deliver services of high quality and reliability. That depends heavily on the design of the spacecraft bus in combination with a need to have an economical system. We have considered only two such constellations: ORBCOMM, which is designed to provide nearly continuous service for data communications and paging, and Globalstar, a single-hop voice communications network that is tightly tied to the PSTN. As discussed in Chapter 11, the Iridium system has a particularly interesting feature in that calls can be relayed from satellite to satellite, as in a daisy chain. That means Iridium can serve a substantial quantity of users even when a significant fraction of its satellites are nonfunctional. All in all, the requirements for non-GEO spacecraft appear to vary greatly.

Spacecraft complexity, power, and mass differ greatly among the various GEO and non-GEO missions. GEO spacecraft generally are the most powerful, to deal with the added path loss and to provide adequate capacity to exploit a particular orbit position and footprint. As we move lower in orbit, the power and capacity demands on the spacecraft decrease. That is a particularly good thing because it compensates, to some extent, for the need for more spacecraft for continuous coverage. When we examine the bus designs used for a large LEO constellation of 60 or more satellites, we perhaps find that the individual spacecraft are small compared to the typical GEO design. That trend is countered when satellite capacity must increase to match a potential megamarket for wideband services.

8.2 Spacecraft Configuration

Commercial aircraft all take essentially the same shape to fly at high speed and carry sufficient passengers and cargo to meet market demands. However, commercial communications spacecraft can and do take on a variety of shapes and configurations because the external forces in space are much weaker. There is, of course, the need to package the spacecraft for delivery to orbit by an Earth-launched rocket. Once in space, a spacecraft is unfurled into one of three different configurations currently in use. Referred to as the simple spinner, the dual spin, and the three-axis, the configurations and their attributes are discussed in this section.

This discussion is meant to give the reader some appreciation for the major tradeoffs in communications spacecraft design. Additional common features include power generation and storage, attitude sensing and control, propulsion for orbit control, telemetry and command, thermal control, structural design, and integration with the launch vehicle. Those criteria are addressed in relevant sections of this chapter on spacecraft bus subsystems.

8.2.1 Simple Spinning Spacecraft

The original configuration to be employed for application as a geostationary communications satellite was the simple spinner. While not currently in production for commercial service, the spinner (shown in Figure 8.3) is a very stable and reliable design. The spinner is said to be unconditionally stable, meaning that the spacecraft will stay erect and even correct itself if disturbed by an external torque against the spin axis. The body and the major components are arranged to provide maximum rotational inertia about the spin axis. That produces a drum shape more akin to that of a tuna can than to that of a pencil.

The payload repeater and the spacecraft bus are housed in the spinning section, while the antenna and the feed are despun. As a result, the transmit and

Figure 8.3   The Palapa A simple spinning spacecraft. (*Source:* Photograph courtesy of Telekomunikasi Indonesia.)

receive microwave signals must pass through a waveguide rotary joint contained in the motorized assembly that physically connects the two sides.

All the functions of the bus are contained in the spinning body. Electrical prime power is obtained from silicon solar cells mounted on the circumference of the drum. Batteries for energy storage are attached to the internal structure to give proper balance and are located to receive a nearly constant temperature environment. While spacecraft stability is guaranteed by its rate of spin, the despun antenna keeps its beam properly pointed by virtue of an active control system that is integral to the bus. A propulsion system provides pulses of thrust to maintain the proper orbit position (stationkeeping) and spacecraft orientation. All the facilities are discussed in detail in applicable paragraphs of this chapter.

8.2.2 Dual Spin Spacecraft

The dual spin spacecraft design came out of continued research and development at Hughes Aircraft Company and is the only type of spinning satellite in production at the time of this writing. The preceding discussion of a simple spinner generally also relates to the dual spinner but with certain important distinctions. To get an appreciation for the enhancements this configuration offers, examine Figure 8.4. Again, the bus section of the spacecraft spins to provide stability; however, the entire payload (repeater and antenna) is despun. That gives greater flexibility in designing the payload for greater communications capability. The mechanical interface between the bus and the payload, called the bearing and power transfer assembly (BAPTA), allows electrical power to reach the payload. In this case, microwave signals do not have to pass across the mechanical joint because the repeater is physically connected to the antenna feeds.

From the standpoint of dynamic stability, the dual spinner is more complex than the simple spinner. First, the fact that the despun section does not rotate reduces the gyroscopic stiffness of the spacecraft. Second, there is a dynamic interaction between the spun and despun sections that can cause the spacecraft to gradually drift out of proper alignment. That process, called nutation, can be observed in a spinning top as it slows down. The top develops a second sideways rotation at a slower rate than the spin rate. Gradually, the top

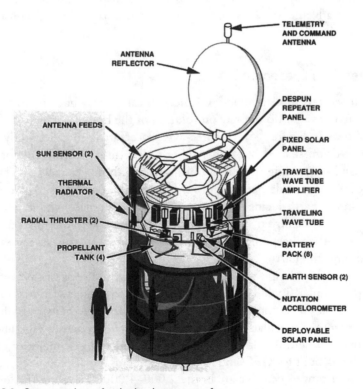

Figure 8.4 Cutaway view of a dual-spin spacecraft.

leans over until it falls on its side. In a similar manner, an uncontrolled dual spinner will lose its vertical alignment and ultimately lean over and go into a "flat" spin, that is, it will be spinning about an axis perpendicular to the spin axis. Said another way, the dual-spin configuration is not unconditionally stable like a simple spinner.

The key to achieving stability in a dual spinner is to introduce a proper degree of nutation damping on the despun section. The damping is not active while the satellite is spinning normally. However, as soon as the angle of nutation (or spin axis tilt) increases to some detectable level, sufficient damping is automatically provided by the onboard control system to cause an equal compensating torque. An understanding of the physical principle is difficult to obtain without a review of rather complex mathematics, which is beyond the scope of this book. The use of damping to stabilize the dual spinner was an important innovation in spacecraft design, producing spinning satellites with expanded capability. Designs produced in the 1990s were long and narrow with telescoping solar panels for increased power capability. The reflector, as shown in Figure 8.4, can be attached to a motor-driven antenna repositioner mechanism that both deploys the reflector and permits adjustment of the beam while on orbit. The advantages of a simple spinner cited at the end of Section 8.2.1 also apply to the dual spinner.

8.2.3 Three-Axis Spacecraft

Several innovations in electronics and mechanical devices have permitted the design of satellites that do not rely on rotation of the body to achieve stabilization. As shown in Figure 8.5, the three-axis spacecraft (also called body stabilized) is not drum-shaped but rather takes on the most convenient form for providing and supporting the communication function. Typically, the body is box-shaped with one side pointing toward the Earth at all times. Antennas are mounted on the Earth-facing side and the lateral sides adjacent to it. The solar panels are flat and deployed above and below the body, with orientation mechanisms provided to keep the panels pointed toward the sun as the satellite revolves around the Earth.

The three-axis design was adopted to deal with the apparent power limitation of the spinner (including the dual-spin) and to make additional mounting area available for complex antenna structures. It took many years for the design to mature, particularly because of the complexity of stowage and on-orbit control. Those difficulties were overcome, and the vast majority of new spacecraft being built for GEO and non-GEO systems are of the three-axis variety. The power ranges available are in the 2,000–4,000W level on the low end up to the 15,000–20,000W level at the high end (at the time of this writing).

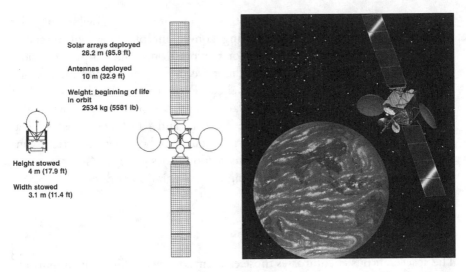

Solar arrays deployed
26.2 m (85.8 ft)

Antennas deployed
10 m (32.9 ft)

Weight: beginning of life
in orbit
2534 kg (5581 lb)

Height stowed
4 m (17.9 ft)

Width stowed
3.1 m (11.4 ft)

Figure 8.5 General arrangement of a three-axis GEO spacecraft.

As power is increased by enlargement of the solar panels, the overall design of the spacecraft becomes complex, particularly because of the added payload equipment and the difficulty of managing the extra heat generated onboard.

Stability of the three-axis is ensured through an active control system that applies small forces to the body to correct for any undesired changes in spacecraft attitude. The most common three-axis attitude-control system ("biased momentum") uses a high-speed gyro called a momentum wheel to provide some stiffness and to act as an inertial reference. A basic control technique is to either speed up or slow down the wheel to cause the body of the spacecraft to rotate in the direction opposite to the speed change, which produces precise east-west adjustment of the platform (depending on the type of mission and the orbit design). The momentum wheel can sense transverse angular motion of the body (nutation), which causes a counterforce on its support. The momentum wheel can be twisted by a gimbal (a motor-controlled pivot), placing a precise dynamic force on the spacecraft. There are at least two momentum wheels to ensure reliability in case of a single failure and to provide additional control options.

An alternative approached, called zero momentum bias, provides separate reaction wheels for each of the three axes. Rather than spinning at a high rate in one direction, the reaction wheel can spin up and down and can reverse its spin and go in the opposite direction. The technique is extremely precise because the axes are independently controlled and adjusted in orbit. Gyroscopic stiffness is nearly zero in comparison to the biased momentum and spinner designs.

The three-axis design has the obvious advantages of allowing the solar panel to reach any size and permitting convenient placement of antennas, which can be either rigidly attached or deployed on orbit. To provide those capabilities, it is necessary to package the spacecraft for launch and then deploy all the appendages once orbit is reached. Furthermore, precise control of antenna beam pointing will rely on a fairly complex system of electromechanical devices and an onboard computer. Thermal and RF design are somewhat more difficult, while assembly of the spacecraft may be easier. In current generations of three-axis spacecraft, the bus section is designed as a separate module to be attached to the payload during final integration and test.

8.3 Spacecraft Bus Subsystems

The spacecraft bus encompasses those elements of the satellite that support the communications payload. As depicted in Figure 6.3, the support is multifaceted, involving the following functions: providing prime power during sunlight and eclipse; pointing the antenna beam(s) precisely at all times; controlling temperature to maintain proper operation of electronics and mechanical devices; providing physical support during launch and preoperational maneuvers; operational ground control of the satellite through TT&C functionality; and maintaining orbital position or parameters during the spacecraft's operating lifetime. Each function involves one or more of the bus subsystems, as described in the following paragraphs.

8.3.1 Attitude-Control Subsystem

Several forces act on the spacecraft that cause the platform and the antenna to move away from optimum pointing. The attitude-control subsystem (ACS) is that combination of elements that senses any change in pointing and then causes the spacecraft to realign itself. This type of function can be found on virtually every space vehicle, including guided rockets, scientific deep space probes, and, of course, GEO and non-GEO communications satellites. Commercial satellites usually have some built-in (autonomous) ACS capability, but manual intervention from the ground often is necessary during certain phases of orbital operations and failure modes.

8.3.1.1 Attitude-Control Loop

Shown in Figure 8.6 is a conceptual block diagram of an ACS control loop intended for autonomous operation. The pointing sensor (located at the upper left) measures the pointing angle with respect to some reference. Having the

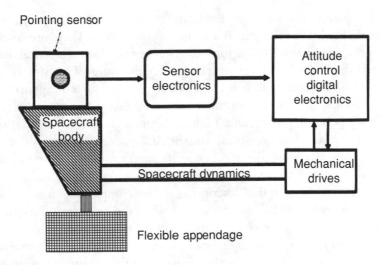

Figure 8.6 Conceptual block diagram of the showing the arrangement of the control loop.

ability to detect changes along two axes (north-south and east-west), the sensor is a critical element because its measurement accuracy has a large impact on over-all ACS pointing accuracy. (Typical sensing techniques are reviewed in a subsequent paragraph.) The ACS loop includes a sensor electronic unit to process the sensor's output signals and produce a properly calibrated error message. Attitude processing typically is carried out in the onboard spacecraft control computer. It accepts the error message for use in determining corrective action. Other functions with the processor tie attitude control to orbit maintenance since thruster firings introduce attitude perturbations. Ground control of the processor can be exercised during spacecraft maneuvers and to update its programming.

The processor control signals are used to activate the electrical windings of the mechanical devices (drives) or other transducers that cause the spacecraft attitude to actually change. In the case of a spinning satellite, the east-west mechanical drive is the BAPTA, which controls the relative spin rate between the antenna and the bus. Three-axis spacecraft typically use several mechanical drives and devices. For example, the speed and the gimbal angle of the momentum wheel are adjusted to cause small changes in spacecraft body orientation. To maintain a stable three-axis platform, those forcing functions must be applied softly because of the small degree of gyroscopic stiffness compared to the spinner. However, modern digital processors are programmed to use adaptive control laws to minimize the extent and duration of transients during orientation changes. The control laws and procedures can be modified after launch by uploading new computer code and data over the command link.

The spacecraft is not a rigid body but actually has several flexibly attached parts, particularly the solar panels and antenna reflectors. Therefore, as shown at the bottom of Figure 8.6, the force for change from the mechanical drives is applied to the body through spacecraft flexible dynamics. That simply means some parts of the body will bend and vibrate when a force is applied to the spacecraft. That, combined with the inertia of spacecraft mass, produces transients in the pointing performance of the antenna beams, which can last several seconds or minutes. It is important that the ACS prevent the peak value of the transient from exceeding the acceptable beam error and that transients are properly damped out without excessive overshoot and duration.

The control loop in this description is now complete because the sensor is mounted on the body and also will move when the attitude of the spacecraft changes. The sensor transmits a new indication of orientation, which in turn is fed into the alltitude control digital processor. The active and adaptive nature of the control loop allows it to continuously adjust the orientation of the spacecraft and to dampen any transients. Failure to do that properly can result in unstable antenna pointing or, in the worst case, physical damage to the appendages. Another critical worry is spacecraft safety in case of a major mispointing event, such as due to failure of a sensor or operator error. The situation can develop when a failure causes the spacecraft to move off proper pointing and even Earth alignment. From there, solar pointing and hence prime power generation could cease. That might combine with an alignment of the satellite with the sun illuminating an area of the spacecraft that cannot tolerate excessive heat (or not illuminating an area that requires solar heat input to maintain adequate temperature). As a result, communication performance is lost, and the spacecraft, if not restored to a safe attitude, could become unserviceable.

The answer to such relatively unlikely occurrences is to preprogram the onboard computer with contingency procedures to put the satellite into a safe condition. From there, ground commands are sent to restore normal operation and to determine the source of the problem. In some cases, the difficulty might be multiplied by improperly programmed emergency procedures.

8.3.1.2 Sensors

The two principle types of attitude sensors are the Earth sensor and RF tracking. Another very precise tracking technique is called star tracking. At certain times in the mission, it is not possible to use the Earth as a reference due to interference from the sun or moon. That usually is overcome with a sun sensor made up of a silicon solar cell and optical barrel.

Depicted in Figure 8.7 and 8.8 are the Earth sensor concepts for a spinning satellite and a three-axis satellite, respectively. Both use a narrow-beam

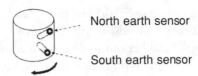

North earth sensor

South earth sensor

(A) Arrangement of north and south earth sensors
on the body of a spinning spacecraft

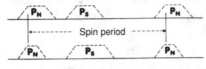

(i) Equal north and south pulse indicate that spacecraft spin axis is aligned on the equator

(ii) Short north pulse and long south pulse indicate that spacecraft spin axis is aligned to the north

(B) Examples of earth sensor pulse timing waveforms

Figure 8.7 Arrangement and operation of dual fixed Earth sensors on a spinning space-craft.

optical device consisting of a small telescope with an infrared detector at the viewing end. The star tracking technique mentioned previously also uses an optical telescope to find and track a bright star. This is an extremely precise but more complicated tracking scheme. For the spinning bus, it is a relatively simple matter to attach two Earth sensors to the rotating part of the spacecraft,

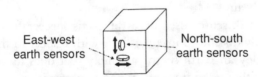

East-west earth sensors

North-south earth sensors

(A) Arrangement of north-south and east-west earth sensors
on the body of a three-axis spacecraft
(double-headed arrows indicated scanning motion)

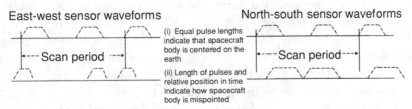

East-west sensor waveforms

North-south sensor waveforms

Scan period

Scan period

(i) Equal pulse lengths indicate that spacecraft body is centered on the earth

(ii) Length of pulses and relative position in time indicate how spacecraft body is mispointed

(B) Examples of earth sensor pulse timing waveforms

Figure 8.8 Arrangement and operation of dual scanning Earth sensors on a tree-axis spacecraft.

allowing them to scan past the Earth. The north- and south-pointing Earth sensors each intersect a cord from the relatively warm Earth's disk. The two cords will be equal in length (and hence the time scans across the Earth will be equal in duration) only if the body is pointed toward the equator. Any miscalibration of sensor hardware usually can be determined by ground measurement and then corrected by recalibration. If the body were pointed north, the duration of the north Earth sensor pulse would be shorter than that of the south Earth sensor, with the precise difference being a direct indication of the amount of misalignment.

The waveforms to the bottom of Figure 8.7 depict the time relationship between the pulses coming from the two spinning Earth sensors. The north sensor is pointed slightly ahead of the south sensor to prevent the pulses from overlapping in time. That allows the two pulses to be sent on the same signal channel. The top line represents the pulses with the body properly aligned, while the set below it corresponds to the condition with the body pointed northward. Pointing along the east-west (rotational) direction is determined by comparing the start of the north pulse with respect to an index pulse generated within the BAPTA (or drive motor for the rotary joint in the case of a simple spinner). A ground-commanded offset is stored in the attitude-control digital processor to provide a reference for comparison with the measured offset between the north and index pulses. Using the Earth pulses, the index pulses and the offset, it is possible to determine attitude along both pointing axes to within ±0.05 degrees, which is adequate to maintain reasonably good pointing of an area coverage beam.

The type of Earth sensor used on a three-axis spacecraft must be different, because the body does not move regularly with respect to the Earth. One approach is to employ an Earth sensor assembly that does its own scanning, producing pulses with the appropriate type of information. Alternatively, a fixed type of infra-red (IR) sensor array can sense motion in both directions using electronic techniques rather than physical motion. In the simplest scanning approach, two mirrors are mounted on a rotating shaft that scans back and forth. The mirrors are oriented north and south just like the respective fixed sensors on a spinning satellite. Infrared radiation from the Earth is reflected by the mirrors onto separate IR sensors. Because the sensors deliver pulses of lengths proportional to Earth chord widths, the technique operates essentially the same as described for fixed Earth sensors on the spinner.

The RF tracking technique, depicted in Figure 8.9, is an accurate system for sensing spacecraft attitude. Some of the same devices used for reception of the uplink microwave signals are applied to tracking. Each axis requires two receive horns and an associated hybrid. Out of the hybrid come two ports, one containing the sum (A + B) and the other the difference (A - B). The sum sig-

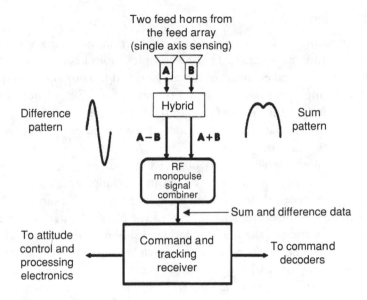

Two feed horns from
the feed array
(single axis sensing)

Figure 8.9 Attitude sensing by the spacecraft antenna and command receiver when tracking an RF pilot signal transmitted from the TT&C Earth station.

nal represents the combination of the two beams, having a combined pattern similar to the one shown on the right. However, the difference signal effectively passes through zero in its response precisely where the individual horn patterns overlap. The difference pattern, which has the appearance of a sinewave, is used to detect a null in signal energy corresponding to proper alignment of the antenna beam. That requires that a reference Earth station be located where the null should be placed and that the station transmit a continuous signal called a pilot carrier.

There remains the problem that the difference signal by itself does not provide a continuous error signal when the beacon enters the null. That is overcome by modulating the A − B response onto the A + B response, which is accomplished in the RF monopulse tracking processor. The modulation then can go through its zero point at the null. Bias and drift in this null are corrected by storing an offset pointing angle in the onboard attitude control processor. It also is advantageous to use a single receiver to perform all commanding, ranging, and tracking. Separation of the attitude-error signal from the command signal usually is accomplished in the command-track receiver, shown at the bottom of Figure 8.9. RF tracking systems have measurement accuracies in the range of ±0.01 to ±0.04 degrees. Performance of the star tracker is of this order, or possibly better.

8.3.1.3   Antenna Pointing

The real measure of ACS performance is its effect on the accuracy and stability of spacecraft antenna pointing. This is a complex subject because it involves the performance of several elements of the payload and the bus and the interaction thereof. Pointing performance is the most critical for GEO missions for which pointing angles are the smallest. In contrast, pointing for a LEO satellite is nearly one order of magnitude less critical; for that reason, bus operation can be somewhat simplified.

Overall pointing accuracy is predicted by identifying each contributor and assessing its impact in the form of a pointing budget. Sensor accuracy contributes approximately 10%–20% of the total accuracy of pointing. A second big contributor is the movement of parts of the physical antenna due to variations in solar heating of the reflector and feed supports. Referred to as thermal distortion, that effect can produce as much as one-half of the total error, depending on the thermal stability of the material used in antenna construction. Another error is that due to the precision with which the antenna and sensor are attached to the spacecraft, a process that requires the use of optical alignment instruments during assembly on the ground. By providing a ground commanding reflector alignment mechanism, the static beam pointing error can be removed almost entirely. Last of all, the transient behavior of the ACS during thruster firing will cause the antenna to be misdirected momentarily. Analysis of those effects involves the individual accuracies, the interaction of the elements, and the operation of the spacecraft in different stages of the overall mission. That places considerable demands on the prelaunch design and test of the spacecraft, the operating principles of the spacecraft when in orbit, and the means to respond to contingency situations that arise from time to time.

Figure 8.10 provides the general view of the spacecraft, indicating the three classical axes of attitude: roll, pitch, and yaw. The axes are based on the view of the spacecraft as it flies over the Earth like an airplane. The roll axis, which typically is in the orbit plane, is pointed in the direction of orbital flight; pitch corresponds to the up-down motion of the nose of the spacecraft along an axis perpendicular to flight and in the orbit plane; and yaw is a twisting motion of the antenna about the axis pointed directly down from the spacecraft (e.g., sideways motion of the nose).

For a typical operating satellite, the overall pointing error of the antenna beam is in the range of ±0.05 to ±0.15 degrees in either the north-south or east-west directions. Motion about the third axis of the spacecraft (yaw) causes the antenna pattern to twist with respect to the Earth and therefore represents a third error-prone direction of motion. A benefit of the standard ACS approach is that yaw error is automatically removed in one-quarter orbit (6 hours later

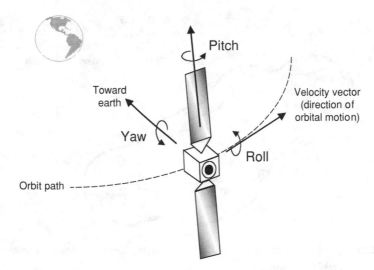

Figure 8.10 Definition of the three axes of attitude change: roll, pitch, and yaw.

for GEO) because yaw error turns into roll error at that point. For coverage regions some distance from the equator, yaw can be nearly compensated for by east-west and north-south ACS control.

Examine the antenna coverage of Mexico in Figure 8.11, which is based on the shaped beam antenna design described in Chapter 7. The pattern provides 30 dB of gain (the black contour) over the Mexican landmass and reaches 33.5 dB at the peak. The approach commonly used to evaluate antenna pointing error is to employ circles or boxes with dimensions corresponding to the maximum pointing error anticipated. Because Mexico is considerably north of the equator, yaw error has been broken down into east-west and north-south components, which are already included. In the example, the ACS pointing error is assumed to be ±0.10 degrees in both directions, which makes the error box 0.2 degrees on each side. The conservative (worst case) approach is to use boxes at the test points (circles can be used under the assumption that north-south and east-west errors are uncorrelated). The circles then are placed at critical locations such as key cities or at points along the border of the region or country being served. In Figure 8.11, error circles have been placed at Mexico City as well as two points along Mexico's coast that protrude into the water. The Mexico City circle shows that the gain at that point nominally will be 32 dB but can be expected at times to decrease as much as 0.5 dB (obtained by interpolation between the 32- and 30-dB contours). The lower right circle (located at the tip of the Yucatan Peninsula) shows that even with maximum error, a

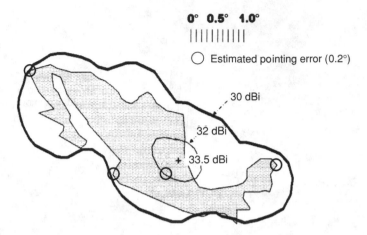

Figure 8.11 Spacecraft antenna gain performance in decibels indicating how pointing error is taken into account.

gain of 30 dB is maintained. However, the error circle at the left (Pacific coast) extends outside the contour; therefore, the gain at times will drop below 30 dB. As an alternative to using boxes or circles, the size of the map of Mexico can be increased by the extent of the maximum pointing error.

This procedure is used to compare the predicted performance of the spacecraft to the specification. While values of gain are shown in Figure 8.11, it is more appropriate to label the contours with EIRP and G/T, determined in accordance with the procedures given in Chapter 4. That then brings into play the combined performance of the communications payload and attitude-control system.

8.3.2 Solar Cells and Panels

All Earth-orbiting communication satellites use panels of solar cells to provide prime power to the payload and the bus. Over the years, performance of silicon solar cells has improved; however, high-performance solar panels are being constructed of GaAs solar cells, which can outperform silicon by a factor of two. Shown in Figure 8.12 is a comparison of the two common panel forms: the cylindrical drum and the flat panel. The dimensions D and L define the effective area illuminated by the sun. The power output of a solar cell is proportional to the intensity of solar radiation reaching the cell. The intensity is maximum with the rays arriving perpendicular to the cell and decreases with the cosine of the angle of the ray with respect to the perpendicular. A condition of zero out-

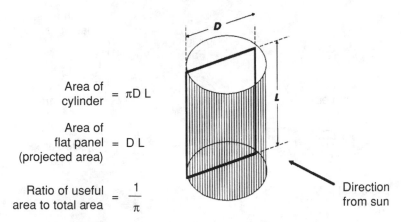

Area of cylinder = πD L

Area of flat panel (projected area) = D L

Ratio of useful area to total area = $\dfrac{1}{\pi}$

Direction from sun

Figure 8.12 Comparison of cylindrical and flat solar panels from the standpoint of the effective area in the direction of the sun.

put occurs when the rays are parallel to the cell's surface. A sun angle off-pointed vertically in Figure 8.12 will likewise produce a loss of dc output, this time affecting both panel types.

In the case of the type of flat panel on a three-axis spacecraft, a drive motor maintains the array in the direction of the sun and all cells receive identical illumination. Alignment in that manner is optimum for a GEO spacecraft; however, for a non-GEO mission, the spacecraft must be maintained with proper attitude for panel illumination even though the orbit is not in or near the equatorial plane. For a cylindrical panel, the cells are connected in series of strings running the length of the cylinder. The output of a string remains at zero as the string rotates from behind to a point at the left visible edge and increases to a maximum as the strings rotate to be in direct line with the sun's rays. The mathematical relationship shown at the left side in Figure 8.12 indicates that a cylindrical panel would have π times as many solar cells as a comparable flat panel that is continuously oriented toward the sun.

The output power from a solar panel, be it cylindrical or flat, varies over the course of a year due to the orbital geometry. A simplified drawing of the geometry is shown in Figure 8.13, indicating that the equatorial plane at GEO altitude is tilted approximately 23 degrees with respect to the ecliptic plane (i.e., the plane of the orbit of the Earth around the sun). That means a satellite experiences the seasons just as we do on Earth. (The figure uses summer and winter as seen from the northern latitudes.) In summer and winter, the sun's rays hit the panel at an angle, reducing the efficiency of the solar cells. The worst case occurs on the solstices, when the efficiency factor is equal to

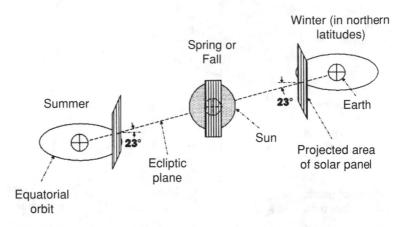

Figure 8.13 Variation of the illumination of the solar panel as the Earth and a satellite in GEO revolve around the sun.

cos(23 degrees), that is, 92%. In the spring and fall, on the other hand, the sun provides better illumination of the panel, and the cells give their maximum output at the equinoxes. It happens that the point of minimum panel output occurs at the summer solstice, when the distance between Earth and sun is greater than in winter.

For non-GEO spacecraft and constellations, the question of solar panel illumination can be much more complicated. Among non-GEO systems, the arrangement of orbits and communication coverage requirements differ significantly, imposing unique constraints on solar power generation. That would not be a problem if the spacecraft was a sphere covered with solar cells to catch the sun independent of attitude. However, flat-panel arrays are both more convenient and more efficient, so some means must be found to provide separate alignments throughout the orbits and the mission. In an inclined orbit system like Globalstar (Figure 8.14), the dual flat panels align with the sun, and the Earth-facing phased array antenna creates 16 overlapping beams. To keep the panels on all satellites in proper alignment with the sun, the beams must rotate about the yaw axis as the satellite passes over the Earth.

In contrast, the polar-orbiting Iridium satellites each keep the orientation of their 48 beams fixed, which means the panels cannot be maintained in a perpendicular orientation toward the sun (Figure 8.15). Without yaw motion, satellites with their orbit plane nearly perpendicular to the sun line experience poor solar illumination and therefore deliver only a fraction of their potential. The panels are oversized so that adequate power is generated for the bus and the payload. A fortunate consequence of this particular orbit orientation is that no power is required for battery charging since the satellite does not go into

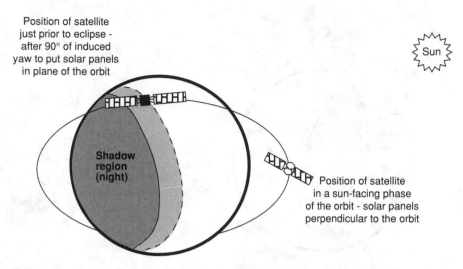

Figure 8.14 Sun illumination for an inclined orbit constellation such as Globalstar.

eclipse during this time period. Optimum alignment, shown in Figure 8.16, brings with it daily eclipse periods and a requirement for rapid battery charging when the satellite is in daylight. Second generation LEO constellations may align their solar panels in two axes to optimize solar output at all times.

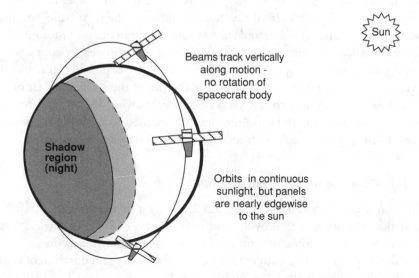

Figure 8.15 Sun illumination for a polar orbit constellation such as Iridium. This orbit has the poorest solar-panel orientation but does not experience eclipse.

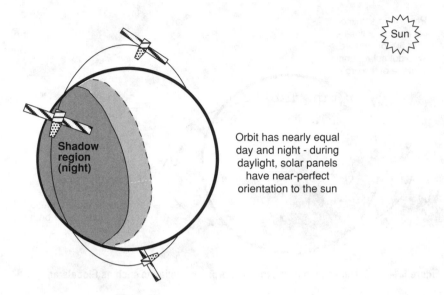

Orbit has nearly equal
day and night - during
daylight, solar panels
have near-perfect
orientation to the sun

Figure 8.16 Sun illumination for a polar orbit such as Iridium; this orbit has ideal solar-panel
orientation but experiences frequent eclipse.

These geometrical factors should provide some appreciation of the basic
requirements for solar panel design. There are, however, many other factors to
consider. The space environment degrades panel performance over time. For the
geostationary orbit region, local electrons impinge on the solar panel and gradually degrade cell performance and solar storms errupt in blasts of protons whose
energy is particularly damaging to solar cells. LEO satellites experience less electron degradation. By proper cell and panel design, it is possible to reduce the
degradation to the range of 15–20% over the life of the satellite. Most of that
power loss occurs in the first few years of operation, while the loss in later years
tapers off. Mechanisms and drive motors that are used to deploy and orient the
panels must be very reliable with redundancy built in where possible.

8.3.3 Battery Design and Configuration

The solar array described in Section 8.3.2 is only a part of the spacecraft electrical power system that provides continuous power to all electrical loads.
During the eclipse period, the sun is concealed by the Earth, and the spacecraft
must derive power from a battery system. The frequency and duration of eclipse
as a function of orbit altitude is graphed in Figure 8.17. In a LEO constellation, some fraction of the orbits will subject their satellites to eclipse periods,
which is a natural consequence of entering and exiting the side of the Earth

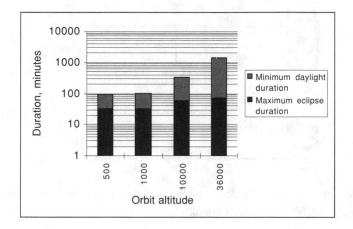

Figure 8.17 Maximum eclipse duration as a function of altitude.

experiencing nighttime. There is a polar orbit called sun synchronous that avoids eclipse entirely by adjusting the orbit plane to be perpendicular to the sun. In the Iridium system, for example, one or two orbits are in effect sun synchronous while the rest are not. MEO and GEO satellites experience eclipse much less frequently and also have seasons in which there is no eclipse at all.

8.3.3.1 Electrical Power Bus

Figure 8.18 is a simplified block diagram of a typical electrical system incorporating dual independent electrical busses. Separate busses provide redundancy and give assurance that a major failure (such as a short circuit) could eliminate only half the system. Alternatively, a single power bus architecture can be used for the primary electrical source with the split to diverse systems occurring downstream. To understand the construction of the electrical system, examine the arrangement of the upper half of Figure 8.18 (i.e., the part corresponding to bus 1).

Prime power is generated in the solar arrays and delivered to the bus. The limiter immediately to the right of the array is provided to drop the voltage during periods of maximum solar illumination (e.g., the equinoxes in GEO missions) or minimum power demand in the spacecraft. Switching in or out of the limiter is accomplished automatically by sensing the voltage on the bus (see the dashed line). The downstream electronic units further regulate their internal voltage, consistent with the specific circuit requirements. Alternatively, the primary voltage can be tightly regulated on a central basis to simplify the design of the powered units in the spacecraft.

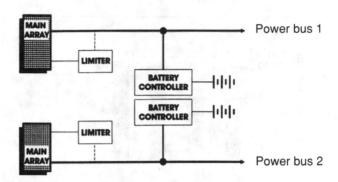

Figure 8.18 Spacecraft electrical power system, including solar arrays and batteries with dual independent power busses.

In Figure 8.18, the symbol composed of alternating short and long lines represents the battery composed of electrochemical cells used to store energy for eclipse operation. Nickel-cadmium cells have been used successfully for many years, but the more efficient nickel-hydrogen (NiH_2) cells are now standard in commercial satellites. Illustrations of a typical NiH_2 cell and battery system are provided in Figures 8.19 and 8.20, respectively [5]. The basic operation of this type of cell depends on the chemistry of gaseous hydrogen and water. The hydrogen gas stored in the pressure vessel increases in pressure as the cell is charged. When the cell subsequently is discharged, the nickel-active material is reduced to its original preoxidation state and the hydrogen gas is oxidized to form water. The cell pressure gradually decreases to its low value as the hydro-

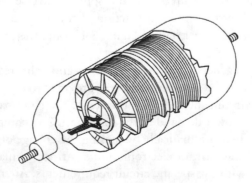

Figure 8.19 Standard NiH_2 dual stack independent pressure vessel.

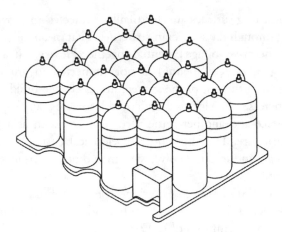

Figure 8.20　Typical NiH$_2$ battery design.

gen gas is consumed. The energy storage efficiency and lifetime of this class of cell currently are the best in the industry.

　　Battery charge is maintained by directing excess solar array current to the battery system during sunlight. A battery controller switches the electrical bus from the array to the battery at the start of eclipse and back at the conclusion. The action can be automatic or by ground command, depending on the particular approach taken in the design of the system. The battery controller also is used for maintenance of the batteries, which consists of delivering a light "trickle" charge most of the time and periodically commanding a rapid discharge into resistors for cell reconditioning. The overall process of battery maintenance is a complex subject, particularly in light of the criticality of energy storage for the entire mission.

8.3.3.2　Electrostatic Discharge

The spacecraft electrical system provides conduction paths to every electronic and electrical unit in the spacecraft and can allow potentially damaging electrical transients to propagate through the system. Low-level transients generated aboard the satellite by normal operation of the payload can be rendered harmless by proper power supply design. However, it is possible for spikes in the range of hundreds to thousands of volts to occur from electrostatic discharge (ESD), a phenomenon not unlike lightning.

　　ESD is the result of a buildup on external surfaces of the spacecraft of a static charge from electrons traveling in space. In a well-designed and constructed satellite, all conducting surfaces are electrically connected to provide a common ground running through the spacecraft structure. That means all points inside the spacecraft rise in negative voltage potential together. However, if any external surface (conducting or nonconducting) is not connected to ground, its voltage potential differs from the rest of the spacecraft. As change builds up and potential difference increases, an electrical arc could leap across any gap that separates the external surface and another (grounded) element such as a piece of metal waveguide or structure. The large surge of electrical current that would ensue could pass from ground into the power system and be carried into sensitive electronic equipment.

The effect of ESD has been experienced many times in the past, particularly on new spacecraft designs. A seemingly random shutdown of an operating transmitter could be a symptom of ESD as could a "glitch" (a wild variation) in the telemetered output of a sensor. In some cases, ESD has been known to permanently damage digital electronic equipment and power supplies. Therefore, it is vital that ESD be carefully taken into account by the use of appropriate materials and by proper grounding.

8.3.4 Liquid Propulsion System

The onboard propulsion system uses the principle of expelling mass at some velocity in one direction to produce thrust in the opposite direction (Newton's first law). As with any rocket, there are principles and practices that govern how safely and effectively propulsion is achieved. Three basic types of mass can be ejected: sold fuel (e.g., a controlled explosion), liquid fuel (which decomposes into a gas jet), and ion propulsion (which is based on accelerating charged particles with an electric field). We begin with the liquid type of propulsion system, the most common.

The majority of commercial communications spacecraft control their orbital positions by using small liquid fuel rocket engines. Propellant options include a cold gas like nitrogen, which is low in cost and performance, a single combustible propellant like hydrazine that produces hot gas thrust when exposed to a catalyst, or two liquids (fuel and oxidizer) that produce combustion on contact. Low levels of thrust are needed for small corrections in orbit, while high levels are needed for major orbit changes or corrections. Large amounts of thrust required for major orbit changes can be provided either by solid rocket motors or by large liquid fuel engines. The general arrangements of spinning and three-axis spacecraft propulsion systems are compared in Figure 8.21.

The launch vehicle provides the boost to place the spacecraft into an initial orbit from which adjustments can be made using the onboard propulsion

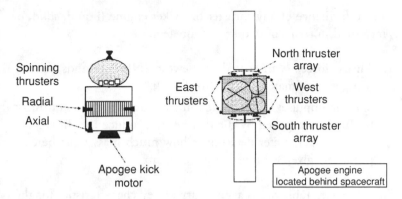

Figure 8.21 General arrangement of propulsion systems for spinning and three-axis space-craft configurations.

system. In the case of most GEO missions, that is an elliptical orbit with perigee at approximately 300 km and apogee at near-synchronous altitude (36,000 to 37,000 km). The configurations in Figure 8.21 assume the use of a solid fuel apogee engine that is permanently integrated into the spacecraft and fired by ground command.

In the case of the spinner, there is a need for four liquid thrusters of approximately 22N (5 pounds), all mounted on the rotating section. Either one of the axial thrusters when fired continuously will push the spacecraft upward, which is useful for north-south stationkeeping. The spin axis can be moved to adjust spacecraft yaw attitude by pulsing one axial thruster in synchronization with the spin rate. Because the spacecraft is acting as one gigantic gyro, the pulsing must be done 90 degrees ahead of the desired spin axis change. Pulsing of a radial thruster also must be synchronized with spin to move the spacecraft in the plane of the orbit. To prevent misalignment of the spin axis, the direction of thrust passes through the center of mass of the spacecraft. Note that only two liquid thrusters are required (one axial and one radial) to satisfy requirements for attitude correction and stationkeeping.

The fact that a three-axis satellite does not spin to provide stability leads to the need for many more thrusters and a much lower level of thrusting. As shown in Figure 8.21, the typical three-axis uses 12 thrusters arranged in arrays on the north, south, east, and west faces. Low thrust levels are needed to minimize disturbance of spacecraft attitude during thrusting. In comparison, the high gyroscopic stiffness of a spinner requires the use of relatively large thrust levels to correct spacecraft attitude.

The performance of any thruster or rocket engine (liquid, solid, or electric) is measured in terms of two key parameters:

1. Thrust force, F, measured in Newton (N) or pounds (force), there being approximately 4.45 N/lb (force);
2. Specific impulse, I_{sp}, measured in seconds, is the ratio of thrust force to the mass expelled by the thruster to produce the thrust. This performance parameter determines how much mass must be ejected to produce a given orbit velocity increment.

For reference, Table 8.2 is a summary of key characteristics for the types of thrusters described here. They break down into chemical thrusters, electric augmentation and ion thrusters, and solid-fuel rocket motors. The range of I_{sp} values is from under 100 sec on the low end for cold gas up to 2,000 sec or more for the different forms of electric and ion propulsion. As is usually the case in high technology, the lower performance approaches are simple and very reliable; as we move toward the higher performing systems, the complexity and cost increase dramatically. However, the trend in large satellites is toward the higher performing types of thrusters that also provide relatively low thrust force levels. Table 8.2, summarized from [1], is for illustrative purposes and should not be used as a design guideline.

Table 8.2
Basic Comparison of Onboard Thrusters Used for Various Commercial Satellite Missions

Type	Propellant	Energy	Typical I_{sp} (sec)	Thrust force (N)
Cold gas	N_2, NH_3, He	Gas pressure	100	0.01–50
Solid fuel rocket motor	Dry mixture of fuel, oxidizer, binder	Chemical	300	10–10^6
Monopropellant	Hydrazine	Exothermic decomposition	200	0.05–25
Bipropellant	MMH and N_2O_4	Chemical	350	1–10^6
EHT	Hydrazine, NH_3	Exothermic decomposition and electrothermal	300	0.05 to 3
ARCJET	NH_3, N_2H_4	Electric arc heating	800	0.05–5
Ion propulsion	Hg, Xe, Ce	Electrostatic	3000	10^{-6}–0.5

8.3.4.1 Reaction Control System

The liquid fuel tanks, fuel lines, valves, and thrusters make up the reaction control system (RCS), which is an integral part of the bus. Figure 8.22 shows a conceptual drawing of the spinner's RCS, which uses a single liquid fuel called hydrazine. With the tanks having been pressurized prior to launch and with the spacecraft spinning on orbit, fuel exits the tanks and remains in the lines. The drawing is not accurate in that the lines are actually connected to the tanks on the side facing the exterior (solar panel) so that rotation forces the fuel out of the tanks.

The system shown in Figure 8.22 is split into redundant halves, either of which could support the mission with one axial and one radial thruster and a pair of tanks. An interconnect latch valve, which normally blocks connection between the two halves, can be commanded open to allow fuel from one side to move to the other. It normally is used during the mission to equalize fuel load and pressure, but it also can allow access to half the fuel in the event of a thruster open failure. Note the use of an isolation valve in the line between the radial thruster and the tanks, which can be closed if a thruster fails in the open position. Tanks and lines typically are made from welded titanium and have been thoroughly tested on the ground to preclude even the slightest possibility of a leak.

In the case of a three-axis satellite, the flow of fuel in the tank is controlled by use of a device in the tank rather than the force of rotation. One approach is to use a rubber diaphragm separating the fuel from the pressurant gas, allowing the gas to push the fuel out of the tank, much the way we squeeze a tube of tooth-

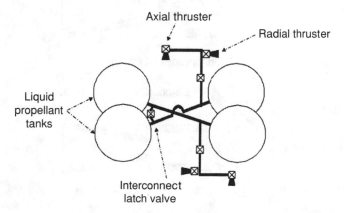

Axial thruster

Radial thruster

Liquid propellant tanks

Interconnect latch valve

Figure 8.22 Simplified diagram of a spinning reaction control subsystem employing monopropellant hydrazine fuel.

paste. Another more common technique employs vanes and spongelike surface-tension devices to direct liquid fuel toward the tank exit by capillary action.

Conventional liquid monopropellant systems use hydrazine fuel alone, which on contact with a catalyst decomposes into its constituents. The process releases energy, resulting in high-pressure gas at the thruster nozzle. Another important innovation in spacecraft RCS was the introduction of bipropellant technology, that is, the use of a separate fuel and oxidizer. That is now the standard for GEO satellites. Monopropellant and bipropellant thrusters are compared in Figure 8.23. Having separate fuel and oxidizer produces greater thrust for the same weight of fuel, which allows a bipropellant system to yield longer life (or conversely save fuel weight for the same life). Of necessity, bipropellant systems have more tanks, lines, and valves; therefore, there is some penalty in cost and increased complexity.

An improvement in monopropellant performance has been achieved with the electrically heated thruster (EHT), which marginally increases thrust by heating the propellant with an electrical winding. Thrusting with the EHT has to be done in such a way as to use excess electrical power; otherwise, there would be a weight penalty for the additional solar cells. The benefits of bipropellant and EHT technology outweigh the costs for larger spacecraft and with missions that extend beyond 10 years.

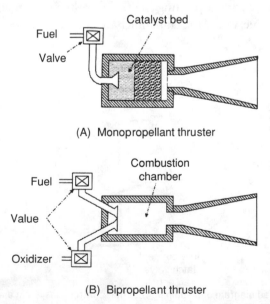

(A) Monopropellant thruster

(B) Bipropellant thruster

Figure 8.23 Propellant thrusters employing monopropellant and bipropellant technologies.

8.3.5 Electric and Ion Propulsion

Electric propulsion, which depends on power derived from the solar panels to provide most of the thrust, is being used in commercial missions, particularly at GEO. The specific technologies under this category include the following:

- ARCJET generates intense heat using between 1 kW and 20 kW of prime power to produce a wide-field, high-energy exhaust plume using ammonia as the elemental fuel. The engine incorporates a nozzle to control the exhaust plum (while the other propulsion technologies, discussed below, do not). I_{sp} available is in the range of 500 to 800 sec, and thrust level is approximately one order of magnitude below (1/10) that of hydrazine.

- Pulsed plasma thruster (PPT) utilizes Teflon propellant to provide over 1,000 sec of I_{sp} while operating at power levels of 100–200W. Thrust can be conveniently adjusted by varying the pulse rate. PPTs have demonstrated thrust levels of only three orders of magnitude below (1/1000) standard hydrazine thrusters.

- Hall thruster (HT) uses a low-pressure propellant gas that supports an electric discharge between two electrodes. A radial magnetic field generates a Hall-effect current that reacts with the magnetic field to place a force on the propellant in the downstream direction [6]. These devices are developed and produced in Russia and deliver I_{sp} of the order of 1,600 sec.

- Ion propulsion is not a new technology; devices in this class have existed since the early 1960s. Ion propulsion carries very little fuel, relying instead on acceleration of a plasma to very high velocity. I_{sp} up to 3,000 sec at a power of 1 kW represents the potential. Current devices deliver about 2,500 sec and are capable of operating throughout a typical GEO mission of 15 years. Thrust level is quite low, requiring that the thruster be operated for extended periods at a time to provide sufficient velocity deltas.

The basic arrangement of an ion thruster is shown in Figure 8.24. That particular model, developed by Hughes Electron Dynamics, ionizes elemental Xenon gas propellant to form a charged plasma. Thrust is produced by accelerating the plasma in an intense electric field, causing gas molecules to be expelled at high velocity. The thruster nozzle can be directed along two axes to facilitate precise pointing of the entire spacecraft. Typically, a pair of thrusters and control systems ensure redundancy. An ion thruster of this type first flew on PanAmSat 5, launched successfully in 1997. Subsequent missions use ion

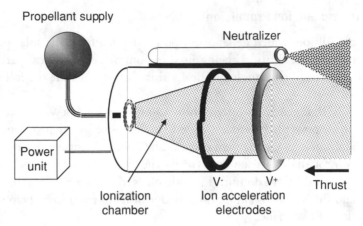

Figure 8.24 The principle of ion propulsion.

thrusting for stationkeeping, attitude control, and augmenting the launch rocket for raising the final orbit. The latter is attractive because the I_{sp} of ion propulsion is substantially higher than any chemical system and so leaves more dry mass once final station is reached (or, conversely, reduces the amount of fuel required to reach orbit). The consideration here is the extra time (typically months) needed to achieve final orbit because of the low thrust force afforded by ion propulsion compared to chemical systems.

Electric propulsion in its various forms shows great promise for communications satellites, the majority of which depend on some form of stationkeeping. In GEO systems, being able to maintain orbit position throughout its life is critical to a satellite's operating success. There are significant benefits to reducing the weight of stationkeeping fuel, which represents a substantial part of the mass to be delivered to orbit. Even in the absence of stationkeeping, electric propulsion could improve the price and/or the performance of satellites that must be raised from the initial launch vehicle altitude to the final orbit. There are many considerations in the use of electric propulsion, the most significant of which are the lifetime of the thruster itself, the possibility of contamination of various critical spacecraft surfaces from thruster effluent, and the need for high voltage and power to operate the thruster itself. In particular, the wide plumes of the engines result in cant angles of 45 degrees or more, reducing the achieved I_{sp} and complicating the orbit and attitude control methodology to a degree. All those factors can be and are being considered in the development of this attractive technology. Chemical propulsion systems using liquid fuel and solid fuel (to be discussed next) will continue to be applied in near-term satellites, particularly for LEO applications, in which stationkeeping is minimal.

8.3.6 Solid-Fuel Rocket Motors

Solid-fuel rocket motors are filled with a relatively hard, rubbery combustible mixture of fuel, oxidizer, and binder. When ignited by a pyrotechnic device in the motor case, the mixture burns rapidly, producing intense thrust lasting 1 min, more or less. Figure 8.25 shows a cross-section of a typical small solid motor. This type of rocket motor delivers an I_{sp} of approximately 300 sec and is suitable for major orbit changes, such as for perigee or apogee kick. Redundancy is applied to the igniter, as shown. The case is constructed of titanium, although high-strength fibers can be used when wound in the proper shape. Attached at the exit is a nozzle assembly made of mixtures of carbon and other materials. During firing, the nozzle gets red hot from the heat of the expelled gasses and is eroded by the particles in the combustion products. Nozzle design is as much an art as a science, and the importance of quality control and testing cannot be overemphasized. Failures of such nozzles in the past have left satellites in unusable orbits.

As shown in Figure 8.21, a single solid motor can be integrated into the spacecraft and the empty case carried for the remainder of the satellite's lifetime. Other solid motors can be attached at the bottom and discarded after use. There has been a recent trend to use relatively large liquid bipropellant motors that may or may not be part of the RCS because such a motor can be restarted and fired at successive apogees. Solid motors will, however, still find application during launch and orbit change simply because of their efficiency and simplicity.

8.3.7 Tracking, Telemetry, and Command

The communications payload and spacecraft bus, although designed to serve for a period of many years, require ground support and intervention even under

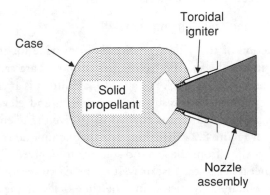

Figure 8.25 Typical solid fuel rocket motor to be part of the spacecraft for use in apogee kick or perigee kick.

normal circumstances. Consequently, there is need for a two-way system of radio communication to allow ground computers and support personnel to determine the status of the spacecraft and to control the various subsystems of the payload and the bus. That is the purpose of the TT&C operations, which, as illustrated in Figure 8.25, consist of both ground and space elements.

The TT&C ground station and its associated SCC provide a sophisticated facility consisting of RF equipment, baseband and digital processing equipment, computers, and operator display consoles. (All aspects of the TT&C ground facilities are described in Chapters 9 and 11.) The RF equipment is in the form of a fairly typical Earth station with the exception that the antenna has the ability to track the satellite to help measure its position. Commands generated by operators or computers are sent to the satellite on the command uplink. Onboard the satellite, the commands are decoded and activate some aspect of satellite operation. The telemetry downlink carries what is called housekeeping data, which are measurements of temperature, electrical voltage, RCS tank pressure, and other important parameters. Telemetry data are received by the ground facility, processed in the computer system to provide a current picture of satellite health, and stored for evaluation of long-term performance.

Most communications spacecraft employ two different antenna systems for TT&C. As shown in Figure 8.26, an omnidirectional antenna (discussed later on in this section) is used during transfer orbit, at which time the main reflector is not pointed toward the Earth. The high-gain (reflector) antenna subsequently is used because it permits the TT&C station to reduce its power significantly. In the case of RF tracking, the high-gain antenna includes the feedhorns, which simultaneously receive the command carrier and generate pointing error signals for use by the ACS. LEO spacecraft may rely on omnidirectional antennas throughout their mission due to the close range and rapid motion relative to TT&C ground stations.

8.3.7.1 Satellite Tracking and Ranging

One important function of the TT&C station is to provide for the measurement of satellite position. The specific techniques employed are called tracking and ranging, illustrated in Figure 8.27. For a GEO satellite, it is a relatively simple matter to fix its position by measuring the azimuth and elevation of the Earth station antenna when its beam is aligned to receive the telemetry signal. Note also at the top of Figure 8.27 that there is only one unique value of range (the distance to the satellite from the ground station) for a given GEO satellite once we know whether it lies to the east or west of the Earth station's longitude. The precise range is obtained by measuring the time delay at the TT&C station between transmission of a reference signal (like the pulse in the bottom drawing) and its reception after being relayed by the satellite. Taking into

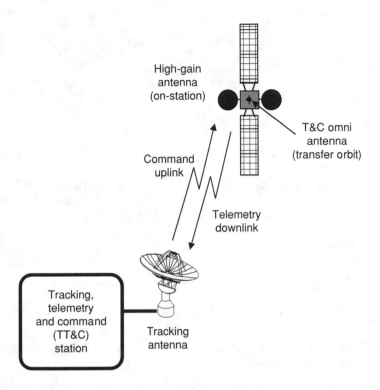

Figure 8.26 TT&C operations from the ground using the omnidirectional and high-gain antennas of the satellite.

account many factors in the link, the range can be calculated from the delay because the speed of propagation is fixed at the speed of light. The amount of time delay through the satellite and Earth station equipment, which must be removed from the measurement, is determined by ground tests and by long-term analysis of ranging data. The combined accuracy of range measurement for GEO satellites typically is 25m or less.

Tracking and ranging of non-GEO satellites apply the same measurement techniques. Because the associated ground stations require multiple tracking antennas to ensure continuous communication, it becomes a relatively simple matter to include tracking and ranging as added functions. The data subsequently are relayed by suitable communication means to the SCC, where the satellite performance and location are determined. Position accuracy requirements depend on the particular mission but could be relatively simple compared to GEO because all ground stations are required to operate with moving satellites.

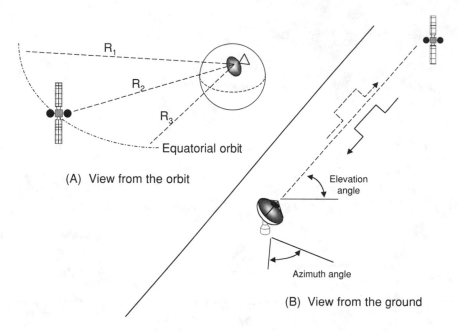

(A) View from the orbit

(B) View from the ground

Figure 8.27 A single TT&C Earth station measures the range and direction to the satellite so the satellite's position can be computed.

8.3.7.2 Spacecraft Telemetry System

The telemetry link between the spacecraft and the ground station is used to provide data concerning the status and health of the payload and bus subsystems. To get a feel for how the data are gathered onboard the spacecraft and transmitted to the ground, examine Figure 8.28, which is a simplified block diagram of the telemetry system. At the left are examples of some of the critical elements of the spacecraft: a battery, a fuel tank, and an RF power amplifier. Measurement of a key parameter of each is made through a transducing device, which senses the parameter and produces a corresponding output voltage. Battery voltage is taken directly and adjusted to a standard level range with a set of voltage-dropping resistors. A pressure transducer is connected into the fuel lines to provide information about the amount of fuel and gas remaining. The device outputs a corresponding voltage, which also can be telemetered. The last example in Figure 8.28 is an SSPA for which we might want to telemeter the output power and GaAsFET gate current. A power meter detects a small amount of RF output power taken from a 20-dB coupler (that steals only 1% of the total output of the amplifier, equivalent to an output loss of 0.04 dB) and again produces a corresponding voltage to be telemetered.

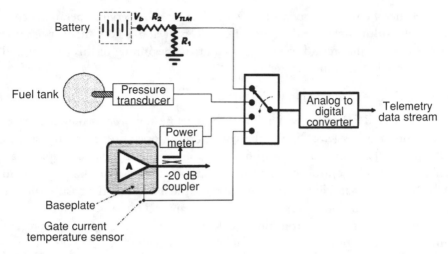

Figure 8.28 The spacecraft telemetry subsystem makes analog measurements and converts them to digital data before transmission to the ground.

The simple system depicted in Figure 8.28 requires the transmission of four channels of telemetry. On a real spacecraft, several hundred separate channels could be involved, some being simple on/off indications and others being complex analog measurements, like the four previously discussed. Therefore, TDM is used to combine the channels into a single stream for downlink transmission. While some of the early satellites transmitted the analog samples directly (i.e., pulse amplitude modulation), all modern commercial telemetry systems encode the samples using PCM. That is the same technique used in digital telephone and data communication systems (described in Chapter 5).

The output at the right of Figure 8.28 is, therefore, a multiplexed stream of digitized telemetry samples repeated in intervals of typically 1 sec or multiples of 1 sec. The frame structure allows some parameters to be measured more frequently than others. In addition, it might be possible to add parameters when required for mission support. That could include the ability to increase the frequency of measurement of critical parameters, particularly if an anomaly is detected. All the telemetry information is transmitted to the ground, where it is recovered by the baseband and digital processing equipment in the TT&C station.

It was mentioned previously that reception of non-GEO satellite telemetry must, of necessity, be intermittent. That limitation is overcome by storing vital telemetry data onboard and downloading when in view of a suitable ground station. Systems engineers need to determine the mission requirements for telemetry coverage as part of the specification of in-service operations concepts and, ultimately, the design of the spacecraft and TT&C network. Some approaches in that regard were suggested at the beginning of this chapter and are reviewed further in Chapter 11.

8.3.7.3 Spacecraft Command System

The command link is received onboard the satellite and processed by the spacecraft command system. Unlike the continuously operating telemetry system described above, the command link need only operate when the spacecraft must be actively controlled. The one exception is the RF tracking system in which the command link is continuously on the air. While the basic structure of the command link is reasonably secure, many TT&C systems incorporate various forms of added protection, such as command encryption to prevent unauthorized parties from taking control of the satellite.

Shown in the simplified diagram of Figure 8.29 is an example of how the command system might control some elements of the spacecraft (in this case, the payload). Each unit to be controlled has associated with it a control driver, which can be as simple as the single transistor circuit shown at the top. The control driver requires two inputs before it will activate the unit under control. First, the COMMAND line from the spacecraft command decoder must go "high" (meaning that a positive voltage must be applied to the base of the transistor). Then, while COMMAND remains high, the EXECUTE line must be grounded so the transistor will conduct. Under those circumstances, the output to the unit under control will change from the high voltage, V, to near ground potential, activating the unit. The reason the dual process is used is to provide an extra safeguard against operator mistakes and equipment anomalies.

Just how the COMMAND and EXECUTE signals are processed is depicted in the lower part of Figure 8.29. Each function to be controlled has assigned to it a unique digital command word, like a post office box number. Those words are maintained in a list that is unique to the particular type of spacecraft. To protect against unauthorized access, the words can be encrypted prior to uplink transmission and subsequently decrypted in the spacecraft. Commands are selected by the spacecraft control operator, who sends them over the command link to the satellite. In the spacecraft, the specific command is identified by the command decoder, which has a direct and unique connection to each control driver. Three typical examples are shown at the right. The S switch can be commanded between its two positions using its control driver. Activation of the TWTA requires two separate ON commands, first for the heater and then (after a predetermined warmup period) for the high voltage.

The standard procedure for commanding a unit ON is as follows. First, the uniquely defined command word is sent to the spacecraft, where it is

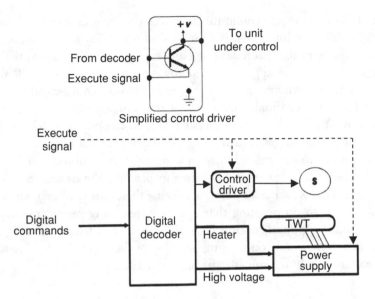

Figure 8.29 Simplified block diagram of the spacecraft command subsystems showing how the COMMAND and EXECUTE functions are used.

received, decoded, and retransmitted to the ground via telemetry for verification. The ground station computer automatically checks to see that the received command word is the same as that which was sent. At that point, the operator can be sure the spacecraft is ready for the EXECUTE, independently of the COMMAND. Only that unit for which the proper command word was sent will respond to the EXECUTE. Autonomous onboard verification of the command (with retransmission to the ground) is accomplished through a rigorous data integrity check in the command decoder similar to the CRC sequences used in computer communications. Three-axis spacecraft also use stored commands, which are either maintained in computer memory onboard or uploaded ahead of the desired function. That involves command sequences needed to alter spacecraft configuration or even orbit.

8.3.7.4 TT&C Omnidirectional Antenna

The TT&C system on a spacecraft may use two different antennas during the mission lifetime. For most of the initial operations in transfer orbit and before the main communications antenna is functional, the TT&C omnidirectional antenna provides all access to the spacecraft. Similar comments are appropriate for non-GEO satellites, which could, because of shortened path length, employ omniantennas for an entire mission.

An ideal omniantenna would function like a point source of light, radiating equally in all directions. That is a physical impossibility for a real antenna, due to blockage from the spacecraft body and its appendages, coupled with the difficulty of designing an appropriate microwave structure. Figure 8.30 shows some more practical antenna patterns currently used on commercial satellites: the toroidal and the cardioid. Placement of the omniantenna on a three-axis spacecraft can, at times, be difficult. That is due to the potential RF blockage from reflectors, solar panels, and other appendages. The solution has been to incorporate two or more cardioid or horn antennas that combine signals in the command receiver or telemetry transmitter to produce the desired near-complete coverage. That is particularly critical because three-axis spacecraft are somewhat more prone to off-pointing than spinners, requiring careful monitoring and fast response by ground controllers. In the absence of adequate coverage, the spacecraft may have to rely on preprogramming of the control processor to try to put the spacecraft into a suitable attitude for TT&C operation.

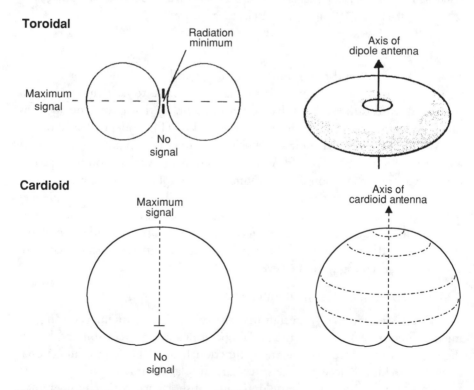

Figure 8.30 Radiation patterns of near-omnidirectional antennas.

8.3.8 Thermal Control

One of the little appreciated yet vital areas of spacecraft design is thermal control. Space can be a harsh environment in terms of temperature, both hot and cold. Active components would prefer to operate at or near room temperature (22°C), while external appendages, like reflectors, feeds, solar panels, and thrusters, can expect to experience temperature extremes of −100 to +100 C. Therefore, the thermal control system represents a common denominator for all operating elements of the spacecraft. Unexpectedly high temperatures cause great concern for such components as moving bearings and solid state electronic devices, which may have been designed and tested for a more nominal temperature range. Failure of the thermal control system can be caused by inaccurate analysis of heat distribution followed by insufficient ground environmental testing. Therefore, the criticality of proper analysis and testing of the thermal control system cannot be underestimated.

To get an appreciation for the elements of the thermal control system, examine the cross-section of a hypothetical spacecraft configuration shown in Figure 8.31. The picture does not represent a particular spacecraft design but shows the key components and their arrangement. There are two fundamental approaches to thermal design: passive thermal control and active thermal control. Passive thermal control means that the desired temperature range is maintained using appropriate thermal surface coatings (paint or deposited material such as aluminum or gold), blankets, and mirrors. Those techniques are supported by simple conduction and radiation, augmented by thermostatically controlled electrical heaters.

Active thermal control depends on mechanical devices of one sort or another that can enhance or directly alter the thermal properties of the spacecraft. The primary example is the heat pipe, which uses a gas-liquid thermal cycle to move heat from a warm place (the interior) to a cooler place (exterior of the spacecraft). High-power GEO three-axis spacecraft in the 4-kW and above class rely on heat pipes embedded in the equipment-mounting surfaces. The absorptance and emissivity of the outer surfaces of the north and south panels of three-axis spacecraft are enhanced through the use of optical solar reflectors, which are specially designed quartz mirrors. The east, west, nadir (Earth-facing), and zenith surfaces are usually covered with multilayer insulation blankets. Satellites designed for GEO service at more than 5 kW are now designed as closed cages of heat pipes.

Reviewing the elements of a passive design in Figure 8.31, the heat producing components (TWTs and other electronics) typically are attached to a heat-conducting surface such as an aluminum honeycomb equipment shelf. The primary source of external heat is the sun (shining on the spacecraft from

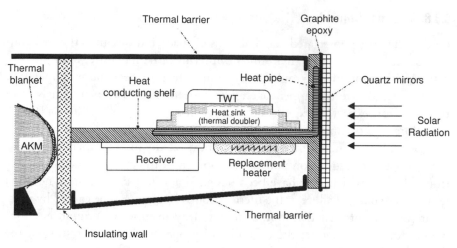

Figure 8.31 Overview of passive and active thermal control techniques, which rely on the conduction and radiation of heat.

the right), which must be precluded from raising the temperature in the compartment above acceptable limits. In addition, the heat generated in the compartment by the electronics must be transmitted to space, or it, too, will cause the internal temperature to rise. Both those necessities can be accomplished with a quartz mirror radiator located on the side facing the sun. The mirror acts like a filter, reflecting visible and ultraviolet radiation away and transmitting infrared radiation from inside to outer space. Mirrors have to be placed carefully on the spacecraft because any contamination or clouding of the surface during the life of the satellite reduces effectiveness. Generally, the temperature in the spacecraft gradually rises on a yearly basis, which must be taken into account in the overall design of the thermal control system.

If the spacecraft contains an apogee kick motor (AKM), it is necessary to provide considerable thermal control measures for it. As shown at the left of Figure 8.31, the AKM is surrounded by a thermal blanket. Electrical heater wires can be placed under the blanket if needed. The purpose of these measures is to keep the motor and propellant temperatures above a prescribed minimum prior to firing. To protect the rest of the spacecraft from the heat of the motor firing, an insulating wall and thermal barriers are placed between the motor and the sensitive components.

A similar design consideration exists for the batteries, which need to be kept cooler and within a tighter temperature range than other equipment. On GEO satellites, the prevailing design is to give the batteries their own isolated thermal environment, usually at the aft end of the spacecraft.

In the thermally controlled compartment, the electronics receive and transmit heat during their normal operation. It is the purpose of the thermal control subsystem to achieve thermal balance in the desired temperature range during seasonal variations and when units are individually turned off and on. The TWT shown in Figure 8.31 produces localized heat, which can be distributed more evenly by metallic heat sinking using thermal doublers (that technique, unfortunately, adds considerable weight to the spacecraft). The receiver mounted below it, therefore, is heated by the TWT, aiding in its temperature control. Thermal conduction is improved by installing heat pipes under the hot portion of the TWTA. If and when the TWT is not operating, the heat necessary to warm the receiver is provided by a replacement heater located in proximity to the units. Replacement heaters, operated either automatically or by ground command, are necessary to maintain an overall "bulk" temperature for the spacecraft.

High-power DTH satellites can be designed to enhance heat rejection by employing TWTs with direct radiating collectors. The collector end is exposed to space by mounting of the TWT on an externally facing panel of the spacecraft. As mentioned in Chapter 6, the collector develops about half the total dissipation of the TWT, so direct radiation is beneficial in terms of reducing thermal control system complexity and weight.

This discussion of the primary aspects of thermal design is intended only as an introduction. An important aspect to such study is the creation of analytical thermal models for use in predicting temperatures in the spacecraft at various critical points and times in the mission. The models must be verified by ground testing in a simulated space environment. Once in orbit, it is difficult to compensate for errors in analysis or design, although it happens all too frequently. That is why it is important to provide a thermal design with adequate margin and to use telemetry to measure the temperature of critical elements after the satellite is operating in orbit. Satellite operators need to keep a close eye on temperature trends through alert monitoring and careful recordkeeping. The data should be fed back to the spacecraft designer to help improve future models.

8.3.9 Structural Arrangements

The last area to discuss in spacecraft design is the structural arrangement or physical support system for the spacecraft. Figure 8.32 shows the common arrangements for the spinner and the three-axis. To provide rotational symmetry, the spinner uses a central cylinder or truncated cone. That type of structure, also called monocoque, is rigid and strong along its axis, which is the direction of maximum force during launch of an expendable rocket. The monocoque can

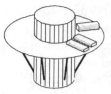

(A) Thrust tube and shelf in a spinning spacecraft

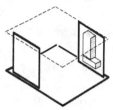

(B) Structural panels in a modular three axis spacecraft

Figure 8.32 Basic structural arrangements used with spinning and three-axis spacecraft configurations.

be replaced by a trusswork to provide greater strength in cross-directions. Attached to the central structure is one or more equipment shelves, which allow placement of payload and bus electronic components. The cylindrical solar panel is attached around the circumference of the equipment shelf. In the case of a dual spinner, the spinning and despun sections are each constructed in this manner and brought together for testing as a system. The sections can be separated for shipment.

The box structure at the bottom of Figure 8.32 often is used for many three-axis satellites because it facilitates the attachment of solar panels and antennas. A monocoque still can be placed at the center to handle launch loads, although flat panels and struts may be preferable to maximize mounting area and volume. Equipment is mounted on panels, which are integrated with the spacecraft. It is now a practice to configure the structure in separate payload and bus modules.

Structural design of the spacecraft must consider all phases of the mission, ranging from buildup during manufacture, on-ground testing, transportation to the launch site, launch and final orbit injection, deployment on station, and operation for the lifetime of the satellite. A major consideration is that the spacecraft must fit properly within the launch vehicle shroud, as will be discussed in Chapter 10. All components must be placed appropriately in the spacecraft and be capable of operating properly under all expected conditions. The structure also functions as a shield cage to protect sensitive electronics from space radiation and ESD. Another area of complexity is the continuing drive to reduce weight because that is what determines lifetime (due to fuel availability) and cost (associated with launching of the satellite or constellation). There is a need to find lightweight materials, both metallic and nonmetallic, that can be properly fashioned into the structural components of the vehicle and withstand the rigors of space flight.

References

[1] Larson, W. J., and J. R. Wertz, *Space Mission Analysis and Design,* Torrance, CA: Microcosm and Dordrecht, The Netherlands: Kluwer Academic Publishers, 1992.

[2] Cenker, R. J., M. Halverson, and R. Nelson, "Design and Operation of a Non-Stationkept LEO-Constellation." *AIAA 16th Internat. Communications Satellite Systems Conf.,* Washington, D.C., February 25–29, 1996, Collection of Technical Papers, American Institute of Aeronautics and Astronautics, New York, paper no. 96-1082, 1996.

[3] Yarbrough, P. G., "Operation Concept for the World's First Commercially Licensed Low-Earth Orbiting Mobile Satellite Service," *AIAA 16th Internat. Communications Satellite Systems Conf.,* Washington, D.C., February 25–29, 1996, Collection of Technical Papers, American Institute of Aeronautics and Astronautics, New York, paper no. 96-1049, 1996.

[4] Smith, D., "Operations Innovations for the 48-Satellite Globalstar Constellation," *AIAA 16th Internat. Communications Satellite Systems Conf.,* Washington, D.C., February 25–29, 1996, Collection of Technical Papers, American Institute of Aeronautics and Astronautics, New York, paper no. 96-1051, 1996.

[5] Coates, D. K., et al., "The Design and Manufacture of Nickel-Hydrogen Batteries for Spacecraft Power Systems," *AIAA 16th Internat. Communications Satellite Systems Conf.,* Washington, D.C., February 25–29, 1996, Collection of Technical Papers, American Institute of Aeronautics and Astronautics, New York, paper no. 96-1023, 1996.

[6] NASA, "Overview of Electric Propulsion Devices and Power Systems," http://trajectory.lerc.nasa.gov/aig7820/projects/e.

9

Earth Stations and Terrestrial Technology

Earth stations provide access to the space segment, interconnecting users with one another and with the terrestrial network. In the first decades of this industry, Earth stations were large, both in physical and financial terms. Individual ownership of a satellite terminal was within the reach only of major organizations and perhaps a few wealthy individuals. By the late 1980s, low-cost TV receive-only (TVRO) terminals cost as little as $1,000; many people who lived in remote areas were pleased to be able to receive commercial television programming through this means. During the 1990s and past the year 2000, we have the wide-spread adoption of low-cost user terminals (UTs) that provide access to two-way as well as one-way services. Interestingly, the electronic complexity of the modern UT is no less than what formerly took up several equipment racks in the large Earth stations.

The overall RF and baseband performance of the Earth station has been covered in previous chapters. In this chapter, much more will be said about the various Earth station architectures in relation to the types of services that are provided. Included are digital TV transmit and receive, data communications, and fixed and mobile telephony. A discussion is provided of the TT&C Earth station and satellite control facility, both of which are needed to implement and operate the space segment. They are not typically used for communication services, but their design and operation are no less important to meeting the technical and financial objectives of a satellite mission. Tradeoffs in design of the various types of Earth stations are also described. The terrestrial "tail" that connects the

Earth station with the outside world is reviewed, as are considerations for implementing the building and other support facilities.

9.1 Basic Earth Station Configuration

For the purposes of this discussion, an Earth station can be broken down into the following major elements: the RF terminal (including the antenna), the baseband and the control equipment, and the user interface. In addition to these electronic systems, every fixed Earth station must be supported by a physical facility capable of housing the equipment, supplying reliable electrical power, and maintaining the temperature and humidity of the electronics within acceptable limits. These latter functions are similar to the ones provided by the spacecraft bus (while the functions in the previous sentence are comparable to those of the communications payload). The key difference here is that the majority of operating Earth stations can be serviced by maintenance personnel. That allows the larger Earth stations to be designed from building blocks that are, for the most part, easily repaired and changed.

The previous remarks obviously apply to the traditional type of permanent Earth station such as what is used to uplink digital TV. Modern UTs, such as advanced TVRO terminals and MSS phones, are self-contained and some are battery powered. Just like consumer electronics, UTs of this class do not require servicing except for preprogramming (such as a user would do with a VCR) and replacement of batteries. Repairs are performed by taking the unit to a repair facility or, in the case of the ultimate consumable electronic item, simply throwing it away as one would an obsolete PC or cell phone. VSATs, however, fall in between the two extremes, and it is common for repairs to be made by roving technicians who are called on to bring a terminal back into service.

The major elements of a generic Earth station are illustrated in the block diagram in Figure 9.1. The dotted box is the RF terminal, encompassing the antenna, the HPA system, the LNA, and upconverters and downconverters. In the figure, the LNA and the HPA are connected to the cross-polarized ports of the antenna feed. (The operation and performance of those components were covered in Chapter 4.) Where an Earth station is used as a major hub of a network or is required to provide reliable service, redundant, or backup, equipment (as shown) is included in the RF terminal.

The RF terminal provides an intermediate frequency (IF) interface to baseband equipment that performs the modulation/demodulation function, along with baseband processing and interface with the terrestrial tail. The specific configuration of the baseband equipment depends on the modulation and multiple access method employed (covered in Chapter 5). In Figure 9.1, base-

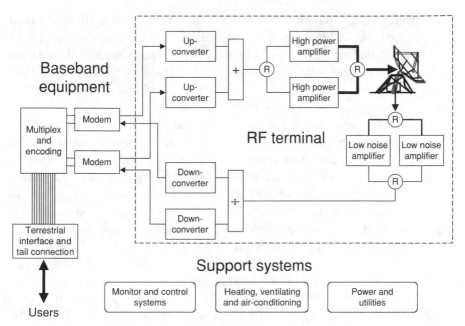

Figure 9.1 Major elements of an Earth station used to provide two-way digital communications links.

band equipment consists of digital modems and a time division multiplexer. Multiple lines interconnect those facilities because there would be several independent data channels. Also, equipment redundancy in some form could be included. The last electronic element of the station is the terrestrial interface along with the terminal end of a tail connection of some kind. Interfaces must be properly engineered and specified so that communication services can be carried over the satellite link without disrupting the operation of the overall network. Unfortunately, it is all too common for that interface not to be defined adequately or correctly ahead of installation, causing incompatibility problems that take a considerable amount of time and money to correct. Problems of that type often happen because of subtle differences in local interface requirements due to different network standards in use in different countries and by different users.

The tail connection itself would be composed of a fiber optic cable or microwave link, extending to the point of service such as an office building, telephone switching office, or television studio. Tail connections deserve more attention then they usually get because they affect quality and reliability in a direct and significant way. Poorly engineered or installed tails introduce noise, distortion, and, in the case of voice, echo. Furthermore, the high reliability of the satellite-to-Earth station path is easily compromised by the susceptibility of buried cable tails to breakage, deterioration, and electrical discharges (due to power line faults and lightening). These remarks about the terrestrial interface and tail do not apply to VSATs and UTs, because they typically are self-contained and may not need to extend the interface to external devices (if they do, the interfaces again need to be properly considered).

The power and the building facilities for the Earth station (analogous to the bus of the spacecraft), often get overlooked but are important nevertheless. Prime ac power for the station usually is obtained from the local electric utility company. It is common to utilize battery backup for critical parts of a station that provide vital communications service. To do that, a multiple sequence of ac-to-dc-to-ac conversion is followed. For example, in a typical uninterruptable power system (UPS), commercial power (or a diesel generator in the case of a remote installation) is converted in a rectifier to dc and tied to a lead-acid battery system capable of powering the essential electronics (called the critical load). Nonessential loads, like office lighting and air conditioning, are not protected in this manner. The stable ac required to power the critical load is obtained by passing the stable dc through a dc-to-ac inverter. Critical loads that obtain their power in this manner are isolated from the commercial power source, protecting them from voltage drops and power surges. In addition, the batteries are available to supply power automatically in the event of a short disruption, say, 30 minutes. That usually is sufficient, since backup diesel generators are started automatically if the disruption lasts more than a few minutes.

The function of the UPS is available in a small package suitable for a VSAT-type installation. It consists of the ac-dc-ac electronics conversion system plus battery backup. A diesel generator is not appropriate because of the desire for compactness and low cost. However, it would be entirely feasible to provide enough backup battery power to supply the terminal for many hours. That usually is sufficient time for commercial power to be restored or to connect a portable emergency generator.

A properly engineered facility also provides heating and cooling to maintain the equipment within its prescribed operating temperature and humidity range. Electronic racks often are mounted on top of a raised floor (called computer flooring) so that conditioned air can be conveniently fed from below; that also allows cables and waveguides to be neatly routed between racks. When done properly, a large Earth station facility looks attractive and allows easy and convenient maintenance of the electronic equipment. A small Earth station such as a VSAT, on the other hand, can be treated like telephone or data processing equipment installed in an office building. Building modifications usually are needed to access the outdoor RF terminal, often roof mounted, and reliable power with UPS provisions may still be required.

Special consideration should be given here to the RF terminal, which often is left outside in the elements. The RF equipment could be contained in a waterproof box similar to the cabinets used by telephone companies for street installation. As long as the door is closed, this type of container can withstand almost anything nature can throw at it (except a massive earthquake or hurricane). The problem can come if the equipment must be serviced during bad weather, particularly heavy rainfall and wind. The high voltages and ac power that may exist in the box should not be exposed to the elements or to unprotected human beings. The solution is to utilize an equipment shelter large enough for the service person to work inside and be protected from the weather. Also, at least two people should work on high-voltage power supplies and amplifiers, with one person observing and ready to disconnect the power in case of threat to life.

Monitoring and control (M&C) functions of an Earth station allow local and remote control operation. In modern satellite networks, M&C capabilities are advancing rapidly and involve sophisticated networks of their own. Earth stations that incorporate digital switching and TDMA require a centralized system of control to ensure that the proper channel and data packet routing are established. That can be done once or twice daily in a preprogrammed fashion or may be changed on user demand (i.e., demand assignment). The operation and status of RF and baseband equipment also can be determined, and in the event of failure, redundant equipment can be switched by remote control. In principle, M&C is to an Earth station what TT&C is to the satellite itself.

9.2 Performance Requirements

The performance requirements of the Earth station mirror those of the satellite. That is to ensure that there is a satisfactory RF link between the ground and the space segments under all expected conditions and for the range of required services. In addition, the Earth station determines the baseband quality and much of the end-to-end communication performance of the services being provided. We first consider the key parameters of EIRP and G/T as they relate to the ground.

9.2.1 Transmit EIRP

The Earth station EIRP simply is the value that applies to the uplink. As with the satellite (downlink) EIRP, it is obtained by multiplying the antenna gain (as a ratio) by the transmit power at the input to the antenna. In decibel terms, it is the antenna gain plus the transmit power in decibels relative to 1W (dBW).

As shown in simplified block diagram in Figure 9.2, the EIRP also considers any RF loss between the HPA and the antenna. The loss factor, l_t, is a number greater than 1, and the ratio $1/l_t$ is actually the transmission factor (e.g., a gain less than 1). Major Earth stations typically are designed for reliable uplink operation with low outage due to rainfall. Therefore, the required uplink EIRP margin is substantially higher than that of the downlink. That may require the use of vacuum tube HPAs capable of transmitting several kilowatts or at least a substantial fraction of 1 kW. The antenna gain will, of course, depend on the uplink frequency and the size of the antenna. While the behemoth 30m antennas of the early INTELSAT system no longer are being constructed, it is common to find antennas in the 7m to 13m class for service at C, X, Ku, and Ka bands.

VSAT networks are designed to not require vacuum tube HPAs at the remote sites, favoring instead lower power SSPAs. EIRP for a hand-held UT is going to be low because of low transmit power and low antenna gain. In non-GEO systems, the satellite can be literally anywhere in the sky, so the link should tolerate a significant degree of off-pointing of the UT antenna. The approach that seems to make the most sense is to provide a broad-beamed low-gain antenna that, at the least, provides nearly uniform radiation in all directions. The expected elevation angles toward the satellites as they pass overhead, as well as the variety of physical orientations that the user might impose, must be considered.

High-power uplinks bring with them a concern about radiation hazard to humans and other living things. That is one of the reasons major Earth stations

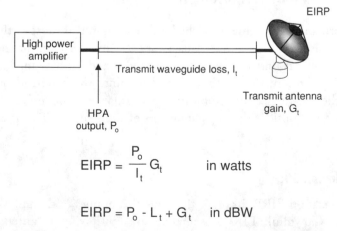

$$EIRP = \frac{P_o}{l_t} G_t \qquad \text{in watts}$$

$$EIRP = P_o - L_t + G_t \qquad \text{in dBW}$$

Figure 9.2 Definition of Earth station transmit EIRP, indicating the primary contributors to performance.

are not located in highly populated areas or are shielded from such exposure. Careful consideration must be given to that aspect of safety. Low-power uplinks in VSATs and MSS terminals may pose less of a risk but still should be considered potentially harmful to humans. Hand-held UTs expose the user to direct radiation, but power is kept to levels as low as typical cell phones. This is an area of active research and is still controversial with regard to the operation of common cell phones, which transmit similar power levels. In all cases, appropriate research and testing are prudent.

9.2.2 Receive G/T

The figure-of-merit of overall Earth station performance in the downlink is the G/T. G/T is the ratio of the antenna gain (itself as a ratio) to the total system noise temperature (in Kelvin, which is with respect to absolute zero) of the receiving Earth station, converted to decibels per Kelvin. The contributors to G/T are indicated in Figure 9.3. Again, l_r is a factor greater than 1, and $1/l_r$ is the transmission factor (e.g., less than 1). Higher antenna gain improves sensitivity as does lower noise, hence the use of a ratio. The G/T of the station is used in link budget calculations (discussed in Chapter 4) to compute the ratio of carrier to noise. In general, Earth station downlink performance can be adequately measured and optimized by consideration of the G/T figure.

The figure of merit (G/T in decibels per Kelvin) of a C-band Earth station is plotted in Figure 9.4 as a function of antenna diameter for three common levels of LNA noise temperature: 50K, 80K, and 100K. We assume that

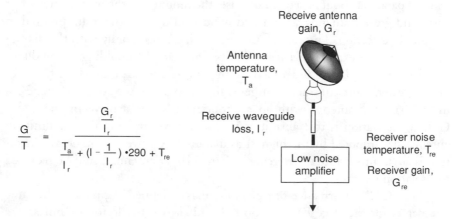

Figure 9.3 Definition of Earth station receive G/T, indicating how noise contributes to performance.

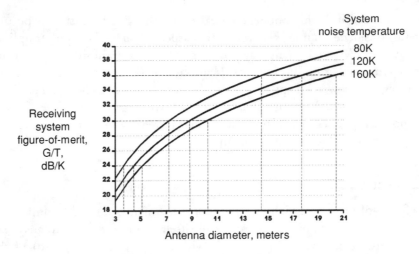

Figure 9.4   Earth station receiving figure of merit (G/T) as a function of antenna diameter
and system noise temperature.

the antenna temperature in each case is a typical value of 50K, common for parabolic reflectors at elevation angles greater than about 20 degrees. The other factors that make up G/T consist of the noise contribution and loss of the connecting waveguide, as indicated in Figure 9.3.

Larger diameter antennas are expensive to build because of the structural strength needed to maintain reflector alignment, particularly in high wind. Also significant is the effect on gain of surface accuracy, which is the accuracy with which the surface fits a true paraboloid (e.g., the surface obtained by rotating an ideal parabola about its axis). Likewise, the mount and the motor drive must withstand greater antenna weight and wind load to keep the beam pointed in the proper direction. The cost of an antenna increases rapidly with the diameter, following an exponential relationship. For example, doubling of the diameter from 5m to 10m (a 6-dB increase in gain) can increase cost by a factor of 10 or greater. Cost consideration generally encourages the system designer to use the smallest antenna with an appropriate LNA to achieve the necessary G/T. If the station is to transmit as well as receive, there is a lower limit on antenna size imposed by the highest available HPA power. Another consideration is adjacent satellite interference, where a larger antenna could be required for isolation reasons.

The G/T performance of small antennas is defined in precisely the same manner as for the large Earth station. Terminals with reflector antennas, like VSATs and TVROs, literally are miniature versions of their bigger cousins but are only able to deliver G/T consistent with their receive gain. A more complex situation exists for the hand-held UT to be used in non-GEO MSS services, since the antenna points in many directions at the same time. Here, the antenna temperature often is over 100K, due to pickup from the ground and local obstructions. The G/T is a negative number and is much more difficult to define.

9.3 Radio Frequency Equipment

A simplified block diagram of the typical RF portion of the Earth station is shown in Figure 9.5. In it, we see many of the components of the standard bent-pipe type of repeater: a receive chain consisting of the antenna, LNA, and downconverter, and a transmit chain consisting of an upconverter, an HPA, and, again, the antenna. It is almost always the case that a transmit Earth station uses the same antenna to receive, since we are establishing communication with a common satellite. The upconverters and downconverters interface with the rest of the station equipment at the IF, which typically is 70 MHz. That allows the baseband equipment (particularly the digital modem) to be designed and optimized for a particular common carrier frequency. The upconverters

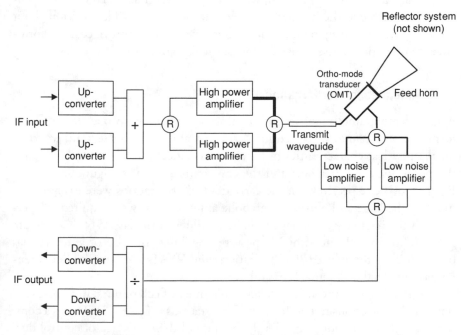

Figure 9.5 Block diagram of a typical RF terminal.

and downconverters then provide the translation to the necessary RF frequencies assigned for the uplink and the downlink, respectively. That is the case whether the station operates at C, Ku, or Ka band. (There is a caution here because while the center of the IF is 70 MHz, the higher end of the typical C-band transponder bandwidth extends to 100 MHz, which is within the standard FM broadcast band. That makes the Earth station somewhat more susceptible to RF interference from a close-by FM transmitter.) An exception to using the standard IF is likely for compact UTs such as applied in MSS because of the potential for economies in the details of the interior design. For example, it might be possible for the modem to transmit directly on the uplink L-band frequency.

The RF terminal often is contained in a separate shelter or cabinet adjacent to the antenna. That reduces cable and waveguide losses, which is desirable from the standpoint of EIRP and G/T. Another option is to package the uplink equipment as an exciter, containing the modulator and the upconverter, and the downlink equipment as a receiver, containing the downconverter and demodulator. The RF terminal then consists only of the antenna, LNA, and HPA. Further, TVRO and VSAT installations generally combine the LNA with the first stage of the downconverter into what is call a low-noise block converter (LNB). That has the nice feature of low-noise amplification and downconversion of all the channels to L band, where they can be transferred over inexpensive coaxial cable to the integrated receiver decoder (IRD). The decision about the RF terminal configuration is up to the designer to improve cost, performance, or operational ease of use for the particular application.

9.3.1 Antennas for Earth Stations

A considerable amount of technical information on the properties and performance of antennas was given in Chapters 4 and 7. In this section, we focus on the physical characteristics of antennas specifically used in Earth stations: the prime focus fed parabola and the cassegrain, shown in Figure 9.6(A) and Figure 9.6(B), respectively. In the early days, Earth stations were easily recognized by the large and obtrusive parabolic antennas that were required to operate with the relatively low-power C-band satellites of the late 1960s (see Figure 1.15). With the advent of higher powered satellites, the size of antennas has been significantly reduced. TVRO stations and VSATs use even smaller antennas, as indicated in Figures 1.16 and 1.17.

Antennas for hand-held UTs resemble those of cell phones, but must support satellite positions rather than land base stations (Figure 1.9). This is a complex subject, as mentioned previously, because of the user's lack of control during link operations. For that reason, there is a general feeling in the industry

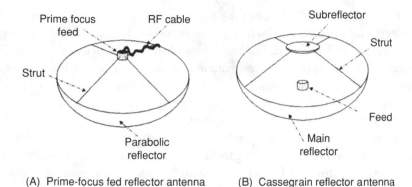

(A) Prime-focus fed reflector antenna   (B) Cassegrain reflector antenna

Figure 9.6 Earth station antenna configurations commonly used for receive-only and two-way communications links.

that some degree of user cooperation is needed to allow communications to proceed with adequate link quality and performance.

9.3.1.1 Prime Focus Fed Parabola

Antennas with a diameter less than 4.5m and particularly those used in RO service employ the prime focus fed configuration shown in Figure 9.6(A). Within this class is the offset fed parabaloid, used extensively for antennas less than about 2m in diameter. The cost usually is the lowest because of its simplicity and ease of installation. To minimize RF loss and noise, a compact LNB package is located directly on the waveguide port of the feed to minimize loss between the front-end amplifier and the feed. The RF cable to connect between LNB and indoor receiving equipment is attached to one of the struts. Unless the electronics can be fitted at the feed, the prime focus configuration is not attractive for transmit/receive applications.

In a variant of the prime focus approach, the feedhorn is at the end of a hook-shaped piece of waveguide that extends from the vertex of the parabolic reflector. The LNB is attached to the end of the waveguide behind the reflector, and the assembly is supported at the feedhorn with either struts or guy wires. That has the obvious advantage of allowing the electronics to be accessed more easily and provides placement out of the way of the main beam. The mechanical rigidity and thermal stability of the bent waveguide system are limited, but small-diameter antennas used for TVRO purposes can get by with that approach.

9.3.1.2 Cassegrain Optics

Many of the shortcomings of the prime focus fed antenna are overcome with the cassegrain system, shown in Figure 9.6(B). Using the folded optics principle of the reflector telescope, the focus of the parabola is directed back by a hyperbolic subreflector to a point at the vertex of the center of the main reflector. Proper illumination is still provided, but the feed and the electronics can be joined at an accessible point in the antenna structure. In large antennas, a hub enclosure is provided at the back for placement of essentially the entire RF terminal in close proximity to the feed. The cassegrain system eliminates the cable or waveguide runs between RF electronics and feed.

As with the prime focus system, there is a variant of the cassegrain that uses a piece of waveguide extending from the vertex of the reflector to the prime focus. However, instead of a bent end with a horn, illumination is provided through a flat or shaped plate, which functions like the subreflector of the cassegrain. The waveguide may now be supported within a structurally sound hollow tube, eliminating the need for struts or guy wires. This type of system has become very popular in nearly all sizes and types of ground antennas because it combines the benefits of the prime focus feed system with those of the cassegrain.

The offset fed geometry, which was reviewed in Chapter 7 for spacecraft antennas, also has been applied to Earth stations. Shown in Figure 9.7 is the common home Ku-band TVRO used in high power DBS. A main benefit of the offset approach is the elimination of main beam blockage by the feed and struts, which tends to improve antenna efficiency and reduce sidelobe levels. It also permits rain and snow to fall off the surface, because the reflector is at a more acute angle relative to the ground. In a particular design, there may be mechanical or cost advantages as well.

Figure 9.7 Typical DBS home TVRO. (*Source:* Courtesy of Sony.)

9.3.2 Antenna Beam Pointing

The relatively narrow beamwidth provided by a fixed Earth station antenna results in some variation in signal strength as either the satellite or the antenna physically moves with respect to each other. As discussed in Chapter 4, the 3-dB beamwidth of an antenna defines the angle for which the gain will be one-half its peak value (or 3 dB down from maximum). For a GEO satellite, the antenna can be fixed to its mount as long as the stationkeeping error is well within the 3-dB beamwidth. If that is not the case, some form of tracking will be needed.

Many antenna mounts are configured like an artillery piece, with adjustments being provided for azimuth (0 to 360 degrees) and elevation (0 to 90 degrees). That makes sense if the antenna must be capable of pointing anywhere in the sky, which is the case for antennas used to track a satellite in transfer orbit or for LEO operations. In moving the antenna between GEO satellites, the azimuth and elevation axis have restricted pointing ranges and must be adjusted in a coordinated way. Another common design for GEO satellites is the polar mount, named for the fact that the main axis of rotation is aligned with that of the Earth (e.g., pointed toward the poles). The reflector then can be moved from satellite to satellite by rotation along one axis. However, even polar mounts may need a slight azimuth adjustment when scanning to the edge of the arc. Polar mounts tend to be simpler in construction, are more compact, and allow greater flexibility in covering the orbital arc.

Gateway and TT&C antennas used in non-LEO systems are perhaps more of a challenge due to the need to precisely track the various satellites as they pass. Fortunately, the closer range allows the operator to use a considerably smaller diameter than is required for GEO service. In Figure 9.8, a typical gateway Earth station for the Iridium system (a very substantial facility by anybody's measure) is shown. The station may have as many as four of these antennas: a primary and a backup for the currently employed satellite and a second pair for the next satellite to be acquired. That ensures continuous service to users as the connection is handed off from satellite to satellite.

9.3.3 High-Power Amplifiers

The performance of an Earth station uplink is gauged by the EIRP, the same parameter that applies to the satellite downlink. As discussed earlier, the EIRP is the product of the HPA output power, the loss of the waveguide between HPA and antenna (expressed as a ratio less than 1), and the antenna gain. The result is expressed in decibels relative to 1W, so it also is convenient to add the component performances in decibel terms. For example, an EIRP of 80 dBW

Figure 9.8 Antennas used at a typical Iridium gateway Earth station. (*Source:* Iridium, Inc.)

results from an HPA output of 30 dBW (i.e., 1,000W), a waveguide loss of 2 dB, and an antenna gain of 52 dB. Since the dimensions of an Earth station antenna are not subject to the physical constraints of the launch vehicle, the diameter and, consequently, the gain can be set at a more convenient point. That could be the optimum, which occurs where the cost of the antenna plus HPA is minimum. In small UTs, particularly of the personal and hand-held variety, HPA power must be held to an absolute minimum to preserve battery life and minimize radiation, so the optimization usually is reversed to rely on the satellite G/T and EIRP to produce a satisfactory link.

Figure 9.9 presents a tradeoff curve for the design of an Earth station uplink for use in full transponder digital video service (in C, Ku, or Ka band).

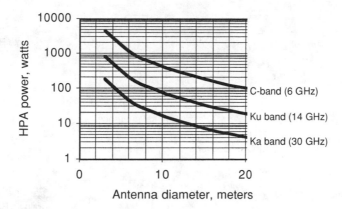

Figure 9.9 Earth station HPA power versus antenna diameter for full transponder uplinking (e.g., EIRP of 80 dBW).

To deliver an EIRP of 80 dBW, the curve gives the requisite antenna diameter and HPA power. An allowance of 2 dB for waveguide loss has been made in the curve. At Ku-band, relatively low HPA powers (less than 200W) are acceptable with antennas 6m in diameter or greater. To be able to employ a more compact reflector of 3m in diameter (typical for a truck-mounted transportable Earth station), the HPA must be capable of outputting 1,000W. There is an additional problem with the smaller diameter antennas, namely, the excessive power also will be radiated through the antenna sidelobes, producing unacceptable interference in the adjacent satellite. Also, these numbers do not allow for Ku or Ka band uplink power control (UPC) which requires 10 dB of HPA power margin (e.g., the power indicated in Figure 9.9 must be multiplied by 10).

The discussion of HPA technology in Chapter 4 provides another perspective with which to evaluate the tradeoff in Figure 9.9. Sufficient antenna size permits the use of lower powered, less expensive HPAs. In FDMA services, it is possible to use an SSPA to achieve the required Earth station EIRP provided sufficient antenna gain is available. Likewise, a klystron HPA could be avoided if the power for video service is less than 600W, which is possible in the example with a 6m Ku-band antenna. There is an important consideration for multiple carrier service in that the HPA would have to be backed off to control RF IMD. That increases the HPA power "size" by as much as a factor of 10.

9.3.4 Upconverters and Downconverters

Downconverter and upconverter equipment, illustrated in Figure 9.10(A) and Figure 9.10(B), respectively, is needed to transfer the communication carriers

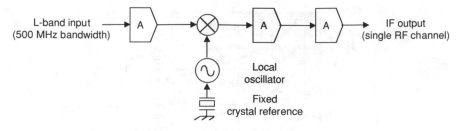

(A) Downconverter block diagram (L-band to IF)

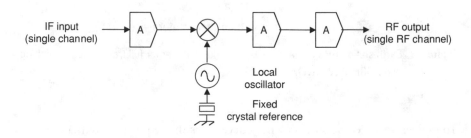

(B) Upconverter block diagram (IF to RF output)

Figure 9.10 Typical upconverter and downconverter block diagrams used in fixed Earth
stations.

between the operating RF frequencies and the IF used by the modems and
baseband equipment. A typical upconverter has a first stage of amplification to
provide adequate gain for the operation of the station equipment. The actual
frequency conversion is accomplished in a mixer and LO combination such as
that shown. A frequency-agile upconverter employs a frequency synthesizer to
generate the LO so any carrier frequency within the satellite uplink band can
be used. Proper filtering is needed to prevent the LO and its harmonics from
reaching the uplink path, a common consideration in the design of frequency
translation stages.

The design of the downconverter is much the same as the comparable
stage of the satellite receiver. Here, the first amplifier stage provides the needed
overall gain and reduces the noise contribution of mixer and IF equipment. A
synthesizer also has be used to provide agility in receive frequency operation. In
small UTs, the upconverter and the downconverter can be driven by the same
synthesized LO to reduce cost and to access the same channel of operation over
the satellite.

9.3.5 RF Combining

In many large Earth stations, there may be a requirement to transmit two or more simultaneous carriers and to receive multiple carriers as well. The simple approach is to use passive microwave components such as the hybrid and filter multiplexers to combine and separate carriers, as appropriate. That principle was addressed in Chapter 6 in connection with the satellite repeater. An example of a passive power combiner is provided in Figure 9.11. On the uplink side at the RF frequencies, microwave filters must be used, one for each transmit frequency. Power combiners of this type look much like the output multiplexer in a bent-pipe satellite repeater, which is not surprising because they perform the same function.

9.3.6 Uplink Power Control

Ku- and Ka-band transmit Earth stations and some MSS UTs as well require automatic uplink power control (UPC) to compensate for link fading. The extreme nature of the fading problem at Ka band makes UPC a standard feature in applications such as TV transmission and telecommunications gateway

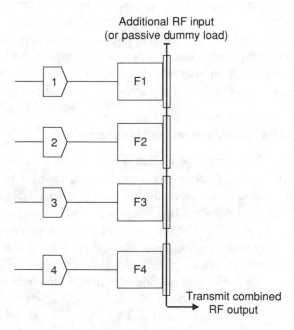

Figure 9.11 Example of a four-channel power combiner for an Earth station uplink.

services. The principle behind UPC is that rain-induced fading is compensated in direct response to the attenuation. That can be a complex process because the uplink itself can be measure precisely only at the satellite, which is something generally not provided. Instead, the uplink Earth station measures downlink power and uses that as an indication of what is happening to the uplink.

In an MSS terminal, UPC may be used to compensate for multipath propagation and attenuation by obstructions. Unlike rain fading, which is relatively slow, multipath and blockage are fast and difficult to track on a precise basis. Instead, the uplink is adjusted in a broad (macro) sense to compensate for longer term variation, such as would occur for a user who is standing still behind a tree or in front of a building. Rapid fluctuations, common when the user is in a moving vehicle, cannot be removed through power control, so sufficient power margin must be included in the link budget.

9.4 Intermediate Frequency and Baseband Equipment

The IF and baseband equipment, illustrated in Figure 9.1, establish the basic service capabilities of the Earth station, including modulation and FEC, multiple access method, and interface to the user or terrestrial network. In many cases, this portion represents the largest investment on the ground, particularly for a major gateway or network control facility. That is because, while the RF terminal usually is fixed for a particular application, the quantity and the design of the baseband equipment must be adjusted to support the bandwidth and number of communication channels the stations serves. This section reviews typical configurations of the major elements of this section of the Earth station.

9.4.1 Modulators, Demodulators, and Modems

The most essential portion of the baseband-to-IF chain consists of the modulator and demodulator or a combination of the two, called a modem. On the receive side, the demodulator detects the incoming carrier, synchronizes the data, performs convolutional decoding (if appropriate), and outputs a clean bitstream for the particular application. The transmit side works in the opposite direction. Threshold performance in terms of error rate and synchronization is determined at this level of the station. Because there must be one modem for each frequency, FDMA stations can contain large quantities of these units, representing a sizable investment. In TDMA, the frequency channel occupied by the carrier is shared on a time basis by several Earth stations. There would need to be only a single modem per Earth station, since it can receive the bursts from the other Earth stations so long as they do not overlap in time. That is, of course, central to the operation of a TDMA network (as opposed to a CDMA network, where stations may transmit simultaneously on the same frequency). The one requirement is that the modem must be capable of wide bandwidth transmission and reception due to the increased data rate to cover the total capacity of the network.

Figure 9.12 is a block diagram of a typical PSK digital modem, such as that used in a TDMA VSAT. Modulation is provided by mixing the data with a 70 MHz carrier, inverting the sinewave when a change from 1 to 0 (or vice versa) is desired. We obtain QPSK by adding a 90-degree phase shift to a second modulated stream of data, thus doubling the use of the uplink frequency channel. Reception works in the reverse direction, using the same mixing process but resulting in the original data being recovered. Additional circuitry is needed on the receive side to remove satellite link noise and resolve the bit pattern. The latter is accomplished by a decision circuit that, simply enough, decides if the received bit is a 1 or a 0. The presence of noise and distortion in the demodulated bits causes the decision circuit to make a bad decision occasionally, a process that gives bit errors. It is the average rate of those errors, the BER, that determines received data quality.

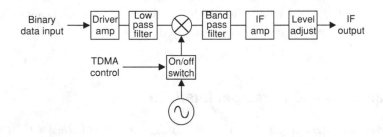

(A) Biphase PSK (BPSK) modulator

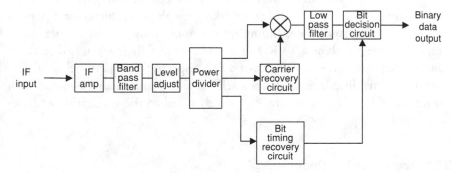

(B) Biphase PSK (BPSK) demodulator

Figure 9.12 Functional elements of the typical PSK modem.

The typical satellite modem (and terrestrial modem, as well) also contains the convolutional coding and decoding circuitry that reduces the BER by factors of 100 to 10,000, depending on the strength of the coding technique. Other functions include burst synchronization for TDMA, baseband processing such as compression/decompression, and preemphasis/deemphasis.

9.4.2 Multiplexing and Packet Processing

It is common practice to use the outbound and inbound data bitstream for multiple functions, such as multiple information channel transmission, network control, and facility monitoring. The convenient way to do this is with TDM or packet data transfer. Both rely on inserting different blocks of data into the common stream to and from the satellite. In TDM, the individual streams or information channels are inserted according to prearranged timing, usually periodic and of fixed length. That works for a standard format without compression. On the other hand, many applications use dynamic bandwidth allocation, so fixed timing cannot be assumed. In that case, it is better to use some form of packet transmission, which is a standard feature in most telecommunications networks, satellite and terrestrial. The equipment usually is designed specifically for satellite network use, but in some cases, conventional multiplexers and packet switches are employed.

9.5 Tail Links and Terrestrial Interface

The terrestrial interface can be called the "business end" of the Earth station because it is where the actual user services are made available. In larger stations, the terrestrial interface can be quite complex, as suggested in Figure 9.1. A gateway station looks to the outside world like a telephone exchange or a portion of a broadcasting studio, depending on the application. On the other hand, simple terminals such as a TVRO or a VSAT render their services directly to consumer equipment, like a TV set or a PC, respectively. A hand-held UT may have no external interface since it is a self-contained unit. The requirements of the terrestrial interface, therefore, vary widely based on the application and user facility provided.

9.5.1 Terrestrial Tail Options

A terrestrial tail may be needed to connect the communications Earth station to one or more remote user locations. The distance to be covered can range from a few meters to hundreds of kilometers. In C-band satellite systems, the Earth station is often isolated from a city to reduce RFI difficulties, in which case an elaborate tail is required. On the other hand, terrestrial interference is not present in most Ku- and Ka-band systems; therefore, tails can be relatively short. An exception is the case where a large Earth station (i.e., a teleport) is shared by several users each of which must be reached by local terrestrial transmission links, often provided by local telecommunications carriers. Site diversity may be needed at Ka band and higher to provide reception when one location is severely affected by rain. The tail link between the diverse sites can represent a considerable investment in fiber optics or microwave towers due to the requirement to transfer a large bandwidth without loss of service.

A few examples of the tail configuration for a major Earth station are illustrated in Figure 9.13. A cable interfacility link (IFL) connects each of two RF terminals (lower left of the figure) to the main Earth station building. Within the building can be found the baseband and interface equipment appropriate for the types of services being provided. It is assumed that all the traffic (voice, data, and video) is to be transported to the nearest city. These transmission requirements are met in the example with a single-hop terrestrial microwave link, equipped with sufficient receiver-transmitter units to carry the video, voice, and data traffic. The customer location or locations access the first

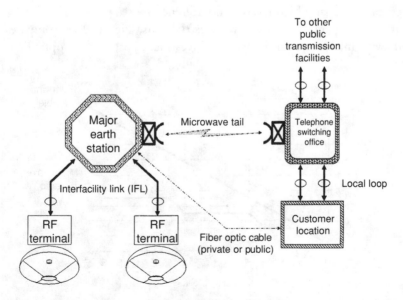

Figure 9.13 Use of terrestrial tails between Earth station facilities, switching offices, and customer locations (distances not to scale).

telecommunications office through either public or private local loops consisting of fiber optic cable and conventional multiple pair cable.

The telecommunications office and local loops can be bypassed (the dotted line between the customer location and the major Earth station) with a private tail fitting the requirement. The reason for bypassing can be based on economics, it possibly being less expensive to own the tail facility rather than leasing capacity from the local telecommunications utility. The private bypass, which permits the user to maintain control of the resources, eliminates some of the time necessary to implement new public facilities. Reliability can be greatly improved with the path diversity of using both the public access lines and the private bypass link.

The two principal types of terrestrial tails for bulk transmission purposes are line-of-microwave and fiber optic cable. Both are effective and reliable and can be economical when applied properly. Short distances between buildings, such as for the IFL within an Earth station site or when a VSAT is connected to one or more users, are best traversed with fiber optic cable. It has become a practice in large Earth station design to rely on fiber optic cabling because of its low-noise performance and its ability to reduce radiated electromagnetic interference and conducted electrical surges. Another option for short tails is single-hop microwave communication. Available in capacities of a few megabits per second to over 1 Gbps, single-hop microwave systems are relatively inexpensive to install and operate. They, too, have the benefit of not transferring electrical surges and high levels of interference.

Focusing on large-capacity terrestrial tails, Figure 9.14 compares fiber optic with microwave on the basis of investment cost. A fiber optic link is rel-

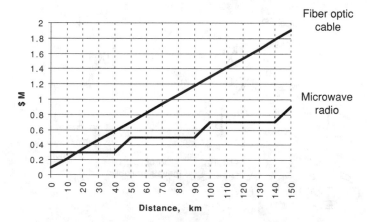

Figure 9.14 Comparison of fiber optic and terrestrial microwave links on the basis of investment cost for a terrestrial tail (excluding land and right-of-way).

atively costly per kilometer of construction, particularly in a metropolitan area. The black line of constant slope is the cost of the cable, electronics, and installation for a basic point-to-point link, assuming that the right-of-way is already available at essentially no cost. Most of the expense is for the repeater and terminal electronics and for installation, which is labor intensive. Fiber optic cable itself is a small part of the cost, so it usually is wise to facilitate future expansion and repairs by including more fiber pairs than are necessary for the current demand.

The cost of putting in a high-capacity terrestrial microwave system, including receiver-transmitter units, towers, and antennas, is shown with the stair step, where the step occurs at the assumed repeater spacing of 50 km. In comparing the two technologies, it is clear that fiber optic is attractive for relatively short tail lengths, that is, less than the extent of one microwave hop. On the other hand, the capacity of the fiber optic link can be made many times greater. That is because the cable contains extra fiber pairs that are available for expansion (and that were included in the original cable at little extra cost). In addition, transmission rates on a given fiber pair can be increased as newer optical modems and electronics are introduced.

It should be kept in mind that any real-world situation should be examined in detail before the tail technology is selected. That is because installation and right-of-way costs can vary widely, as can the cost of the equipment itself. Figure 9.14 illustrates a valid framework for making such a comparison. New approaches, both through purchase or lease, constantly appear on the market, so the tradeoff will change over time. Consequently, an investigation of this type should be conducted prior to commitment to a particular approach.

9.5.2 Terrestrial Interfaces

Earth stations are a means of providing access to a satellite communication network of some type. An important part of their design deals with the interface between the station and the user, illustrated as the last box at the lower left of Figure 9.1. It was emphasized at the beginning of this chapter that an improperly designed or installed interface will seriously degrade the quality of service. Often, access to the satellite network will be blocked entirely, particularly in the case of digital traffic and signaling. This section reviews general criteria for those interfaces as they relate to voice, data, and video information.

The most common interfaces used in communications services carried by both satellite links and terrestrial systems are shown in Figure 9.15, in simplified form for clarity. Each line and arrow indicate an independent signal path, required to properly interconnect the communication equipment with the user. Because this is an overview chart, many of the physical details, such as voltage

Telephone (private line or switched PSTN)

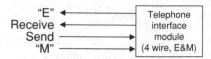

Data transmission (serial digital interface)

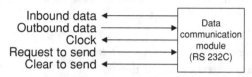

Broadcast television (analog format)

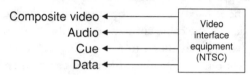

Figure 9.15 Overview of terrestrial interface alternatives for telephone, data, and broadcast television service.

level, timing requirements, and connector type, are missing. Detailed information can be obtained from technical specifications of equipment and international standards such as those promulgated by the technical committees of the American National Standards Institute (ANSI), the Institute of Electrical and Electronics Engineers (IEEE), and the ITU.

9.5.2.1 Telephone Interface

A common analog telephone interface is illustrated at the top of Figure 9.15, usable either for private line or switched service. Separate directions for send and receive (i.e., four wires) normally are provided, since the conversion to two wires for the subscriber loop is accomplished by a telephone hybrid located in the telephone switching equipment (not shown). Circuit control is exercised by "supervisory signaling" over the E (for "ear") and M (for "mouth") wires. The telephone switch at the near end alerts the distant end that a call is about to come through by transmitting an inaudible signal over the M lead. At the distant end, the signal arrives over the E lead. Unlike the receive and send lines, which are each a pair of wires, the E and M leads are single wires that indicate either a voltage different from zero (e.g., 5V) or zero volts (i.e., ground potential). When an Earth station is connected to a terrestrial tail at voice frequency, the E lead of the station is connected to the M lead of the tail and vice versa.

Many of the interfaces encountered in telephone communications are digital in nature, operating in the various hierarchies used in North America, Europe, and Japan. They are specified in the applicable ITU recommendations and national standards. (A summary of the hierarchies can be found in Tables 5.1 and 5.2.) The digital type of interface is most efficient when connecting a digital PBX or data processing equipment to TDM multiplex at the Earth station.

9.5.2.2 Data Transmission Interface

Data transmission links at digital basebands come in various forms to consider the information format, timing references, and control signals. That is what is termed the physical layer in data communications. According to the Open Systems Interconnection (OSI) model, a total of seven layers is needed for a data communication application to be fully functional [1]. Without the physical layer at the bottom of the protocol stack, no communication is possible. Once the physical layer requirements are met (connector specification, signal characteristics, synchronization and timing, and basic "hand shaking"), the link layer and higher can begin to function according to the particular protocol or protocols.

The information shown at the center of Figure 9.15 is an example of one arrangement for low-speed data communication equipment using an interface such as the common RS-232 standard. The inbound and outbound data lines are essentially the same as their counterparts in telephone. Since data communications require precise timing, a clock output is provided to synchronize the receiving side (the "slave") with the sending side (the "master"). If the illustrated end were the master, the clock would be sent in the opposite direction.

The request-to-send (RTS) and the clear-to-send (CTS) amount to a form of supervisory signaling. When the transmitting end wishes to send data, the RTS line is activated much like the M lead. The receiving end, if operating and not occupied by some other task, responds by activating its CTS line. Because digital communication equipment has more capability than historical telephone equipment, the CTS and RTS are bidirectional.

Numerous other interface arrangements have been specified and are available for data transmission. There has been much effort to develop international standards under ISDN and OSI so that digital systems will interconnect with one another and be capable of passing the maximum amount of data. Other layer 3 and higher protocols, namely TCP/IP and ATM, are being promoted as the foundation of new generations of data communications applications. They are being addressed by developers of Earth stations and the networks they serve.

9.5.2.3 Television Interface

The last example in Figure 9.15 presents the terrestrial interface for analog video reception, where the transmit side would have exactly the same lines. Composite video is the term for the complete color TV baseband signal with its luminance (black and white) and chromanance (color) components. It is customary to measure the time waveform of the output in terms of the peak-to-peak voltage swing of the luminance signal, which extends from the tip of the sync pulse (called "blacker than black") to the maximum possible white level. The value is specified for the particular video format that has been dictated for the country of operation. Other such standards apply to the chromanance (color) signal and to the line and frame repetition rates. All these interface parameters for video are specified in engineering documents available to the domestic broadcasting industry. The composite video signal can be viewed only on a video monitor and is intended to be transmitted from a VHF or UHF broadcast station after modulation using the vestigial sideband (VSB) technique common in terrestrial broadcasting.

The audio portion of the broadcast video signal is sent separately because broadcast stations have individual video and audio transmitters that are RF combined on the same antenna tower. (Normal home TV receivers have the capability to separate the audio from the video.) At the audio interface in Figure 6.15, the baseband channel covers a bandwidth of 20 Hz to 10 kHz, providing a high-fidelity program channel. It is important that audio levels be set according to a common interface standard to prevent wide changes in volume as the user switches between television channels. An additional audio channel called a cue channel is provided for network control purposes and is not intended for broadcasting. As discussed in Chapter 5, each audio channel is modulated on a separate FM subcarrier and placed above the video baseband using FDM.

Digital video and audio transmission is typically according to a standard such as MPEG 2 and the appropriate digital hierarchy. The resulting data stream would interface either bidirectionally or in one direction, depending on whether it is a two-way video link or one way using broadcasting. The interface operates much the same as a digital data interface.

The last interface at the bottom of Figure 9.15 provides a one-way (point-to-multipoint) data broadcast channel extracted from the baseband. The common approach to adding data to analog video signal is within the vertical blanking interval (VBI). Alternatively, digital modulation such as QPSK might be used on a subcarrier. The interface of the demodulator would be that illustrated for data transmission at the center of Figure 9.15. Although only one direction of transmission is possible, the digital interface still would provide all the interface lines and functions with the demodulator exercising appropriate control of data flow.

9.6 Earth Station Facility Design

The last practical consideration to be covered is the physical installation of the Earth station, that is, the design of the building and the arrangement of equipment therein. It is instructive to take as an example the layout of the site and building for a major Earth station, shown in Figure 9.16. Many of the same principles can be applied to smaller stations, including VSATs. The drawing assumes operation with GEO satellites with the station located north of the equator. Consequently, the antennas are aligned along the southern perimeter of the site to afford good visibility of the orbit arc. This arrangement of the antennas is usually the most flexible if there is sufficient space between the reflectors to allow independent pointing of each main beam to any orbit position. Care in placing the foundations must be taken so that antennas never touch each other during repointing and that the beams do not overlap one another to a significant degree. The latter minimizes the interaction of the electromagnetic fields around the antennas, including the effects of diffraction. With proper isolation between antennas, the beams will be properly formed, giving the desired cross-polarization isolation and control of sidelobes.

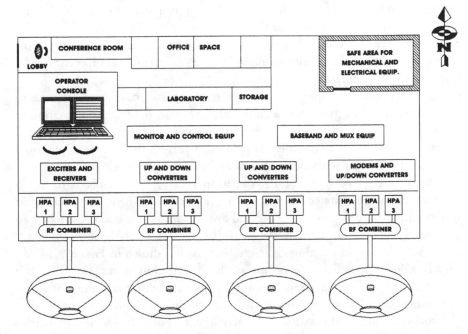

Figure 9.16 Site plan and equipment arrangement for a typical major Earth station.

The next important aspect is the location and arrangement of the RF equipment. As described in a previous section, the LNAs or LNBs should be mounted directly on the receive port of the feedhorn to minimize noise and loss. That also can be done for the HPAs as well if the power level is in the range of 5W to 150W, because compact SSPAs and TWTAs can be mounted near or on the reflector. Power levels in the hundreds to thousands of watts require high-power TWT and klystron amplifiers, respectively, which must be protected against the elements. Operation in two or more transponders at the same time is accomplished by connecting the HPAs in the building to an RF combiner of the type described in a previous section. The HPAs and RF combiner should be located as close as possible to the antenna to minimize waveguide losses. That may require that this portion of the RF terminal be contained in an equipment shelter away from the building but close to the antenna.

The baseband and control equipment can be located anywhere in the building since baseband and IF cable lengths are not critical to the performance of the Earth station. The main considerations here deal with allowing station personnel to operate and maintain the equipment. The appearance of the equipment in the room also is important both to the people who work there and to guests who may come to visit. Equipment that could be hazardous to humans, such as batteries, high-voltage electrical equipment, and diesel generators, should be isolated from the general flow of people, as shown in the upper right of the drawing. A major Earth station also deserves to have a front lobby to control access to the facility and office space for administration and management personnel. Test, calibration, and repair of electronic equipment can be done in a small laboratory.

The importance of providing adequate space between adjacent antennas to minimize RF interaction has been stressed. Figure 9.17 illustrates another important aspect in antenna placement dealing with the protection of personnel on site and the local community. The far-field pattern of the ground antenna contains the familiar main beam and sidelobes. However, the near-field region can produce elevated levels of RF radiation that are potentially harmful to humans and other living beings that enter the near field. The normal practice is to fence in the site to prevent people and animals from ever reaching such close proximity; in crowded metropolitan areas, however, that may be impractical.

A shield wall of metal or concrete such as that shown in Figure 9.17 can greatly reduce the radiation exposure in the surrounding area. The top of the wall must reach above the highest point on the antenna for any angle of elevation it may take. Usually, the worst case occurs when the antenna is pointed at the horizon, which also maximizes the absolute level of radiation along the particular azimuth. A high wall minimizes the diffraction over the top, which is the same consideration in using natural or man-made shielding to prevent RFI.

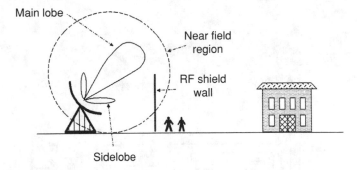

Figure 9.17 Protection of personnel and buildings from local RF fields around the transmitting Earth station antenna.

Analytical techniques using wave and diffraction theory can predict radiation levels in the vicinity of the antenna and the wall, provided that the detailed structure of the antenna and the wall are included in the analysis.

9.7 Major Classes of Earth Stations

A unique class of Earth station is employed for each type of communication service being rendered. In addition, the size and the complexity of a station depend on the quantity of traffic and, if appropriate, the variety of services offered. This section briefly describes some of the more common station configurations in use for domestic and international communications. To start off, the TT&C Earth station and its associated satellite control center are presented to give the reader a feel for the scope of the ground facility used to operate the space segment. The rest of the section covers stations ranging in size from the largest class (used for television uplinking, bulk data transmission, and MSS gateway service) to the smallest (VSATs, TVROs, and tabletop and hand-held MSS UTs).

9.7.1 TT&C Ground Facilities

The TT&C ground facilities of a satellite operator cover a variety of functions and capabilities, as depicted in Figure 9.18. In some systems, all the functions are centralized at one location to minimize capital and operating expenses. That approach occupies the least land and building space and can economize on equipment. Additional savings accrue from minimizing expenses for running the buildings and by being efficient in the use of personnel. For example, a single

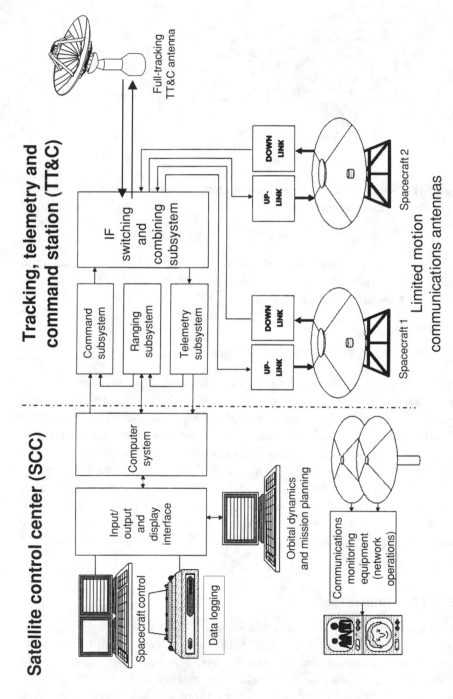

Figure 9.18 General arrangement of TT&C ground facilities for a space segment consisting of two geostationary satellites.

maintenance team can service all the electronic equipment. A second approach splits the facility between a TT&C Earth station and an SCC, as indicated by the vertical broken line in Figure 9.18. Using a tail link such as microwave or leased private lines, the SCC is established at a distant point, such as the head-quarters of the satellite operator. That is advantageous where much of the busi-ness is carried out in a central city and the TT&C is located remotely to avoid RFI and environmental hazards. It also is the case in non-GEO systems because a given TT&C station can view only a few satellites at any time. This section reviews the design and capability of each half in the TT&C facility. More detail on the operation of these facilities is provided in Chapter 11.

9.7.1.1 TT&C Earth Station

The TT&C station has essentially the same capability as the generic Earth sta-tion given in Figure 9.1, incorporating the RF terminal, baseband equipment, and terrestrial interface (appropriate to a separated SCC). The antennas employed for GEO satellites are usually 10–13m in diameter (whether C, Ku, or Ka band) to provide the maximum possible link margin because transfer orbit and potential abnormal on-orbit conditions need to be accommodated. In a typical TT&C station for GEO operations, there is one limited motion antenna pointed at each operational satellite (two are assumed in the illustra-tion), while the full tracking antenna is used for transfer orbit operations and for testing purposes during the life of the satellites. A TT&C station used for a non-GEO constellation must, of necessity, rely on tracking antennas, which also function for communications. An example of a TT&C station with full and limited motion antennas is shown in Figure 9.19. All antennas aid in track-ing the satellites' positions by providing azimuth and elevation readouts to the data processing system from their respective mounts. The full tracking and lim-ited motion RF terminals interface with the rest of the station at intermediate frequency.

The baseband equipment is shown at the upper center of Figure 9.18. In the command subsystem, the actual digital commands are taken from the incoming data stream and modulated on the IF carrier using either FSK or PSK, depending on the design of the spacecraft command receiver. Other spe-cial features can be incorporated in the command subsystem, such as providing the RF pilot carrier modulation used by a spacecraft antenna tracking system. For commanding purposes, the command carrier need be on the air only when it is sending commands and for ranging. That allows the uplink chain to be used on another satellite or for other purposes when commanding is not being performed.

Figure 9.19   A typical TT&C Earth station with a full-motion antenna for tracking and testing as well as several limited motion antennas used for full-time TT&C links and communications services. (*Source:* Photograph courtesy of Japan Satellite Telecommunication, Ltd.)

The telemetry subsystem receives the telemetry carrier at IF from the downlink and demodulates the actual stream of telemetry data. The process is straightforward and directly analogous to the reception of data communication in a standard Earth station. The third function of the baseband equipment is for ranging, which is the process outlined in Chapter 8 for measuring the position and velocity of the satellite. The ranging subsystem generates a baseband ranging signal, which the command subsystem modulates on the uplink carrier. After passing through the TT&C subsystem of the satellite, the ranging signal is received at the TT&C station, demodulated, and the baseband compared with what was transmitted. The measured time delay and frequency provide the basis for computing the range and the range rate. In an alternative ranging approach, the ranging baseband signals are modulated on an independent carrier and uplinked to a frequency within one of the operating transponders of the communication payload. Referred to as transponder ranging, the technique eliminates the need to switch the command link between the commanding and ranging functions but consumes some of the communications bandwidth of the satellite repeater. In yet another approach, the signal rides along with a communication carrier, thus reducing the wasted bandwidth.

The interconnection and switching of command, telemetry, and ranging links between each other and among different satellites is facilitated by the IF switching and control subsystem. For example, during transfer orbit operations, the TT&C link to the satellite is established through the full tracking antenna. A switchover to the assigned limited-motion antenna normally is accomplished with the IF switching and control subsystem after the satellite is located in its assigned orbit slot and has been fully deployed and tested. The TT&C station must be capable, however, of switching the full-motion antenna back in line for testing purposes or in the event of an emergency situation, which could happen at any time in the operating lifetime of the satellite.

In a global TT&C network for non-GEO operation, several TT&C stations transfer their data to a central location, such as the SCC. Coordination and switching of signals from literally dozens of satellites at a central facility can be a complex task, requiring special computer equipment and software.

9.7.1.2 Satellite Control Center

The SCC is the brain of the satellite operation, providing the computing power and human intelligence necessary to operate and control a system of several satellites and TT&C Earth stations. Depicted in the left half of Figure 9.18 is the configuration of a typical SCC. The baseband signals between the TT&C and the SCC can interface directly with the computer system. Additional pieces of specialized digital processing equipment may be placed between computer and baseband in the integrated system. An example of such a device is the command generator used to format and modulate command transmissions for the uplink. In an emergency situation, the command generator, which itself is a small computer, allows commands to be entered directly from its front panel. A complementary device called the telemetry decommutator accepts the telemetry TDM data stream and demultiplexes the telemetry channels. On a manual basis, individual telemetry channels can be examined on its front panel. Another type of special device is the spacecraft simulator, usually a computer workstation programmed to behave like an operating satellite (from a bus standpoint). The simulator is used to prepare for critical maneuvers and for training operators to deal with normal and contingency situations, such as a satellite temporarily losing attitude pointing control.

The single box labeled "computer system" in Figure 9.18 contains sufficient computing power and redundancy to support all the requirements of the SCC, including reliability. The computer system performs real-time functions of command generation, telemetry reception and processing, and ranging. Additionally, several SCC activities are handled by the computer system in the background. Orbital dynamics deals with the determination of the satellite orbit and the planning of orbit correction maneuvers necessary to maintain satellite position. Using ranging data as an input to the orbital software, the orbital dynamics personnel generate maneuver plans, which are presented to the spacecraft controllers (i.e., the people who "fly" the satellites) for action by the command system. The orbital dynamics personnel and spacecraft controllers access the computer system with the workstations shown at the upper left of Figure 9.18. In the photograph in Figure 9.20, spacecraft controllers can enter spacecraft commands through workstations in front of them. The computer system keeps an ever watchful eye over the telemetry data and compares the reading to previously stored threshold alarm values. The printout device in Figure 9.18 logs every action taken by the controllers and records unusual spacecraft data identified by the computer. Hard copy formats have largely been replaced by computer mass storage and graphical workstations that can access any data when needed.

The last function of the TT&C ground facility is the monitoring of the communication transmissions to and from the satellite. This particular example is for a bent-pipe type of satellite, where all uplink transmission to the satellite can be monitored in the downlink. Prior to the start of service, the communication payload is thoroughly tested and every major component checked out from the TT&C Earth station, using automatic test equipment associated with the full tracking antenna. With commercial communications services being provided over the satellite, it is important that the satellite operator have a system to continuously monitor the entire downlink spectrum of each satellite. That is needed to ensure that all RF carriers are set to their prescribed power levels and are located at their assigned frequencies. If the satellite employs dual polarization to achieve frequency reuse, the purity of polarization of every carrier also must be checked prior to start of service and monitored during operation. Most of the transmissions to the satellite do not emanate from the TT&C station; therefore, the communications controllers must have telephone access to the uplinking Earth stations no matter where they may be located.

If the communications control console is located at the TT&C site, the limited-motion antennas can receive the downlink spectrum and deliver it to the SCC for inspection with spectrum analyzers and video monitors. Figure 9.18 indicates the other approach, in which the communications control is colocated with the SCC some distance from the TT&C Earth station. In such cases, one or more separate RO antennas can be provided but only if the satellite footprint covers the location of the SCC. Satellites that provide frequency reuse using narrow spot beams, such as applied at Ku and Ka band, do pose a bit of a problem in downlink monitoring since a separate Earth sta-

Figure 9.20 The satellite control console where the JC Sat satellites are controlled and monitored though the video display terminals. The Yokohoma TT&C station is used as the primary operation center for the JC Sat systems. (*Source:* Photograph courtesy of Japan Satellite Telecommunication, Ltd.)

tion is required in each independent footprint. Some approaches to this problem are discussed in Chapter 11. The communications console used to monitor transmissions from the JC Sat satellites is shown in the photograph in Figure 9.21. Note the use of video monitors to augment RF measurements

Figure 9.21 The downlinks of the JC Sat satellites are monitored on a 24-hour basis by communications controllers at the JC Sat network operations center. (*Source:* Photograph courtesy of Japan Satellite Telecommunication, Ltd.)

and the availability of telephone communications for contacting uplinking Earth stations and other satellite operators.

Monitoring of communication transmissions for digital processing satellites and non-GEO constellations used for MSS services is considerably more complex than the preceding discussion would indicate. That is because the downlink from the satellite comes in many separate beams and would involve reformatting of information onboard the satellite. The inclusion of intersatellite links adds another level of difficulty since the signals cannot be observed on the ground. Monitoring of services, therefore, depends on special facilities onboard each satellite and the relay of data either through the telemetry system or by a special downlink for that purpose. On the ground, the data form part of the overall network management system used by the satellite operation.

9.7.2 TV Uplinks and Broadcast Centers

Earth stations used for commercial video transmission services generally are large in class because of the high bandwidths and extensive signal routing and switching capabilities. A video uplink provides the origination path for TV as well as the capability to receive signals through terrestrial means and retransmit them to the satellite. In addition, a broadcast center adds a whole range of other technical and production activities and so mirrors aspects of a TV studio. A typical broadcast center is the core of a DTH network, including the ability to retransmit and originate as many as 200 independent video channels. The term *teleport* has been adopted to refer to a major Earth station that is open for business to serve the needs of a number of customers. Any of the above classes of video Earth stations has the full range of facilities discussed at the beginning of this chapter and shown in Figure 9.1, including a full complement of redundant equipment.

The configuration of the electronic systems of a basic video uplink station with the capability to transmit four simultaneous TV carriers is shown in Figure 9.22. There are separate interfaces for the video (color picture) and stereo and audio services. Video and audioswitching and distribution are provided prior to the uplink equipment to allow operators to redirect specific programming to the various uplink channels without changing the physical connections. That is often under computer control according to a prearranged schedule. Four of six video exciters provide the RF input to klystron HPAs. On the receive side, the antenna delivers the downlink band to video receivers that are used to monitor the uplink originated channels and to acquire programming from other sources (sources that potentially include other satellites, if additional antennas are available). The design of the overall broadcast center is as much science as art, relying heavily on the designer's experience and understanding of the transmission

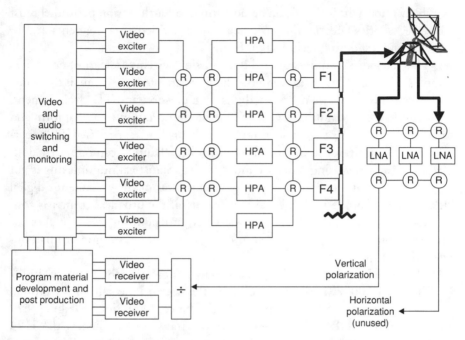

Figure 9.22 Example of a major communications Earth station with the capability to provide full duplex analog or digital video transmission.

and business needs. A station that delivers digital video is much the same in its basic configuration. However, the analog format of the video signal may, in fact, not exist in the station. Conversion from analog to digital could be performed remotely at the studio and the interface be entirely digital in nature. Modulation for the uplink would be with QPSK, but the remainder of the uplink portion of the station would not change.

9.7.2.1 Video Equipment Chain

For analog video, the baseband covering a frequency range of 0 to 4.2 MHz (e.g., for the NTSC standard) is accepted from the terrestrial interface by the video and audio monitoring and switching unit, shown at the upper left of Figure 9.22. Along with the video comes one or more channels of high-quality audio (monaural or stereo, possibly with additional channels for other languages) as well as data channels that support the management of the TV network. The data also could be a data broadcasting service, as discussed in Chapter 3. The monitoring and switching equipment is used in both the transmit and receive directions to route the baseband signals between uplinks, downlinks, redundant equipment, and TV test equipment. Video quality is extremely important in television broadcasting, so Earth station personnel must have the facilities and training to make standard video measurements and to help in troubleshooting technical problems.

Digital video compression and transmission systems are part of new video uplink stations and broadcast centers. The basic architecture is much the same, with the inclusion of the necessary digital compression and TDM equipment. The latter is used to multiplex video and audio information in digital form and to multiplex channels together for multiple-channel per carrier (MCPC) operation. The quantity of simultaneous digital video channels can be nearly doubled on each carrier using statistical multiplexing. Statistical multiplexing is not new to telecommunications, but it only recently was made available for digital video transmissions. Another important feature of the broadcast center is conditional access to control delivery of the video programming. That makes use of encryption technology to prevent pirating of the programming or unauthorized viewing.

For video service, the HPA has an output power in the range of 400W (TWTA) to 1,500W (KPA), depending on the size and the gain of the ground antenna and the sensitivity of the satellite. The configuration in Figure 9.22 shows additional HPAs that operate in other transponders; hence, the HPA outputs are combined in an RF multiplexer. These very high power levels demand the lowest possible RF loss and adequate cooling of the waveguide filters of the combiner. Redundancy switching is accomplished with waveguide rotary "baseball" switches (discussed in Chapter 6 in Section 6.4.2.4) to cross-connect online and standby HPAs. The resistor symbols indicate high-power dummy loads that absorb the output of an HPA when it is not switched into the operating chain and antenna.

The downlink equipment mirrors the uplink with the only difference being in the configuration of the LNAs. A separate operating LNA is provided for each polarization, although the station in Figure 9.22 is not using both simultaneously. When receiving a backhaul video point-to-point feed, either polarization may come into use. The LNA connected to the vertical antenna port provides sufficient gain to overcome the loss of a passive power splitter that feeds the entire frequency range to the video receivers. Each receiver can be tuned to any transponder in the downlink frequency range, including that of the uplink for monitoring purposes.

9.7.2.2 Video Routing and Switching

The quantity and the arrangement of the video routing and switching segment depend heavily on the particular application of the station or broadcast center. The basic devices are sold to cross-connect a particular quantity of video and audio signals. For example, an eight-by-eight router can connect each of eight input video/audio ports to any one of the eight output ports. It is akin to a telephone cross-bar switch, with the added complexity that multiple signals are switched simultaneously and the device is computer controlled to support a dynamic programming schedule common to the TV industry.

In a broadcast center, the routing and switching function is central to the operation of the entire network; hence, its design and installation are critical to success. The function requires a large capacity in terms of the number of simultaneous video/audio channels (numbering in the hundreds). That capability cannot be implemented as a single unit, so there typically is a bank of switching/routers to support the operation under any expected situation. Both computer and human control are involved: the computers are programmed ahead of time with the requirements of the network, and humans are always present to observe that the system is working correctly. Numerous backup modes are provided, including dual feeding of the same (important) video channels, and human intervention may be needed at any time to ensure that the right programs are being transferred to the right channels. More recently, computer automation has been added to reduce the amount of manual labor involved and to improve the overall reliability of switching operation. Mistakes at this level can be detrimental to the quality of service and, ultimately, the revenue derived from programming and advertising.

9.7.2.3 Program Acquisition and Playback

Whether we are talking about a basic video uplink or a comprehensive broadcast center, there is always a need to obtain the programming input from some source. Program acquisition in this context means that the video signal is originating outside the station from one or more remote sources. An uplink that serves a TV network acquires its programming from the originating studio through a terrestrial link. As discussed earlier in this chapter, tail links use either line-of-sight microwave communication or fiber optic cable. Fiber optic cable transmission usually is preferred because of the wider bandwidths for either analog or digital video formats and because of the absence of signal fading (except, of course, for a backhoe fade). It has become popular to acquire programming from a satellite link, on the same satellite or on other satellites. If we are talking about receiving from other satellites, there must be sufficient TVRO antennas to provide the desired access. For that reason, a large uplink station, such as that shown in Figure 9.23, includes a considerable quantity of transmit and receive antennas.

The other potential source of programming is from video tape, which is either developed at the site or delivered to it (a process called "bicycling"). Tapes are received, cataloged, and stored for subsequent playback. In some installations, the video is transferred from the incoming tape to a master tape reel,

Figure 9.23 The DIRECTV broadcasting center in Castle Rock, Colorado. (*Source:* Courtesy of DIRECTV, Inc.)

which is what is actually played on air. There is a trend to use digitized format for tapes as well as satellite transmission, using the standard format of CCIR Recommendation 601. Some facilities use hard disk video servers to contain prerecorded advertising and even programming. The advantage of video server technology is that it precludes mistakes in setting up the transmission and allows content to be selected for particular transmissions (e.g., the insertion of promotions and advertising targeted to a particular audience).

9.7.3 FDMA Digital Communications Service

The traditional type of Earth station used for telephone trunking and wideband data transmission often employs the FDMA approach. There are two ways to implement FDMA: single-destination carriers and multiple-destination carriers. In the single-destination approach, there are two carriers for each bidirectional (duplex) link, thus connecting a pair of Earth stations. We would expect each carrier to contain multiple voice and data channels using TDM or possibly statistical multiplexing. Time assignments on the respective carriers are also paired so that each voice or data channel is duplex as well.

The multiple-destination type of carrier contains traffic not for one station but rather for the group of stations that are to be networked. If there are five stations in the network, each station transmits one carrier that contains multiplexed traffic for all other stations. In total, there is a requirement that five carriers access the satellite. In comparison, using single-destination carriers and five stations, there would have to be a total of 20 carriers to provide full interconnectivity between all stations.

A digital communication subsystem capable of transmitting four T1 (or E1) carriers is shown at the left of Figure 9.24. Each T1 has its own modem, tuned to a separate frequency in the transponder. Redundancy is provided for the modems on a five-for-four basis using an integrated switching system. The T1 channels interface with the terrestrial network through a TDM multiplexer that combines the four streams into one 6-Mbps (T2) channel. The transponder contains carriers for each station in the network. Operation in a single transponder allows the use of one operating upconverter and one operating downconverter, with full redundancy provided to maintain high reliability.

Because Earth station transmissions must be on different frequencies, FDMA can be complicated to arrange and even more difficult to change once the network goes into operation. One inherent advantage, however, is that the

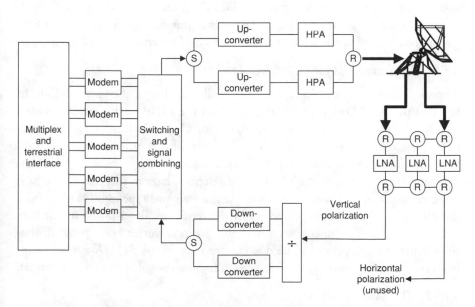

Figure 9.24 Example of a major communications Earth station with the capability to provide full duplex digital transmission FDMA service.

transmit power of the station can be tailored to its traffic requirement (i.e., the total number of carriers and channel capacity). Another advantage is that station transmissions are independent of each other (because they are on separate frequencies) so they can be mixed in with other unrelated carriers in the same transponder.

9.7.4 TDMA Earth Station

TDMA Earth stations were first developed to use an entire transponder in the most efficient manner, transmitting at 60 to 120 Mbps to squeeze together the maximum number of voice channels possible [2]. Later on, narrowband TDMA was introduced to allow the use of small, lower cost RF terminals. As discussed in Chapter 5, there is some overhead in terms of guardtime and burst synchronization, so the actual throughput is somewhat less than the bandwidth would support. Also, it usually is desirable to use convolutional coding to reduce the error rate for sensitive applications like digital video and corporate data transmission; therefore, throughput is further reduced by the coding rate. Still, TDMA is widely accepted in satellite communications because its mesh capability provides flexibility as well as its ease of integration into digital networks. Also, the digital elements of a TDMA terminal can be configured for different transmission or services and can be reprogrammed at any time.

Figure 9.25 presents a simplified block diagram of a typical TDMA Earth station. This discussion relates to the type of major Earth station used for full-transponder TDMA, while the approach for VSATs is covered in a later section.

9.7.4.1 RF Equipment for TDMA Service

The configuration of the RF terminal is similar to that of the video uplink. In full-transponder TDMA service, the G/T and the EIRP of this type of station usually demand an antenna of 10m or 7m for C- or Ku-band service, respectively. The reason for that is that TDMA establishes a true mesh network in which stations can transmit directly to one another through the same transponder. That is in contrast to the VSAT hub (described later in this chapter), which is the center point of the star network. For the standard type of TDMA station, the HPA must have an output power in the range of 400–1,000W to saturate the transponder. The upconverter and the downconverter are typical off-the-shelf items. An exception for the downconverter is if the TDMA network supports transponder hopping, in which case a frequency-agile LO will be required.

9.7.4.2 TDMA Terminal

Moving to the left half of Figure 9.25, the TDMA terminal section consists of an operating modem and a backup, baseband equipment, multiplex, and M&C

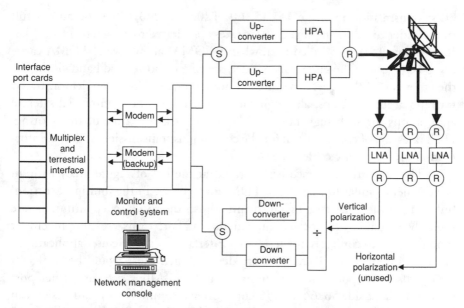

Figure 9.25 TDMA Earth station for full-transponder operation with the capability to provide a variety of digital communications services on a full mesh basis.

equipment. It is customary to purchase the terminal as an integrated subsystem as part of an integrated network from a TDMA manufacturer. The computer workstation connected to the M&C equipment is used to monitor the network and to program the particular terminal for the required traffic arrangement or routing. Usually, one station acts as the network master and is used to create the overall traffic routing pattern. Called the network map, the routing pattern is stored in each TDMA terminal to direct individual TDM communication channels to the proper destination. Depending on the time of day or the traffic demand, the network map is modified by appropriate programming of the central M&C computer. Central control for the network can be exercised from a location not at an Earth station since the computer terminal can be connected through a voice-grade line (terrestrial or satellite).

The design and control of the TDMA modem are critical to the operation of the terminal. This type of modem is intended for burst operation, that is, where the carrier must be turned on and off in rapid succession. The demodulator section has the difficult task of locking onto the incoming carrier at IF and synchronizing with its modulated digital information stream. That dual process must be accomplished for each received burst before the useful data actually can be recovered. Burst modem design has matured over the last two decades so that extremely efficient and reliable units are now available. Typical burst transmission rates are 15, 45, 60, 120, and 240 Mbps; because a full transponder could be employed, those rates are referred to as wideband TDMA. Rates below 15 Mbps (called narrowband TDMA) allow several TDMA carriers to be placed in the same transponder, sharing its power and bandwidth. On the other end of the scale, experimental burst modems have been demonstrated for rates around 1 Gbps, although there currently are no commercial satellite applications for such high rates and bandwidths at the time of this writing. High rates are, of course, useful for the continuous transmission of digital information on fiber optic cable systems.

The terrestrial interface is illustrated at the left of Figure 9.25, with an arrangement similar to that of the TDM multiplexer. In the multiplex section, buffering of the burst-rate digital stream adjusts the speed to a continuous rate typically matching the local digital hierarchy. Port cards are plug-in circuit modules used to configure the terrestrial interface for specific user applications. For example, if only T1 channels are desired, then there would be sufficient port cards to support a specific number of 1.544-Mbps streams. Other port card options include 56, 64, or 256 kbps (often a common port card design can be "strapped" for the desired channel rate); ADPCM or PCM voice; DS3 (45 Mbps); or other specialized interface conditions. Adaptive features such as reprogrammable rates and statistical multiplexing also are available. The type of statistical multiplexing used for voice is called digital speech interpolation (DSI), which transmits digitized voice samples over the satellite only when there is actually an active talker on the line.

9.7.4.3 Carrier Hopping

The capacity of a given TDMA carrier, which is transmitted sequentially from several Earth stations, is limited by its bandwidth and power. When the sum of the channel requirements of the individual stations is greater than the capacity of the carrier, additional carriers must be provided. Carrier hopping is the technique whereby stations can switch between several such carriers to pick out traffic destined to them. Carrier hopping in wideband TDMA must be done between transponders; hence, the RF terminal is equipped with a frequency-agile downconverter. Narrowband TDMA usually is done in the same transponder, eliminating the need for multiple downconverters. However, the demodulator is capable of changing its frequency in the IF range.

9.7.5 VSAT Network and Remote Terminal

VSAT technology brings the features and benefits of satellite communications down to an extremely economical and usable form. Corporate users can implement VSAT networks to obtain services that otherwise would require an extensive terrestrial infrastructure. The favorable economics result from sophisticated digital hardware and software technology in support of the data communication protocols common in business and government applications. The network employs a hub station to form a star network, with the host computer connected to the hub. Such a network, indicated in Figure 9.26, can provide transmission rates of up to 64 kbps per remote terminal, but lower average data rates are more common. Alternatively, VSATs with direct connectivity can be implemented to create a full-mesh network with direct remote-to-remote communication. The advantage of that approach is that the data rate can be increased to 2 Mbps or more, although the transmit power and, potentially, the antenna diameter would have to be increased. The star and mesh arrangements have slightly different technical requirements, but it is possible to integrate the two where needed.

The use of Ku band (in the FSS portion of the spectrum) simplifies Earth station siting by eliminating the need for terrestrial frequency coordination. The higher power satellites permit more services to be carried, including analog or digital video, using antenna diameters in the range of 1.2–1.8m. Implementation of a VSAT network is practical at C band with somewhat larger antennas, which is mainly driven by the concern over adjacent satellite interference. The antenna size also is determined by the required EIRP, indicated in

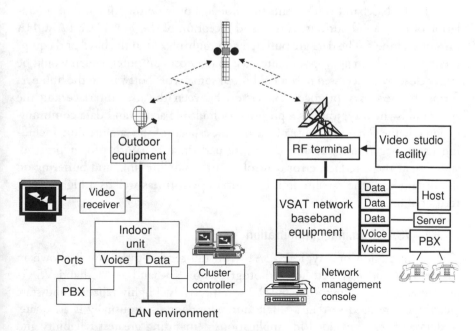

Figure 9.26 Architecture of a typical VSAT star network.

Figure 9.9. The following sections discuss Earth station types that are used to implement star and mesh networks.

9.7.5.1 Hub and Network Control Station

The hub station of a VSAT star network is similar in size and capacity to the TDMA terminal. That is not surprising because most VSAT start networks employ the TDMA access approach for the inroutes (from VSAT to hub) and a relatively high-capacity TDM outroute (from the hub to the VSAT). As shown in Figure 9.26, the hub consists of an RF terminal, a set of baseband equipment (including the modems, multiplexers, and encoders) and network M&C equipment. The RF terminal is designed and constructed in accordance with the principles covered previously. It provides a wideband uplink of one digital carrier per separate network and receives the narrowband channels from the remote VSATs themselves. The specific nature of these carriers depends on the architecture of the particular vendor and the parameters selected by the network operator. Generally, the outroute operates at a multiple of 64 kbps (up to 512 kbps) and the inroute at either 128 or 64 kbps. The EIRP of the VSAT can thus be held to a low value, with the hub antenna being the exclusive receiving point. As a result, the HPA of the VSAT can be held to under 10W and may be as little as 1W. That imbalance in outroute versus inroute carrier level means there are always more inroute carriers.

In the baseband equipment, the modems provide the function of modulation of the digital outroute carrier and reception of the TDMA and ALOHA inroute channels. The data are buffered and multiplexed in the baseband equipment to provide an appropriate interface to the host computer, which would be either close by or accessed over a tail link of some type. Software in the hub performs the necessary protocol conversion between the user interface and the space link, which typically is a proprietary multiple access and data communication scheme. That is because of the unique aspects of the satellite link, including its propagation delay and error rate performance. With proper protocol spoofing at the interface, error control over the satellite link, and buffering on the distant end, the satellite link is made to perform as well as a high-quality terrestrial link.

9.7.5.2 VSAT Terminal Configuration

The configuration of a typical VSAT with full-service capability is shown in Figure 9.27 and reviewed here. Photographs of Ku-band and C-band VSATs are presented in Figures 9.28 and 9.29, respectively. Highly reliable solid state electronics are used so that a single nonredundant string usually is adequate. Redundancy can be added for applications demanding greatest reliability and where the VSAT might be difficult to reach by maintenance personnel. In situa-

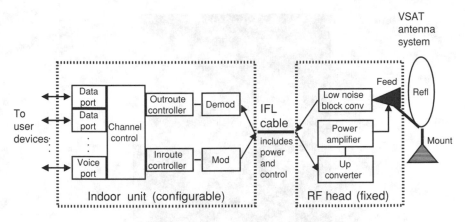

Figure 9.27 Configuration of a VSAT, which combines two-way voice and data with one-way video.

tions in which the PSTN is available at the site, a dial backup facility can be added to the VSAT and hub to provide service in the event of an outdoor unit failure.

The RF terminal of a VSAT is extremely compact and often attached to the antenna itself. As shown at the right of Figure 9.27, the RF equipment is composed of an LNB for reception and an upconverter and SSPA for transmission. The antenna in the illustration is an offset-fed parabola with electronics mounted directly to the feed. For a GEO satellite network, a fixed mount is locked into position, although there still should be provision to repoint the antenna if a change of satellite location is ever contemplated. Non-GEO VSAT networks, which were still being defined at the time of this writing, will require RF terminals capable of tracking at least two satellites at one time.

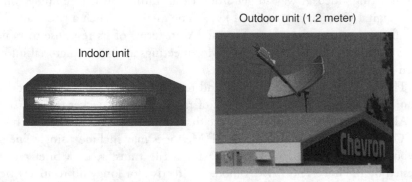

Figure 9.28 Ku-band VSAT employing a 1.2m offset-fed parabolic reflector. (*Source:* Photograph courtesy of Hughes Network Systems.)

Figure 9.29   Photograph of a C-band VSAT installation with a 2.4m antenna.

The remainder of the VSAT indoor equipment, consisting of a compact electronic box about the size of a PC, would be located in the building to which service is provided. As shown in Figure 9.27, the elements are functionally the same as those provided in the major Earth station described earlier in this chapter. The capabilities assumed for the station in the figure included digital voice and data transmission as well as video reception. Because of the low transmit power and small diameter of the RF terminal, a VSAT typically is not capable of transmitting video. A way to get around that limitation is to employ compressed digital video operating at 64 kbps. The quality of such a two-way link is significantly poorer than conventional TV in terms of its response to rapid picture motion but could be acceptable for meetings and static pictorial information (e.g., meeting presentation charts).

The techniques for transmitting and receiving data, both on a broadcast and an interactive basis, involve the use of random access packet transmission (e.g., ALOHA) as well as TDMA. For the first transmission from VSAT to hub, an ALOHA packet normally is uplinked. Messages may include a single line of a keyboard entry; a request to the hub for a file transfer; or a brief report requested by the hub. The TDMA mode is effective for long and relatively frequent transmissions and for voice and video communications, which require a constant bit rate. Switching between random access and TDMA is in response to local traffic demands, activated either by the centralized network management system or automatically when required by the VSAT itself.

The star network structure is assumed for the example in Figure 9.27. If the network is to be used extensively for switched telephone or video teleconferencing service, the configuration of the mesh is going to be far more effective. By allowing direct links between VSATs, the system avoids double hops and multiple hub stations interconnected by expensive terrestrial links. That can be achieved with TDMA depending on the network requirements. In a more or less traditional FDMA demand-assigned scheme, a request for service enters the VSAT over the telephone interface and is transmitted over signaling a network to a hub, where the setup of the link is directed. The hub responds with digital instructions to the VSATs on both ends of the intended circuit to tune their SCPC modems to a frequency pair assigned for the duration of the call. Through that hub, the network management system monitors the use of the network and accomplishes such tasks as data traffic routing and billing. That is similar to the design and operation of an MSS telephone network.

Voice and data interface equipment is needed to make the VSAT appear to be the same as any telephone or data communication circuit. For example, the VSAT network may use a proprietary protocol to carry and process signaling information that the user's PBX does not understand. The voice interface then converts standard telephone signaling into the network protocol. Likewise, data protocols such as TCP/IP are converted to the VSAT network protocol in the data interface equipment. The appropriate protocol and address information are recreated at the distant hub station or remote VSAT. In addition to signaling and protocols, the interface equipment can provide signal conditioning such as bandwidth compression and S/N enhancement.

Video teleconferencing services on a point-to-multipoint basis are well within the capabilities of VSATs. As shown in the upper left of Figure 9.27, a standard video receiver provides a baseband signal to a teleconference unit. The picture can then be carried to a conference room or theater. In addition to the video, the equipment might provide a voice interface for interactive question-and-answer service, using the VSAT to provide the reverse audio link.

This has been only a brief introduction to the types of Earth stations that make up a VSAT star or mesh network. The hardware designs and software technologies that facilitate these systems are undergoing an evolution to much smaller and cheaper installations. The purpose of those efforts is to expand the market into more types and sizes of businesses that can exploit private networks. In addition, consumer designs will appear in response to new satellite systems launched at Ka band immediately after the year 2000. Those prospects are reviewed in Chapter 13.

9.7.6 TV Receive-Only Design

A TVRO Earth station can take on one of a number of possible configurations, depending on the particular application. An installation used for a cable TV head end must be capable of receiving more than one hundred channels at the same time. For a home installation, only one receiver is required (although two or more might be desired); however, that receiver should permit the viewer to change TV and transponder channels conveniently. In multiple-dwelling situations, such as an apartment building, there would be one antenna but several individual receivers for the different users. This discussion reviews the sizing of the antenna and equipment configurations of TVRO stations.

9.7.6.1 TVRO Antenna Sizing

The size of a receiving antenna is determined primarily by the RF power (i.e., the EIRP) of the transmitting source. At the same time, international and domestic regulations limit C-band EIRP because of the possibility of RFI between satellites and terrestrial microwave services that share the band. No such sharing is required for much of the FSS and all of the BSS segments of Ku-band. As a consequence, the power level of C-band satellites is considerably lower than that of Ku-band satellites. That, of course, is the result of the regulatory limit on the design of the satellite transponder and has nothing to do with the physics of propagation. On the other hand, link margin (and hence EIRP) at Ku band must be increased relative to C band to adjust for the rain attenuation in the particular climatic region.

Figure 9.30 presents in graphical form the relationship between TVRO antenna sizing for parabolic reflectors in the range of 45 cm to 3m (shown along the X-axis) and satellite EIRP. A system noise temperature of 100K is assumed. Satellite EIRP values between 30 and 55 dBW (shown along the Y-axis) cover the typical ranges for C band, Ku-band FSS, and Ku-band BSS. We have assumed that the satellite is transmitting a single wideband digital carrier with multiple TV channels that are encoded with the MPEG 2 format (according to the DVB standard). On the basis of power alone, there should be only one curve; however, the factor of rain attenuation forces the curve for Ku band to move up the power range.

The upper curve includes 2 dB of additional power to provide the same link reliability in the Ku-band range (FSS and BSS), which would be sufficient for temperate climates such as in the northeastern United States, Japan, or Europe, and with an elevation angle greater than about 30 degrees. Tropical regions with heavy thunderstorm activity and frequent torrential rains require even greater incremental power margin. The C-band curve begins at a diameter of 1m and an EIRP of approximately 40 dBW because of the international lim-

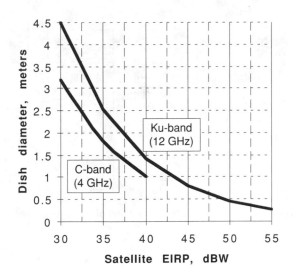

Figure 9.30 Antenna diameter of a TVRO home Earth station as a function satellite EIRP and frequency band.

itations on power flux density. With the rain effect, it takes approximately 42 dBW at Ku band to achieve the same link performance with the 1m antenna. In the Ku FSS range, an antenna diameter as small as 0.5m is adequate with an EIRP of 48 dBW. However, because FSS satellites are not intended for broadcast video services exclusively, the satellites may be spaced so closely together that ground antennas smaller that 1m are subject to unacceptable interference.

The Ku BSS range is available for truly high-power DTH broadcasting, as shown in Figure 9.30. Satellites using the same frequency channels and polarization are to be separated in orbit sufficiently to allow the use of antennas as small as 45 cm in diameter. To achieve that at Ku band with rain attenuation, the satellite must be capable of delivering an EIRP of approximately 50 dBW or higher. The power output of the appropriate transponder amplifier usually is around 130W (depending on the size of the footprint), double that used in the FSS segment of Ku band in North America.

9.7.6.2 Cable Head End TVRO Configuration

The cable head end TVRO, illustrated in Figure 9.31, is used to receive several commercial TV channels for distribution over a cable TV system. Four receivers are equipped in the figure; an actual CATV head end, however, would have 20 to 30 receivers per satellite downlink. Each receiver, in turn, can recover one analog TV channel; for digital video, the receiver provides a multiplexed set of

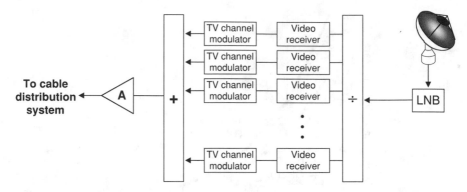

Figure 9.31   Configuration of a cable TVRO Earth station of the type located at a cable head end.

channels. The antenna used to gather the downlink is of sufficient diameter to ensure good signal to noise quality or E_b/N_0 over expected link conditions. While a single LNB is shown, it is more common to find two units: one for the vertical polarization and one for the horizontal. Online redundancy typically is not provided because of the high reliability that solid state LNBs usually provide. In the event of failure, the bad LNB can be rapidly changed out by technical personnel at the head end. The dc power for the LNB usually is carried over the conductors of the same coaxial cable used to bring the RF signals back to the receivers.

The wideband output of the LNB is divided on a power basis to provide the entire spectrum to the video receivers. In analog transmission, each receiver is tuned to a different transponder channel and delivers a video baseband with its associated audio to a TV channel modulator. A power summer and amplifier combine the several TV channels that now lie within the receiving frequency range of normal television sets. Alternatively, the channels could be received in a digital format and either transmitted directly over the cable by digital modulation or, in a hybrid analog/digital system, converted to analog and then modulated as in the conventional analog TV approach. The cable distribution system (Figure 3.13) spreads the ensemble of channels out to cable TV subscribers. Modern cable systems provide each home with a settop box that allows the viewer to select from the full range of TV channels delivered over the cable. Because pay channels may be involved, the settop box has the capability to be programmed to allow the viewer to watch only those channels that have been authorized by the cable company (and for which the subscriber presumably is paying).

9.7.6.3   Direct-to-Home TVRO

In a direct-to-home installation, one video receiver also incorporates the TV channel modulator. Provision is made for convenient tuning of the receiver to any desired channel the viewer wishes to watch. The early C-band TVROs included an antenna with a motor-driven mount and adjustable polarizer. Such receiving systems are made user friendly by microprocessor control, allowing the viewer to key in the desired satellite and channel with a wireless remote control unit. The intelligence in the receiver causes the antenna to move over to the general position of the desired satellite, adjusts the alignment automatically for maximum signal power, and then displays the correct channel on the video monitor.

Modern digital DTH systems, such as that shown in Figure 9.32, use fixed antennas to reduce cost and operational complexity. By operating multiple satellites in the same orbit position and using digital video, the capacity of such a low-cost receive system can be over 200 full-color digital TV channels. The receiver demodulates and decodes the basic bitstream from the downlink

Figure 9.32   A typical DTH installation.

carrier and demultiplexes the individual TV channels. The channel capacity per carrier typically is in the range of 5 to 12, depending on the total data rate, compression parameters, and whether statistical multiplexing is included. The latter has been introduced to nearly double the number of TV channels per carrier by dynamically adjusting the data rate to the motion demands of each channel. Another important feature of the receiver is conditional access, which is discussed next.

9.7.6.4 Receiving Encrypted Programming

The TVRO must include a decoder to recover channels that have been encrypted at the uplink source. Several systems are in use in North America and Europe, each differing in the degree of disruption of the video and audio and in the degree of difficulty of unauthorized receiving of the signal. There has been a trend to adopt the DVB series of standards that were developed in Europe. Even with that standard, there are several proprietary encrypting and conditional access schemes from companies like News DataCom (NDC), Divicom, and Irdeto. The actual transmission system is based on MPEG 2, with DVB providing the supporting arrangement for the multiplexing of added data for control and supervision. (This topic is explored in greater detail in our other work [3].)

9.7.7 MSS User Terminals

Mobile satellite communication is rapidly evolving, as new applications, and terminal types are introduced around the world. The types of terminals that were available at the time of this writing include marine (the forerunners of all MSS UTs), aeronautical, vehicular (primarily for trucks), and tabletop (or mountaintop) MSS UTs. The terminals are supported by the Inmarsat system of GEO satellites (particularly Inmarsat 3 with its spot beams) and certain domestic GEO satellites that mimic Inmarsat. Hand-held UTs became available by 1999 using the first non-GEO constellations.

Examples of each type of UT are shown in Figures 9.33 to 9.36. Most were designed as integrated telecommunications terminals using VLSI and discrete component technology. The Mini-M terminal for Inmarsat 3 is perhaps the most user-friendly GEO unit, being targeted at industrial and high-end individual users (Figure 9.33). In developing regions, the Mini-M has established itself as a viable option for providing a range of vital telecommunications services. The vehicular type of terminal (Figure 9.34) primarily has served to assist trucking lines with locating each vehicle in a fleet, using a GPS receiver integrated with a low-speed data transmitting capability onboard the truck. The

Figure 9.33 MSS phones can be used almost anywhere. (*Source:* Courtesy of NERA.)

Figure 9.34 A vehicular MSS installation for use with Inmarsat. (*Source:* Courtesy of Inmarsat.)

Figure 9.35 A compact maritime mobile terminal. (*Source:* Courtesy of NERA.)

maritime terminal (Figure 9.35) has been scaled down in size but not in capability. Aeronautical terminals are part of the avionics systems of commercial jet-liners, allowing cockpit crews to remain in contact with airline headquarters. The same terminal allows passengers to make phone calls and to transmit and receive fax messages during transoceanic flights.

The approach being taken for hand-held UTs is more in keeping with cell phone technology. They rely on compact circuit modules using application-specific

Figure 9.36 The first hand-held MSS phones, introduced by Motorola in 1999.

intergrated circuit (ASIC) components that operate at low power to conserve batteries. First-generation hand-held MSS UTs are about the same size as second- or third-generation cell phones (Figure 9.36). That is due to the added complexity of operating at L or S band and the fact that the initial manufacturing quantities tend to be not high enough to justify the full miniaturization of the product. In time, the MSS UT will be as small and as handy as third-generation cell phones such as the Motorola Microtac or the Ericsson DH series.

References

[1] Elbert, B. R., and B. Martyna, *Client/Server Computing: Architecture, Applications, and Distributed Systems Management*, Norwood, MA: Artech House, 1994.

[2] Schmidt, W. G., "The Application of TDMA to the Intelsat IV Satellite Series," *COMSAT Technical Review*, Vol. 3, No. 2, Fall 1973.

[3] Elbert, B. R., *The Satellite Communication Applications Handbook*, Norwood, MA: Artech House, 1997.

10

Launch Vehicles and Services

The technologies that make it possible to put one or more communications spacecraft into Earth orbit are the result of rocket science and astrodynamics. Fortunately, we do not have to be rocket scientists to understand the use of rocketry and to work with the people and organizations that provide those services to satellite operators. Rather than attempt to cover those specialized fields, we have taken a practical approach in providing a basic understanding. Needless to say, the necessary systems work quite well, and although significant risks are involved, spacecraft have been placed into orbit, have landed on the moon, and are visiting faraway planets. Thirty or more years of experience with satellite operations have reduced those technologies to commercial practice, although we must never lose sight of the complexities that exist below the surface. For that reason, satellite operators rely heavily on the specialized capabilities of spacecraft manufacturers and the organizations that build and launch the rockets themselves.

In addition to a review of the technologies, this chapter also presents a summary of launch vehicle (LV) systems available at the time of this writing. This chapter should not be taken as the definitive word because the particular set of usable launch vehicles continuously changes in terms of both the particular rockets and their capabilities. Commercial launches were once the exclusive domain of the U.S. rocket manufacturers using the NASA launch facilities at Cape Canaveral, Florida. However, what has evolved is a truly international business, with good capabilities also available from European, Chinese, and Russian launch services organizations. Worldwide, there are 19 different space launch systems from six different countries, flown from 14 different geographic sites [1]. While most of the launches support GEO satellite systems, the advent of non-GEO satellite constellations is making some new and innovative launch techniques more viable for commercial service. The current high cost of launching coupled with the substantial increase in the number of satellites in recent years will encourage the development of new systems and new providers.

The selection of a particular LV system and service provider can be strategic to the success of the overall mission and business. The design and operation of the typical LV are complex and relate as much to technical design as they do to the consistency of the processes used in LV manufacture and operation. The three most important criteria in LV selection typically are the launch mass capability, the reliability or success record of the overall system and process, and the cost of use. This chapter reviews the technology and the systems in use, but it is not a substitute for a thorough investigation of the available options. Any reader needing to make such a selection should base it on current factual information provided by the LV providers, supplemented by other industry sources, such as the spacecraft manufacturers, experienced consultants, and launch insurance brokers.

10.1 The Launch Mission

Spacecraft are designed to be compatible with one or more LVs for placement into Earth orbit. Kepler's second law states that the orbital period scales as the 3/2 power of the semimajor axis. Therefore, the higher the apogee altitude, the larger the period of revolution. A good example is the geostationary transfer orbit (GTO), which has a period of approximately 16 hours. After perigee is raised to GEO altitude (around 36,000 km), the period increases to the prescribed 24 hours.

The launch mission is the sequence of steps that commence when the spacecraft and the LV leave the launch platform (on land, sea, or in the air) until the spacecraft is separated and safely on its way. An overview of that process for a GEO communications satellite is presented in Figure 10.1. Major changes in the trajectory and the orbit of the vehicle are provided by powerful liquid fuel or solid fuel rocket engines, which increase the velocity by the required amounts indicated in the figure. A good reason for first studying the GEO type of launch mission is that it includes other non-GEO orbit altitudes as intermediate steps. A big difference, however, is that LEO and MEO have significantly higher inclinations than would be practical for a GEO mission. That is why, as we will see later, the non-GEO spacecraft are launched toward the north from sites that are more northerly as well.

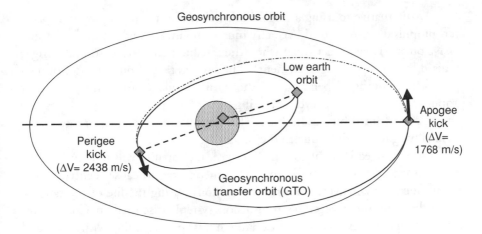

Figure 10.1 Major orbit changes to reach geosynchronous orbit.

For the typical GEO launch mission, the booster lifts off from the launch platform and delivers the vehicle to an altitude of between 150 and 300 km. At that altitude, the vehicle follows a circular orbital path that is used in many LV systems as a parking orbit. The distance is exaggerated in Figure 10.1 because if shown properly to scale 300 km would appear like the thickness of a pencil line. Another rocket stage is then used to kick the vehicle into the elliptical GTO, where the farthest point (apogee) is at or near geosynchronous altitude (e.g., 36,000 km) and the closest point (perigee) is still at the same altitude as the parking orbit. That is because the satellite velocity is not high enough to sustain the altitude at apogee. Once the spacecraft is in transfer orbit, all control is exercised through the TT&C station(s) and the SCC. Because the GTO is elliptical, its period is around 16 hours and the perigee point changes its longitude from revolution to revolution. The Earth's rotation under this orbit also contributes to the changing of the perigee point.

There are fundamentally two different strategies for moving from GTO to the 24-hour GEO. The original and simplest approach is to employ another high-thrust rocket stage to boost the satellite at an appropriate apogee (e.g., one that lies close to the final longitude). The last rocket stage should provide sufficient velocity increment to raise perigee to 36,000 km, thus circularizing the orbit. This stage is fired by ground command and is also used in many GEO missions to correct for orbit inclination of the booster. The latter will result if the launch platform is at a nonequatorial latitude. Both Ariane and Sea Launch tout the benefit of their equatorial launch sites because such sites improve the final usable payload to GEO.

An alternative to using a single high-thrust boost is to employ the space-craft propulsion system to add velocity incrementally, through a number of successive burns. They are activated when the satellite is at apogee, so that perigee is raised by corresponding increments. The process takes from a matter of days to perhaps months, depending on the thrust available from the spacecraft propulsion system. Low-level thrusting from ion propulsion produces the longest period for orbit raising but yields greater efficiency (e.g., more useful payload weight to GEO), due to the higher I_{sp}.

Once perigee is equal to GEO apogee, the satellite will remain in orbit forever because there is insufficient atmospheric drag to cause reentry. Other initial touchup maneuvers and orbit corrections during the life of the satellite are made using the spacecraft's propulsion system (discussed in Chapter 8). Table 10.1 lists the events in a typical launch mission beginning with liftoff and ending with the start of operations in GEO. Considerations for non-GEO orbits are discussed later in this chapter.

10.1.1 The Boost Phase

To reach LEO, the parking orbit, an object must be elevated high enough above the Earth to escape the Earth's gravitational pull. That is done with a ballistic

Table 10.1
Launch Sequence for a Geostationary Orbit Mission Identifying Key
Events and Ground Support Activities

1.	Liftoff of main booster
2.	Parking orbit achieved
3.	Transfer orbit injection
4.	TT&C link established with ground
5.	Reorientation to AMF attitude
6.	Preburn RCS maneuver
7.	Apogee motor firing
9.	Drift to assigned station
10.	Orbit and attitude adjustments
11.	Despin of platform or reorientation
12.	Spacecraft or body deployments
13.	Bus testing
14.	Payload testing
15.	Start operations

rocket stage that lofts the spacecraft, following a path that would return to Earth some distance from the launch site. A second rocket burn pushes the vehicle into Earth orbit. Kepler's first law governs the motion of the vehicle in orbit, wherein the satellite follows an elliptical path with the Earth centered at one focus. The satellite stays in orbit because the force produced by centripetal acceleration is equal to the gravitational pull of the Earth.

Because the mass of the satellite is insignificant compared to that of the Earth, the time period of the orbit is completely determined by the velocity of the satellite. (Pertinent information about this factor is provided in Chapter 1.) Launch vehicles consist of multiple rocket stages to satisfy the demand for fuel. Release of spent stages is beneficial because it reduces the dead mass that the propellant must push into orbit. The second stage of the booster is pointed along the tangent to the desired orbit, and the total thrust increases the velocity of the vehicle to the required value. At a point where the pressure induced by the atmosphere is low enough for safe deployment, the protective shroud around the spacecraft is jettisoned. In the parking orbit, the vehicle makes one revolution around the Earth in approximately one and a half hours. Since that is too fast for conventional TT&C stations to track the LV, operation of the first and second stages is automatically controlled by an onboard computer.

The thin atmosphere that extends to the parking orbit is minute, but there is sufficient drag to cause some slowing of the vehicle. If not corrected in time, the orbit gradually decays and eventually the vehicle reenters and burns up. Because that can take from months to years, there is little risk to the satellite between the time it reaches the parking orbit and when the third stage propels it to GTO. There can be circumstances in which the second stage does not provide its full increment of velocity and the vehicle is then on a suborbital or even ballistic path back into the denser atmosphere. Thus, timely firing of the third stage is extremely important to prevent loss of the mission. In a LEO mission, discussed in Section 10.1.2, this orbit is at or close to the final orbit, so preoperational activities can commence.

There is an issue that relates to the geographical latitude of the launch site. Nearly every launch site is located some distance from the equator, while the GEO is in the plane of the equator. (An exception is Sea Launch, which uses a floating platform that is towed to a location in the Pacific Ocean at the equator.) The parking orbit produced by the second stage may be at an angle with the equatorial plane; this is referred to as inclination of the orbit. This inclination is approximately equal to the latitude of the launch site and must be removed for a GEO mission. This is accomplished using rocket energy at some point prior to putting the satellite into commercial service. Obviously, the closer the launch platform is to the equator, the less additional rocket energy is needed to correct for inclination. The counter to this is that inclination is not a problem, per se, because it can be corrected by subsequent thrust maneuvers from either the LV or the spacecraft propulsion systems.

10.1.2 Non-GEO Missions

Satellite constellations that operate in LEO do not require the use of a transfer orbit. Instead, the LV injects the spacecraft into the appropriate LEO, where they may continue to orbit the Earth. Atmospheric drag is definitely an issue, as are radiation effects from the lower extreme of the Van Allen Belt. Those considerations can be taken into account in the design of LEO spacecraft, whereby sufficient orbit maintenance fuel and electronic device shielding are provided. It is likely that between the constraints imposed by those two factors along with that of investment cost, LEO spacecraft generally are designed for a considerably shorter life than are GEO or even MEO spacecraft.

　　One counter to the cost consideration is the fact that a standard type of LV can place substantially more spacecraft mass into LEO than GEO. More important, LEO spacecraft need to be deployed in such a way as to fill a specific orbit plane. Therefore, LEO spacecraft usually are launched in a bunch of four to eight. In addition, current LEO constellations use polar or highly inclined orbits. That means the launch site can be quite far from the equator with the LV trajectory pointed toward the north or northeast, as opposed to the east.

10.1.3 Geostationary Transfer Orbit

GEO is reached by first placing the satellite into GTO with apogee at or near 36,000 km and perigee at the altitude of the parking orbit (150 to 300 km). The term *perigee kick* is used to describe the action of the last LV stage, which may either be part of the LV or provided separately with the spacecraft. (The mechanical aspects of those alternatives are reviewed later in this chapter.) The LV may perform the perigee kick with a single burn of either a liquid or solid fuel rocket motor. An alternative strategy, employed by the Proton rocket, is to perform two burns: the first to provide the basic perigee kick and the second at apogee to increase perigee altitude. That also helps compensate for the relatively high latitude of the launch site (Baikunor, Kazakhstan, in the case of the Proton).

　　Figure 10.2 shows how the transfer orbit is initiated from parking orbit wherein the perigee kick stage is fired at the point opposite from where apogee is to occur. The mission plan usually specifies that the first apogee after injection into GTO must be in view of a TT&C station at a specific location.

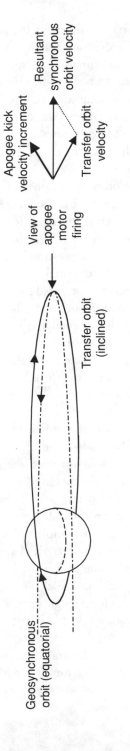

Figure 10.2 Injection from transfer orbit into geosynchronous orbit using the apogee kick maneuver.

Conversely, the firing position may be fixed for the particular launch site and LV, forcing the location of the TT&C station to be in view of the resulting first apogee. The period of GTO is approximately 16 hours, and the satellite is easy to track near apogee when its motion is slowed considerably. Because each successive apogee occurs at a different longitude on the Earth, it is advantageous to have TT&C stations in the eastern and western hemispheres. That can be accomplished without owning additional sites by contracting with a satellite operator in the opposite hemisphere for the use of its TT&C station during the transfer orbit phase of the mission.

The final phase of transfer orbit operations is the injection into geosynchronous orbit. At the right of Figure 10.2 is shown the vector representation of the velocities before and after injection, looking toward the point of apogee motor firing (AMF) from the right. In the figure, the satellite velocity vector in the inclined transfer orbit at the point in question is at the angle with respect to the geostationary orbit plane. To cause the final synchronous orbit velocity to be in the equatorial plane, the spacecraft is oriented so the apogee kick motor fires along the vector direction shown pointed upward. The thrust increment at AMF can correct for inclination if the velocity change is sufficiently greater than that required to produce synchronous orbit. Therefore, the resultant velocity is in the equatorial plane (i.e., horizontal in the vector diagram) and its magnitude is sufficient to circularize the orbit at 36,000 km altitude.

There will always be some inaccuracy in the amount of velocity change, including the precise direction, so the orbits achieved will have some error as well. That is multiplied by the fact that the first key rocket stages are under autonomous control and cannot be guided from the ground. There is a need to use ground-commanded corrections to produce the desired orbit geometry. That can be done only by first precisely locating the satellite using the TT&C station and then determining its orbit parameters. Touchup maneuvers are then performed with the spacecraft reaction control system. For a GEO mission, the particular transfer orbit where AMF is accomplished is selected for operational reasons. For example, it is desirable to fire the AKM at an apogee that is as close as possible to the final orbital longitude of the satellite and thus minimize the time for drift movement from AMF longitude.

The preceding discussion was for a single apogee thrust using a solid fuel rocket motor. However, the majority of spacecraft (including all spacecraft heavier than approximately 2,000 kg dry mass) use liquid propulsion to raise perigee in successive increments. As illustrated in Figure 10.3, the process requires multiple burns of a liquid apogee motor (LAM) that uses either a dedicated bipropellant propulsion system or shares tankage with the reaction control system used for stationkeeping.

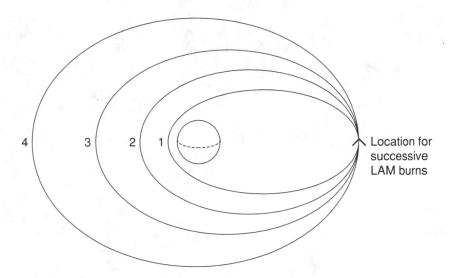

Figure 10.3 Injection from transfer orbit in geosynchronous orbit by successive burns of an LAM.

10.1.4 Drift Orbit for GEO Operation

Following apogee injection, the satellite is in a near geosynchronous orbit and drifting toward the assigned orbit location. For a solid AKM, the detailed parameters of the initial drift orbit are unknown after motor firing, so the first priority of TT&C operations is to determine the orbit as accurately as possible. The next step is to make the necessary velocity corrections to cause the satellite to drift in the desired direction. The LAM approach typically produces a proper drift orbit without injection errors. Figure 10.4 shows how a satellite in drift orbit is actually at a different altitude than the geostationary orbit, which is a consequence of Kepler's second law. That limits the possibility of a drifting satellite bumping into a stationary one. A satellite drifting westward is above the geostationary orbit, while one drifting eastward is below. However, there still remains the possibility of RFI as the drifting satellite passes through the antenna beams of the Earth stations, which point at their respective geostationary satellite longitudes.

Potential problems with RFI in drift orbit are prevented through detailed coordination between satellite operators. Prior to reaching drift orbit, the manager of the mission provides written notification to the operators of every satellite that could conceivably receive interference. Then, during drift, the uplink from the TT&C station and the satellite payload are disabled when

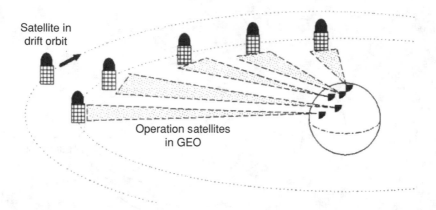

Figure 10.4   Drifting of a geosynchronous satellite from the apogee injection longitude to the final longitude position.

there is inadequate orbital separation (typically 2 degrees of separation at C and Ku bands, but the specific amount should be determined prior to the mission). The harmful interference that could result provides the motivation for cooperation between operators, some of which are potentially competitors in business. Spacecraft deployments and testing could be conducted during drift orbit when the satellite is definitely out of view of operating ground stations, but it is usually best to defer such activities until the satellite reaches its assigned orbit position.

A predetermined amount of reaction control system (RCS) fuel is consumed to stop the satellite from drifting. Final touchup maneuvers with the spacecraft RCS bring the velocity and altitude into alignment with the geostationary orbit. From that point forward, the stationkeeping phase of the mission begins and lasts through the rest of the satellite lifetime. The satellite can be relocated to another longitude by expending two equal amounts of fuel: one to start the drift and the other to stop it when the final longitude is reached. Because the duration of thrusting determines the speed of relative motion, the amount of fuel consumed is roughly inversely proportional to the total time allotted for relocation.

10.1.5   Deployments and In-Orbit Testing

Whether GEO or non-GEO, a communications satellite is delivered to orbit in a somewhat compact condition, consistent with the volume and mounting constraints of the LV. Part of the mission involves preparing the spacecraft for service by deploying appendages like solar panels and reflector antenna systems and thoroughly testing all the active subsystems of the bus and payload. Procedures to calibrate and preload onboard processing and sensing systems also are required prior to start of service. All of that can take as little as a few days or as long as several weeks, depending on the tasks involved and the complexity of the various subsystems onboard the satellite.

GEO satellites probably are the most complex and require more attention during this phase than might the non-GEO satellites. That is because modern GEO satellites tend to be larger in physical and electrical terms to overcome the distance and to provide adequate communications capacity and coverage to justify the required investment. In contrast, some of the smaller LEO satellites are quite simple in mechanical and electrical terms and hence could be placed into service almost immediately after activation in final orbit. That is also driven by the issue of the minimum required quantity of LEO satellites needed to provide the total service from a complete constellation.

To provide some guidance as to the amount of effort involved, Figure 10.5 shows the deployment sequence for a typical three-axis GEO satellite. This particular model requires that the solar panels be deployed early in the mission, before GEO is reached. Reflectors are deployed once the satellite has reached its testing longitude (which could be other than the final assigned longitude if an operating satellite is already in that position). All the subsystems that make up

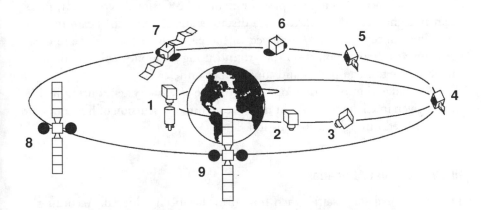

1	Separation from Launch Vehicle	6	Deploy reflectors and spin down
2	Reorient to sun-normal	7	Deploy solar panels
3	Reorient for apogee injection attitude	8	Sun acquisition and bus testing
4	Apogee injection boost	9	Earth acquisition and start of operations
5	Reorient for deployments		

Figure 10.5 Deployment sequence for a typical three-axis satellite (transfer orbit mission summary).

the bus are checked for function and then tested for quantitative performance. Often, the spacecraft manufacturer or operator needs to determine the calibration of sensors and other devices before exercising the entire spacecraft.

Testing of the communications payload can commence once the bus has been tested and configured for operation. Payload testing can be carried out as a brief verification that the payload has withstood the launch mission and provides the basic functions. On the other hand, it is typical to fully exercise all channels of communication and coverage on the ground before entering service. Such a process can be accomplished in a matter of days for a simple bent-pipe type of payload with a single coverage footprint. The only time-consuming aspect is verifying the full coverage of the spacecraft antenna system. Recall that the typical repeater has both an uplink and a downlink band, as well as two different polarizations in each. There likely will have to be tests to determine the cross-polarization performance of all beams. Depending on the demands of frequency coordination, it even might be necessary to verify levels of interference into other countries or regions.

Looking now at more advanced payloads, complexity increases rapidly for greater numbers of individual channels, multiple spot beams, and processing repeaters. As pointed out in Chapter 7, spot beams impose a unique difficulty in that a given testing Earth station may not be able to receive its own transmission directly. The simplest approach might be to move the spacecraft body, allowing each beam to migrate past the main TT&C site. Alternatively, measurement signals could be relayed by a distant station within the beam in question. One concept that was employed in the past was to make separate measurements at different points on the ground using a mobile vehicle containing the necessary test equipment and support personnel.

Consideration of normal operation of the satellite system during its lifetime is given in Chapter 11. Right now, we go into a discussion of how onboard fuel is utilized throughout the mission.

10.1.6 RCS Fuel Allocation

The operational stage starts when final orbit has been achieved. According to the expressed operating plans of the first generation of non-GEO satellite operators, onboard fuel is required to compensate for atmospheric drag. In contrast, a GEO satellite requires frequent correction of orbit parameters to ensure that the desired longitude and low inclination are maintained. Otherwise, the satellite will not remain within the beamwidth of the largest class of ground antenna (e.g., ones lacking automatic tracking). That critical requirement and its consequences for fuel allocation are discussed here.

If the Earth could be represented by a sphere of uniform mass density (or at least composed on concentric spherical shells of uniform mass), the GEO satellite would stay put at this longitude indefinitely. The reality, indicated in Figure 10.6, is that the Earth is not uniform. To that is added the presence of gravitational pull from the sun and from the moon. As a result, the orbit distorts over time so that the satellite appears to move relative to a point on the Earth. First, the orbit becomes somewhat elliptical, a property called eccentricity, and the plane of the orbit inclines relative to that of the equator. Eccentricity is not particularly troublesome and is easily corrected along with the primary orbital adjustments. Inclination, on the other hand, makes the satellite appear to move perpendicular to the plane of the equator, with the period of motion being 24 hours. The bulges in the Earth's mass produce east-west drift along the orbit. With the satellite as shown in line with the bulges, the gravity is in balance and the satellite tends to stay put. Likewise, positions 90 degrees away in the orbit have equal pull from each bulge, and no force acts to move the satellite. Intermediary positions experience a maximum force that tends to drift the satellite toward an equilibrium longitude. Fuel usage for a satellite positioned at the worst longitudes is still a small fraction of the total budget, which is dominated by the required to remove inclination (north-south stationkeeping). The actual physical and mathematical relationships are much more complex than that; however, this simple model can be a way of visualizing the three forms of orbit distortion.

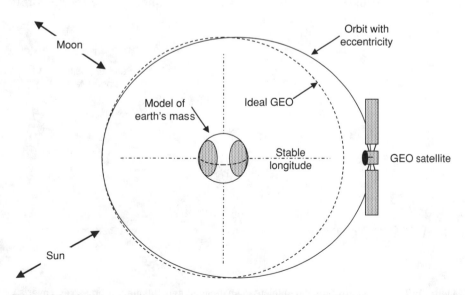

Figure 10.6 Variations in the geostationary orbit in the plane of the equator.

Most stationkeeping fuel of a GEO satellite is consumed to correct inclination. Properties of the orbit are shown in Figure 10.7, which has been exaggerated for clarity. Produced by gravity from the moon and, to some extent,
from the sun, inclination increases by approximately 0.8 degrees per year. For
a given inclination angle, the satellite is below the equator at one point in time
and then the same distance above the equator precisely 12 hours later. When
viewed from the ground, a satellite appears to oscillate in that manner over a
24-hour period. (This topic was covered in Chapter 9 in connection with the
possible need for ground antenna tracking in the event that the beamwidth is
narrower than the north-south motion of the satellite.)

The reaction control system of the spacecraft is used to provide several of
the preoperational velocity increments as well as all those necessary for stationkeeping. Figure 10.8 presents an approximate allocation of RCS fuel for an
entire GEO satellite mission from transfer orbit to end of life. In this example,
the RCS is not required to deliver the bulk of the velocity increment for perigee
kick or apogee kick. One point of clarification is that east-west motion is in the
plane of the orbit while north-south motion is perpendicular to it. From the
Earth station's perspective, those directions usually are not aligned with the
local vertical and horizontal because of the curvature of the orbital arc as seen
from the particular point on the ground (except, of course, for an Earth station
on the equator at the same longitude as the satellite).

Once on station, the orbit is corrected during stationkeeping, introducing velocity changes, which are numerically determined by the specific orbit

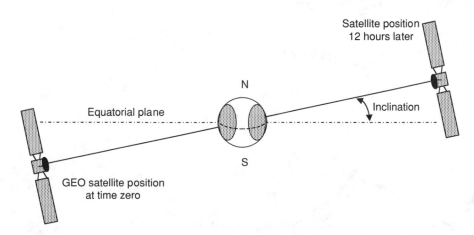

Figure 10.7   Relative position of a satellite in an inclined geostationary orbit as seen over a
24-hour period.

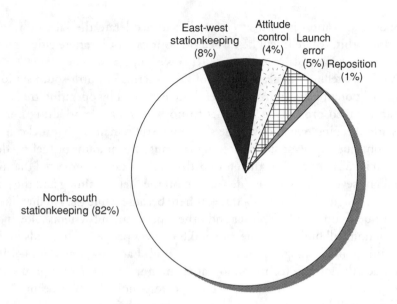

Figure 10.8 Allocation of RCS fuel usage for a typical geostationary satellite over a 15-year period.

parameters. However, the amount of fuel consumed to produce the necessary velocity increments is proportional to the average total mass of the satellite.

As is evident from Figure 10.8, approximately three-quarters of the fuel is required for north-south stationkeeping (correction of GEO inclination). East-west stationkeeping, while critical to every geostationary satellite, requires from zero to a maximum of 10% of total RCS fuel. Both usages are predictable over the satellite's lifetime, and little can be done to extend the mission once geostationary orbit has been reached (except by terminating north-south station-keeping early). One exception is the use of a storage orbit to conserve fuel before the satellite is placed into service. That involves injecting the spacecraft into an inclined geosynchronous orbit so the inclination decreases naturally toward the equatorial plane (which is at the rate of approximately 0.8 degree per year). Thus, an initial "bias" of 3 degrees provides about 3 years of storage potential. A satellite can be brought out of storage prematurely by consuming fuel to remove the remaining inclination.

Most of the attention of SCC personnel is directed toward accuracy in orbit determination and fuel use during routine stationkeeping maneuvers. A small reserve for repositioning provides for moving the satellite to a new longitude if required for some reason. Recall that half that fuel would be used to start the drift and the other half to stop it when the desired position is reached. A final budget allocation is used at end of life to accelerate the satellite into a higher drift orbit, where it will be turned off so it cannot interfere either physically or electromagnetically (e.g., cause RFI) with operating satellites.

A GEO satellite that nears its end of life in terms of north-south stationkeeping fuel potentially can have its useful life extended by operating the satellite in an inclined orbit. That amounts to nothing more than halting north-south stationkeeping operations while some fuel still remains. That fuel can be used to continue east-west stationkeeping. Because the amount of fuel needed per month is substantially less than for north-south, it can be projected that the life would be several years longer (depending on the fuel remaining and the particular orbit longitude). There is a tradeoff here because while the satellite likely will continue to operate (the power and other spacecraft subsystems allowing), the inclination will build up at the rate of 0.8 degrees per year. That adds to the north-south pointing range as seen on the ground. Earth stations would either have to track the satellite (common for large antennas used for TV uplinks and telephone gateway Earth stations) or have correspondingly wide beamwidths (typical for small DTH antennas and mobile user terminals).

10.2 Launch Technology and Systems

The LV situation is an ever changing scene, as new rocket designs replace older counterparts and new LV service providers appear in the marketplace. However, because of the complexities and risks associated with launching satellites, it is often the rule that newer vehicles are merely modified versions of their predecessors. There are always exceptions when new approaches prove themselves in technical and operational terms, as was the case for the Space Shuttle, Pegasus, and Sea Launch. It is likely that future commercial launches still will rely on expendable rockets of improved design, many of which have familiar names like Ariane, Atlas, Delta, Long March, and Proton.

This section reviews the evolution of commercially available launch vehicles suitable for placing a spacecraft into GTO. As discussed in the Section 10.1, that usually involves three stages: a first and second booster and a perigee kick stage. In some systems, the perigee kick stage is part of the LV, while in others the spacecraft provides its own integral perigee kick. The discussion begins with an overview of the technology and evolution of current LVs.

While the details of rocket technology are beyond the scope of this book, the basic alternatives are relatively easy to understand. A liquid-fueled first stage uses separate fuel and oxidizer supplies that burn when combined, producing the necessary thrust. The engine uses pumps to maintain constant fluid pressure, and temperature is controlled by piping the cooler liquids around the hotter parts of the engine. Typical fuel/oxidizer combinations include hydrazine/nitrogen tetroxide (i.e., bipropellant approach described in Chapter 8), kerosene/liquid oxygen (used by the Saturn 5 rocket of the Apollo program), and cryogenic liquid hydrogen/liquid oxygen (used in the main engines of the Space Shuttle).

Three kinds of "hydrazine" are utilized for spacecraft and LV propulsion:

1. Anhydrous hydrazine, N_2H_4, is used in monopropellant hydrazine systems in satellites constructed before 1990 and in current Lockheed-Martin GEO spacecraft. It requires a catalyst to start decomposition, yielding an I_{sp} of around 230 sec. N_2H_4 is somewhat unstable. Burning N_2H_4 with oxidizers, a challenge, is done by the apogee engines in spacecraft dual-mode propulsion systems.

2. Monomethyl hydrazine (MMH), $(CH_3)NHNH_2$, is in bipropellant systems like most GEO spacecraft today (including Hughes product lines). With N_2O_4 (nitrogen tetroxide) as an oxidizer, one can have I_{sp}s of around 300 sec. It is more stable than anhydrous hydrazine; in fact, it can be added to anhydrous hydrazine at the 10% level to stabilize the latter fuel (it removes N_2H_4's monopropellant qualities).

3. (Unsymmetrical) dimethyl hydrazine (UDMH), $(CH_3)_2NNH_2$, is in bipropellant systems for launchers in both the West and the East with various oxidizers. It can be mixed 50-50 with anhydrous hydrazine, the resulting mixture is called Aerozine 50.

Cryogenic propellants must be kept at an extremely low temperature so they remain in liquid form and do not revert to a gaseous state. The tradeoff between high energy cryogenic propellant and propellants that can exist at room temperature is that the latter can be stored for long periods of time (i.e., in a missile silo or in space). Solid fuel rocket motors offer a convenient alternative to liquid because of the simplicity of design and integration and because they can be stored for years at a time. That is the same technology used in the AKM of some spacecraft. (The technical approach to solid motor design is presented in Chapter 8.) The two large rockets attached to the Space Shuttle are solid rocket motors (SRMs) filled with highly combustible propellant material. Solids are attractive for that application because they offer a great deal of boost during critical parts of the launch sequence. Solid rocket motors burn rapidly and at a high temperature, requiring great precautions to be taken with their exit nozzles, cases, and attachments to the vehicle.

The second and third stages of the booster can take the form of solid or bipropellant liquid-fueled rockets. As explained in Section 10.1, the second stage is ignited after the LV is above the denser part of the atmosphere, propelling the vehicle into the parking orbit. Separation of the stages may involve the firing of special small rocket motors or explosives that sever bolts holding the stages together. (Separation in the Proton rocket results from the sheer force of firing the second stage.) Clearly, the sequence of ignition, burn and velocity control, ignition, and separation is vital to the achievement of mission success. Much of the attention and concern of launch operation personnel is focused on this critical sequence, which occurs automatically without ground control.

Preparation of the LV and the spacecraft is another vital area, one that is often overlooked. Many modern LVs are capable of launching two or more payloads at the same time. Even with a single payload, the preparations at the launch site take one or more months since the spacecraft must be checked and then properly integrated with the LV. The launch agency, rocket manufacturer, and spacecraft supplier work essentially around the clock during this period to assemble the LV, install the payload or payloads, check and fuel the system, and conduct various prelaunch tests and rehearsals. The facilities involved are extensive, and there are only a few qualified launch sites in existence. Also, the tracking sites required during each phase of the launch and transfer orbit all must be prepared and checked out prior to liftoff.

This discussion is necessarily brief and of a general nature. Section 10.3 reviews a number of specific LV systems. It is evident that each LV is a unique combination of rocket design and fuel, which should be considered in detail when making a selection for a particular mission. That is because the success of the launch mission requires that each function and step occur properly or at least within acceptable limits. Failure of just one hardware or software component can be all it takes to preclude mission success. Ability to recover from launch failure generally is quite limited, due to the large thrust levels and velocity changes required from the various stages of the LV.

10.3 Typical Launch Vehicles

Before describing the currently available commercial LVs and their capabilities, it is instructive to review some of the recent history. Because of emphasis by the U.S. government on the manned and reusable Space Shuttle, only the Delta and Atlas Centaur were operational for commercial launches during the mid-1980s. Under the plan at the time, all expendable LVs were to be phased out, and production of the Delta, Atlas Centaur, and Titan was nearly halted. Meanwhile, the European Ariane system became available, and some commercial satellite operators took advantage of attractive rates and convenient schedules. Then in 1985, the Challenger shuttle blew up and, after considerable debate within the U.S. government, commercial shuttle launches ceased. The introduction of Ariane is fortunate for the satellite industry now that the shuttle has been shifted away from commercial service, a situation that is expected to persist after the year 2000.

Beginning in the late-1980s, the expendable LV market opened up to even greater competition. Encouraged by the new policies and buying practices of the U.S. government coupled with growth in demand, several commercial LVs have been advanced. Atlas is a key offering from U.S.-based Lockheed-Martin, and Boeing, through its acquisition of McDonnell Douglas, has the Delta II and Delta III to offer. Both Lockheed-Martin and Boeing also are in joint ventures with Russian organizations for the Proton and Zenit rockets, respectively. (Strictly speaking, the Zenit booster is produced in Ukraine and the Block DM final stage is produced in Russia.) The China Great Wall Industrial Corporation, of the People's Republic of China, has significant launch capabilities and is now an important factor in the commercial launch market. Japanese LVs also are capable of placing larger commercial payloads into GTO.

The impression one should take away from this discussion is that the LV lineup changes over the years, yet the particular choices available at a particular time are fixed. The best approach usually is to compare LVs that are based on experienced rockets with good success records. That goes without saying because of the expense and business risk that are involved with implementing and operating a communication satellite system, whether GEO or non-GEO. Those factors and others are reviewed in the following detailed comments on the LVs of the current time frame. Table 10.2 lists various systems (in alphabetical order) and compares their nominal characteristics (as of August 1998).

10.3.1 Ariane

The European commercial expendable LV, Ariane, was developed by the ESA and the French Centre National d'Etudes Spatiales (CNES) and is now manufactured, operated, and marketed by a European company called Arianespace. The basic vehicle is the Ariane 4, which contains first and second stages that use hydrazine–nitrogen tetroxide and a cryogenic third stage using liquid hydrogen–liquid oxygen. Solid and/or liquid strap-ons can be combined with the first stage. The strategy of Ariane was to provide a dual-launch capability to offer a more cost-effective service to commercial customers. The dual-launch capability was proven in U.S. government programs; however, it was applied for the first time in commercial service for Ariane 2. Another important feature of the program is the use of the Guiana Launch Center at Kourou, French Guiana, located at a latitude close to the equator. That reduces the inclination of the

Table 10.2
Characteristics of Typical Commercial Satellite Launch Systems for
GEO Missions (Representative Information as of June 1998)

Launch vehicle	Provider	Launch site	Mass to GTO, kg	Year first launched
Ariane 4	Arianespace (France)	Kourou, French Guiana	4,700	1989
Ariane 5			6,920	1996
Atlas IIAS	International Launch Services (U.S.)	Cape Canaveral, Florida, U.S.	3,719	1993
Atlas IIAR			4,037	
Atlas IIARS			4,264	
Delta II	Boeing	Cape Canaveral	1,870	1989
Delta III			3,800	1998
H-2	Rocket System Corporation (Japan)	Tanegashima Space Center, Japan	4,000	1994
Long March 3B	China Great Wall Industrial Corp. (PRC)	Xichang Satellite Launch Center, PRC	4,000	1996
Long March 3A			2,300	1994
Long March 2E			3,100–3,300	1992
Pegasus	Orbital Sciences Corp. (U.S.)	L-1011 aircraft	180	1990
Proton D-1-e	International Launch Services (U.S.)	Baikunor Cosmodrome, Kazakhstan	4,500	1970
Proton M			5,500	1998
Space shuttle	NASA (U.S.)	Cape Canaveral	5,900	1981
Taurus	Orbital Sciences Corp. (U.S.)	Wallops Island or Cape Canaveral	430	1994
Titan III	Lockheed-Martin	Cape Canaveral	5,000	1989
Titan IV			8,620	1989
Zenit	Sea Launch Co. (U.S.)	Pacific Ocean	5,000	1985

transfer orbit, which offers the capability to save fuel or extend life. The Ariane rocket is a derivative of earlier European rocket programs from the French, British, and German governments. On the other hand, the concept of a commercial company, that is, Arianespace, was totally new and has become the pattern for several others to follow.

The principal commercial vehicles are the Ariane 4 and Ariane 5, the latter having been developed to support the Hermes reusable "spaceplane." Ariane 4, shown in the photograph in Figure 10.9, has the capability to place two satellites into GTO, using a structural casing called the *structure porteuse externe lancement Ariane* (Spelda). GTE was the first U.S. company to choose the dual-launch Ariane system to place its Spacenet and GStar satellites into GEO. For very heavy payloads, such as Intelsat VI, the Ariane 4 became available. The Ariane success record is among the best in the industry, having improved over the past decade through extensive design refinements and improvements in operating procedures.

While Ariane 4 has proved itself a versatile and reliable vehicle, it nevertheless lacks flexibility in terms of the types of spacecraft that can be placed in the lower position of the Spelda. The Ariane 5 rocket is intended to improve that situation by allowing two very similar satellites to be launched simultaneously. Therefore, there should be less of a limitation on which position a given spacecraft can employ.

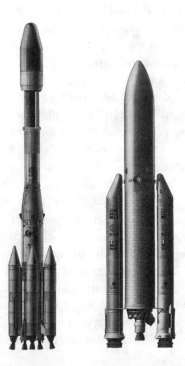

Figure 10.9 Ariane 4 and Ariane 5 rockets. (*Source:* Courtesy of Arianespace.)

Ariane 5 is a new design that can launch two or three spacecraft into GTO. The cryogenic first stage has two large solid-propellant strap-ons and is used for all missions. Above it, the upper composite stage can support Earth and sun-synchronous missions to LEO or, using an attach fitting, missions to GTO. This stage burns MMH and N_2O_4 and is pressurized with helium. Dual- or triple-payload launches to GTO are accomplished with a fitting called the *structure porteuse externe lancements triples Ariane* (Speltra).

At the time of this writing, Arianespace was still working to produce a reliable Ariane 5 system with a track record on a par with Ariane 4. As Ariane 5 gains its position, it is likely that the older Ariane 4 technology gradually will be phased out. In the future, many commercial organizations as well as government satellite operators will continue to use the Ariane vehicle because its capabilities appear to be well matched to the operators' needs.

10.3.2 Atlas

The Atlas rocket is one of a number of LVs made available to the commercial market by Lockheed-Martin (others being Proton and Titan). Having been around for three decades, the Atlas Centaur has played an important role in lifting medium payloads into GTO and other orbits. The liquid-fueled rocket was developed and originally manufactured by the Convair division of General Dynamics. That organization was acquired by Lockheed-Martin in 1994, and Atlas continues to be available through ILS (discussed in Section 10.3.7). The Atlas first stage, originally a ballistic missile, employs a liquid kerosene–liquid oxygen system. The second stage, called the Centaur, was the first high-energy cryogenic liquid hydrogen–liquid oxygen engine developed in the United States. Delivery to GTO is provided by the Centaur stage; the Atlas (acting effectively as one and a half stages) performs a portion of the injection into parking orbit. All launches into GTO are conducted from NASA's Kennedy Space Center (KSC) at Cape Canaveral, Florida.

The Atlas II series was the primary offering at the time of this writing. The vehicle was developed from the existing product line in response to a U.S. government requirement for a medium launch capability to boost military communication satellites to GTO. The upper stage was stretched to increase payload capability, and other improvements were added. Two configurations are offered: The Atlas IIA version includes all those components while the Atlas IIAS adds solid strap-ons to the first stage. In addition, Lockheed-Martin is upgrading the Atlas II for increased payload capability. The Atlas IIAR, due to enter service in 1998, uses a Russian-supplied RD-180 engine and an improved Centaur upper stage.

A new series of rockets called the Lockheed Launch Vehicle (LLV) is in the final stages of development. It is an all-solid fuel system available in two- and three-stage versions. The rocket was first launched in 1997 with a relatively low mass capability. It currently does not have capability to deliver payloads to GTO.

10.3.3 Delta

Referred to as the workhorse of NASA, the Delta rocket has established an impressive track record since the first U.S. launches of geosynchronous satellites. It dates back to the beginning of the U.S. space program, when NASA Goddard Space Flight Center contracted with the then Douglas Aircraft Company for 12 complete rockets. (Current production includes the Delta II and Delta III shown in Figure 10.10.) While the first Delta flight in May 1960 was a failure, its eventual record proved to be the best in the industry. Boeing, who acquired the McDonnell Douglas Astronautics Company, is the manufacturer and now

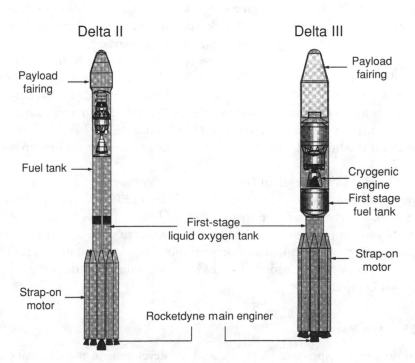

Figure 10.10 The Delta II and Delta III rockets produced by Boeing. (*Source:* Courtesy of Boeing.)

operator of the Delta, which employs a kerosene–liquid oxygen first stage and a hydrazine–nitrogen tetroxide second stage. The Delta II employs up to nine solid rocket strap-ons to the side of the first stage for added lift. To propel the spacecraft from LEO into GTO, a solid rocket motor called the payload assist module (PAM) can be provided. All launches to GTO are conducted from NASA's KSC facility. Non-GEO and polar missions can be initiated from Vandenberg Air Force Base in California.

The newest generation of the series, the Delta III, is a greatly enhanced LV that can compete with the other heavy lift vehicles like Ariane 4 and Proton. Both the U.S. government and Hughes are early purchasers of this system, with first launches occurring in 1998–1999. Following that, Boeing is promoting the Delta IV LV, which is being developed for the U.S. government.

10.3.4 H-1 and H-2

The Japanese government entered the field of launching geosynchronous satellites by employing a version of the Delta under license from McDonnell Douglas. NASDA is the government agency in Japan responsible for launch and other space programs. The N-2 rocket demonstrated a perfect record for its nine launches. The first stage is a kerosene–liquid oxygen stage with solid strap-ons (similar to the Delta), while the second is a liquid hydrazine–nitrogen tetroxide rocket engine built in Japan. Perigee kick is provided by a Thiokol solid rocket motor. Launches take place on the Japanese island of Tanegashima. The upgraded H-1 rocket is built on the N-2/Delta design but with more Japanese technical and manufacturing content. Both the N-2 and H-1 ceased production at Mitsubishi Heavy Industries, to be replaced by the more powerful H-2.

The H-2 LV, the result of Japanese technology development, was built on the H-1 and designed to place larger payloads into LEO and GTO. The first stage employs cryogenic liquid hydrogen–liquid oxygen with a pair of rather large solid rocket motors strapped on the side, as shown in Figure 10.11. GTO is achieved by the cryogenic second stage through two burns of its engine. The first burn places the payload into a LEO parking orbit, and the second performs GTO injection.

10.3.5 Long March

With decades of experience in missile and LV development, the Chinese government has made available its Long March LV to the commercial market. We use the English language version, Long March (LM), instead of the Chinese

Figure 10.11   The H-II and H-IIA launch vehichles (*Source:* Courtesy of NASDA.)

language designation, Chang Zheng (CZ). The contracting authority, China Great Wall Industrial Corporation (CGWC), has provided launch services for several satellite operators, both Chinese and foreign. CGWC is a wholly owned subsidiary of the government-owned China Aerospace Corporation (CASC); the rocket is produced by the China Academy of Launch Technology (CALT), and the launch and tracking sites are operated by China Launch and Tracking Central (CLTC). The first LM launch was of the Dong Fang Hong ("East is Red") 1 satellite, which was placed into LEO in April 1970. Currently available rockets include the LM-2E, LM-3A, and LM-3B. Lift capability is in the low to middle range, making the Long March suitable for many commercial spacecraft such as those used for domestic communications.

In terms of rocket design, the first stage of the 2E and 3B employs four liquid engines using UDMH–N_2O_4 propellant. The second stages of the 2E and 3B differ in that the 2E carries larger fuel tanks. The 3B cryogenic third stage, using liquid hydrogen–liquid oxygen, can place payloads directly into GTO. The first foreign launch of the LM was for AsiaSat 1, an HS-376 satellite that was successfully placed into GTO in April 1990. Subsequently, CWIC provided launches for larger satellites using the LM-2E and LM-3B series, the latter being the newest and more capable rocket. As discussed previously, the LM-3B includes a third stage for direct injection into GTO.

All GEO missions are supported from Xichang because that site is relatively southern and allows overflight over relatively uninhabited Chinese terrain. Accommodations are relatively modern and air service is available through Sichuan's capital of Chengdu. CLTC provides launch services to LEO from a second, more northern site at the Jiuquan Satellite Launch Center in Gansu province (which is on the border of the Gobi Desert). That was the original Chinese launch site and recently has been upgraded to support non-GEO missions using the LM-2F (a modified version of the 2E). Up to four separate launches can be processed simultaneously, two using a more efficient vertical assembly procedure.

10.3.6 Pegasus and Taurus

The Pegasus and Pegasus XL are three-stage rockets that can be lofted by an aircraft. The rockets are designed, built, and offered to the market by Orbital Sciences Corporation (OSC), a relatively young U.S. aerospace company located in the Washington, D.C., area. OSC entered into a joint venture with Hercules Aerospace, which provides the solid rocket motors and the payload fairing. The first user for Pegasus was the U.S. government Advanced Research Projects Agency for six missions. Subsequently, OSC used Pegasus to place its own Orbcomm spacecraft (discussed in Chapter 8) into LEO. OSC operates its own converted Lockheed L-1011 jetliner as the launch platform for Pegasus. The Pegasus has both successes and failures during its initial trials and represents a unique system for launching small spacecraft into LEO.

The basic rocket consists of three solid rocket motors cased in graphite-epoxy composite, produced by Hercules, and sports a delta-wing with a 6.7m wingspan. While the first stage has fixed firing alignment, the nozzle of the second stage can be controlled to adjust course direction. The third stage, like the second, can be vectored, and the avionics for the entire vehicle are located ahead of it. A payload attach fitting is provided on the third stage as well. The rocket motors of each stage burn solid fuel akin to that of the AKM and burn for the order of 1 min.

OSC also offers the Taurus LV, a more conventional rocket that is launched from a ground-based mobile launching system. It is a four-stage system using all solid rocket motors. Thiokol produces the base stage (stage 0), and Hercules produces the remaining three stages. It is intended for all types of missions, including LEO, GTO, and planetary. The unique mobile launching approach allows Pegasus to utilize "austere" launch sites.

10.3.7 Proton

The Proton is the most powerful Russian-produced rocket that was on the commercial market at the time of this writing. The former Soviet Union has launched more payloads than any other country in the world and, during the 1990s, began to provide commercial launch services. The Proton rocket, in particular, is currently the most capable of Russian LVs for missions to GTO. The rocket is offered by International Launch Services (ILS), a joint venture stock company owned by Lockheed-Martin Commercial Services and the Lockheed Khrunichev Energia International joint venture. The latter is a joint venture of Lockheed Martin, Khrunichev State Research and Production Center, and RSC Energia. Khrunichev produces the first three stages and Energia produces the fourth.

The first three stages of the Proton, built at the Khrunichev plant in Moscow, are used for all missions. It can place large payloads into LEO and uses a UDMH and N_2O_4 liquid-propellant system. The first stage is unique among commercial LVs in that it employs six fixed strap-on liquid engines that include integral fuel tanks. The center contains a single tank of oxidizer. Each rocket engine is gimbaled to allow for vectoring of the thrust. Stage 2 employs four gimbaled single-chamber liquid-propellant rocket engines. The third stage contains one fixed single-chamber engine and a control rocket with four gimbaled nozzles. The combined system has been in use for 40 years and has a particularly good success record.

For placement in GTO and other high-energy missions, the Proton can add a fourth stage. At the time of this writing, the operating version of that stage is the Block DM and is provided by the Russian Space Complex (RSC) Energia, located in Moscow. In this particular configuration, the nomenclature for the rocket is Proton D-1-e. The Block DM is intended for multiple-burn operations, meaning it can be fired for a period and then restarted for added flexibility in mission design. It is equipped with one liquid bipropellant engine (e.g., UDMH and N_2O_4) and two "micro" engines for guidance and control. Beginning around 1999, Khrunichev will begin to offer their own fourth stage, called Breeze M.

All Proton launches take place from the Baikunor Cosmodrome, located in the Republic of Kazakhstan. The facility covers an extensive piece of flat real estate and provides for all launches over land. The launch vehicle is transported from Moscow over land by rail and is integrated at the launch site. ILS is also the marketing agent for Proton as well as other LVs made available by Lockheed-Martin.

10.3.8 The Space Shuttle

The Space Transportation System (STS), the official NASA name for the Space Shuttle, was developed for the purpose of inexpensively placing payloads into Earth orbit. The original contract was awarded to Rockwell International in 1972 after 3 years of predevelopment and consensus building within the U.S. government. While failing to achieve its objective for economy, there can be no doubt that the Space Shuttle was and remains an outstanding accomplishment in space hardware development. Sadly, the twenty-fifth STS mission and the tenth launch of the shuttle *Challenger* on January 28, 1986, resulted in disaster and the deaths of its crew of five astronauts. The ensuing debate in Washington over the risks associated with launching commercial payloads with shuttle resulted in a dramatic change in U.S. policy. From then on, the Space Shuttle was not available for commercial launches and expendable LVs became the only alternative. In spite of that situation, we provide here a brief overview of the basic capabilities of the system as a reference.

The main engines of the Space Shuttle employ cryogenic liquid hydrogen–liquid oxygen propellants contained in the large external tank to which both the orbiter spaceplane and the twin SRMs are attached. Only the Saturn rocket of the Apollo program had greater lift capability. The shuttle actually delivers its payloads to LEO (in the same manner as the first two stages of a conventional rocket), and a perigee stage is relied on to take the spacecraft to GTO.

The orbiter of the STS performs several functions, including those of a payload carrier and an airplane. An external tank delivers the propellants to the main engines of the orbiter. In addition, the orbiter has maneuvering engines that are fueled by hydrazine–nitrogen tetroxide. The cargo bay of the orbiter can hold the payloads to be delivered to orbit; other payloads can remain fixed to the orbiter for return to Earth. An added feature of the STS is its demonstrated ability to perform LEO repair operations and to recover spacecraft for return to Earth. However, the orbiter is restricted to LEO and cannot itself reach GEO.

It has long been the assumption that manned launch missions are more reliable than unmanned from the standpoint of reaching LEO. The basis for that is that the vehicle is designed to include many safeguards and backup systems that would be uneconomic for expendable LVs. In addition, the payloads themselves must be protected in such a way that there is little chance of explosion in the shuttle bay. An added benefit is that the human crew can perform functions unavailable in expendable LVs. The primary role of the system now appears to be the support of scientific research, experiments in manned missions, and operations of orbiting space stations such as Mir and Freedom.

From time to time, the space shuttle has been enlisted to help with the recovery or repair of commercial payloads. That capability was demonstrated by the recovery of Westar 6 and Palapa B2, which both failed to reach GTO due to defective perigee kick stages. Another impressive support activity was the rescue of the Leasat satellite; astronauts literally hot-wired the electrical system to bring the disabled satellite back to life. Finally, from a commercial standpoint, the shuttle was used to replace an inactive perigee kick state that was attached to an Intelsat VI spacecraft.

10.3.9 Titan

The military side of U.S. launch operations has relied heavily on the Titan III launch vehicle, developed from the Titan II intercontinental ballistic missile. The lift capability and success record are both good, making the system attractive for larger domestic and international communications satellites. Lockheed-Martin, the manufacturer, developed the upgraded Titan 4, which provides a massive lift capability comparable to that of the Space Shuttle. Commercial launches with the Titan were provided to satellite operators such as JC-Sat in Japan and INTELSAT.

The first stage consists of a hydrazine–nitrogen tetroxide fueled rocket engine augmented by two large solid strap-on motors. To reach LEO, the second stage employs the same hydrazine–nitrogen tetroxide combination. The final phase to reach GTO can be provided by any of several different perigee systems. For example, in a dual launch arrangement similar to Ariane, each spacecraft can employ either a solid rocket motor or an integral propulsion system. The IUS and the TOS are two solid rocket systems for perigee kick, manufactured by Boeing and Lockheed-Martin (under contract to OSC), respectively. It is also possible to use the Centaur upper stage for that function, particularly for NASA's deep space missions. Generally, the Titan is considered too expensive for commercial launches.

10.3.10 Zenit

Sea Launch Company was formed by Boeing Commercial Space Company in 1995 in partnership with Kvaerner of Norway, RSC Energia of Russia, and NPO-Yuzhnoye of Ukraine. This commercial venture provides launch services to GTO from a converted oil drilling platform that is towed to the equator in the Pacific Ocean. Kvaerner modified a drilling rig to use as the launch platform, constructed the assembly and command ship, and now manages all maritime operations. The home port is in Long Beach, California, in close proximity to the first customer, Hughes Space and Communications, which purchased services for 10 GEO missions [2].

The Zenit-3SL liquid propellant rocket comprises the first two stages of Sea Launch. Produced by Yuzhnoye and first flown in 1985 from the Baikunor Cosmodrome in Kazakhstan, Zenit uses liquid oxygen and kerosene in both stages, the first containing four thrust chambers and gimbaled nozzles and the second containing a single main engine and another aft-mounted engine for three-axis control. The two-stage system is horizontally integrated, self-erecting, and self-fueling, which facilitates prelaunch operations onboard the ship and towed platform.

The third and final stage, used to propel the payload from LEO to GTO, is a modified version of the familiar Block DM produced by RSC Energia of Moscow. As noted previously for the Proton, the Block DM is a restartable upper stage using a liquid oxygen and kerosene main engine. A single gimbaled nozzle provides directional control. The vehicle is itself a spacecraft that provides its own three-axis control using a separate reaction control system consisting of five side-mounted engines.

The ship was designed and built in Glasgow, Scotland, in 1996 exclusively for this function. Besides providing the facilities necessary to process the LV and payload and to control the launch, the ship tracks the initial ascent of the LV and is the principle means of transportation from Long Beach to the aquatic launch site. The rocket is assembled and fueled on the ship, then transferred to the platform with a crane. Following launch, the rocket is tracked by another down-range ship and the Russian ground tracking stations.

10.4 Launch Interfaces

From Section 10.3, it should be clear that there is a wide variety of LV arrangements and capabilities. That is reflected to some extent in the interface with the spacecraft to be launched. Many modern spacecraft designs incorporate features that permit the use of different LVs, allowing the satellite operator to obtain the best possible arrangement for launch. Beyond the physical interfaces between the spacecraft and the LV are the management interfaces, which require a continuing process of coordination among the manufacturer of the spacecraft, the purchaser, and the launch service providers.

10.4.1 Physical Launch Interfaces

A communications spacecraft that is to be launched must be designed and tested in such a way that it can be properly attached to the LV. In addition, the spacecraft must withstand the expected acceleration, shock, and vibration of rocket flight. An important consideration is that the spacecraft fit within the dimensions of the LV. The electrical connections between spacecraft and LV, which must be severed at time of separation, are primarily for the purpose of activating separation and the spacecraft itself. In general, the spacecraft is self-powered by its internal batteries during ascent.

Figure 10.12 presents in simplified form the physical interface for two typical expendable LVs: the Delta and the Ariane. The fairing is the outer shell of the LV that contains the spacecraft, constraining its height and width. In the illustration at the left, the spacecraft is shown attached to the perigee kick motor with an adapter ring. The use of an adapter ring allows the spacecraft and LV manufacturers to build their respective attachment systems ahead of time without having to bring the two parts together prior to integration at the launch site. Even with that technique, integration tests are still performed on the first sample of any new combination of spacecraft and LV.

The interfaces in the Ariane are compatible with the dual launch feature. In Figure 10.12(B), one spacecraft is mounted on top of Spelda, within which the second spacecraft is contained. Once in transfer orbit, the casing frees both spacecraft for subsequent injection into GEO by their respective apogee engines. Arianespace uses an enlarged 3m diameter fairing, which makes possible the launch of relatively large spacecraft. The trend in other expendable LVs is now toward such enlarged fairings because of the flexibility afforded to the spacecraft buyer.

As discussed earlier in this chapter and illustrated in Figure 10.1, the spacecraft requires two major orbit changes to reach GEO from the parking

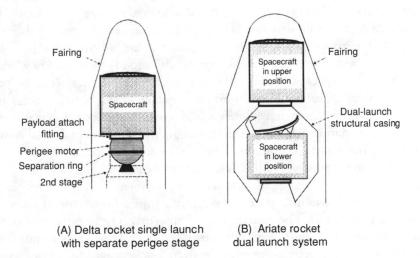

(A) Delta rocket single launch
with separate perigee stage

(B) Ariate rocket
dual launch system

Figure 10.12 Payload integration arrangements for the Delta and Ariane expendable launch vehicles.

orbit. The second change, at the apogee, usually is provided by a rocket engine internal to the spacecraft, such as a solid AKM or a high-thrust liquid engine. Perigee kick can be provided either by the third stage of the LV or by a separate rocket engine attached to the spacecraft. The spacecraft buyer would not need to be concerned with the perigee kick stage if its function is provided by the LV agency (as is the case with Ariane, Atlas-Centaur, and Proton).

10.4.2 Management Interfaces

While not all-inclusive, this section provides a brief overview of the plans and activities carried out by the organizations involved with the launch of a commercial communications satellite. The first step in arranging a commercial LV is to enter into a launch services agreement (LSA) with a launch agency such as Arianespace, Boeing, or ILS. Because the LSA is a legal document, it usually is prepared and negotiated like any other contract. In addition to laying out the schedule for launch and payments to the launch agency, the LSA delineates major responsibilities, including those shared by the buyer of the services, which could be either the satellite manufacturer or the intended operator of the satellite. There is a technical appendix to provide specifications for the rocket and its interfaces with the payload. One of the most important specifications is the maximum payload weight the LV is capable of placing into parking orbit or GTO, as appropriate for the particular LV. Another LSA is required for a transfer stage obtained separately from the LV service.

While the technical appendix to the LSA covers physical interfaces and requirements, it still may be necessary and desirable to prepare a document after the launch agency is under contract to precisely define every aspect of the integration of the payload with the LV. Often, most of the information is provided by the spacecraft manufacturer since that organization is the most knowledgeable about the physical and electrical characteristics and constraints of the payload. However, all parties have a great deal at stake in the launch process and therefore need to take an active interest in the preparation of the management plan. From a contractual standpoint, the requirements of the plan need to stay within the limits of the LSA so the buyer is not subject to increased cost and to ensure that the capabilities of the LV are not exceeded.

While the technical specifications in the documents and agreements cover the physical interfaces, there is a vital need for a time schedule for all the activities leading to liftoff. The manufacture of the spacecraft and the LV are the responsibilities of the respective contractors, but delivery of the hardware from the factories must be consistent with the overall project schedule. Prior to the start of launch site activities, compatibility tests and rehearsals usually are carried out to verify that things will fit together and that the launch itself will go smoothly. Those activities require management attention by the three key parties to the agreements.

A detailed schedule is generated for the activities at the launch site. The spacecraft contractor working at the site performs a series of checks and tests to prepare the spacecraft for integration with the LV. Meanwhile, the launch agency and the LV manufacturer assemble the LV and begin the process of payload integration. In the case of the Space Shuttle, integration is done in a special building off the launch pad. After the complete STS is assembled and the payloads integrated, the entire vehicle is moved to the pad on the mobile launcher, a gigantic tractor. Most Western expendable LVs are integrated with their payloads right on the launch pads; Russian and Chinese LVs and Pegasus are integrated horizontally.

The program plan must deal with the availability of support services and facilities at the launch site since the spacecraft contractor usually will not have its own facility nearby. The launch agency that operates the site may provide those services as part of the cost of the launch itself; otherwise, a separate agreement must be entered into with whatever organization happens to have the ability to provide that support. Critical activities at the site include electrical system checks, fueling of the spacecraft RCS, and the attachment of any external rocket motors, all of which are hazardous.

While many of the responsibilities can be delegated to the spacecraft manufacturer, the ultimate responsibility for the mission direction lies with the buyer of the launch service, often the prospective satellite operator. The mission director, working for the launch services buyer, is positioned at the launch control center and is consulted by the launch agency and its launch team. The launch director (also called the test director, a throwback to the days of missile testing) works for the launch agency and commands the team that executes the actual launch. The launch director, however, will not agree to launch without the consent of the mission director.

Following liftoff and after the spacecraft separates from the LV, all control reverts to the buyer and the subcontractors. This is the transfer orbit phase (described in detail at the beginning of this chapter). As part of the program plan and preparation, the mission direction and TT&C support around the world should have been tested by rehearsals and simulations well ahead of the actual mission. It is also a good idea to have contingency plans to deal with the unexpected. For example, a TT&C site in a remote part of the world may experience an equipment failure just prior to the time of need. Contingency plans can be incorporated into the mission analysis performed by the orbital dynamics group so that tracking and commanding can be accomplished in a number of different ways. In any case, it is important to have the full operational and analytical capability available at all times during the initial phases of the mission.

10.5 Risk Management in Launch and Operation

Placement of communications satellites into service is inherently a risky business. The odds are that a given launch will be successful, but it is not a surety. Therefore, satellite operators must plan for all reasonable contingencies so their customers can continue to obtain communications services. We use the term *risk management* to include those steps taken during the design, manufacture, launch, and initial operation of the spacecraft that make up the space segment. Risk management must be embedded in the operations of all program participants (spacecraft manufacturer, satellite operator, launch vehicle manufacturer, launch service provider, et al.) and should not be considered a peripheral task [3]. It would not be sufficient simply to cover the direct financial risk with an insurance policy, because that alone cannot ensure the service will be available (only that the cost of replacement is covered).

An important first step is to make sure the spacecraft and, in the case of non-GEO systems, the constellation are properly designed from a system and reliability standpoint. As a rule, a spacecraft that comes from an established product line from one of the leading manufacturers probably is more inherently reliable than one that represents a unique design. This can be overcome through extensive design analyses and testing to uncover potential reliability shortcomings, but it comes at considerable extra expense. Careful consideration of redundancy and likely failure modes must be made early in the design of the spacecraft. The customer should maintain oversight during manufacture and test phases of the program. It is common for customer personnel to be located in the spacecraft manufacturer's factory to attend design reviews and witness major subsystem and system tests. That instills in everyone the importance placed on reliability and risk management.

In summary, the risks that apply to commercial space systems relate to cost, schedule completion, launch, and on-orbit operation. We are particularly concerned here with launch and on-orbit risk considerations. This section provides some guidelines and examples of how risk management is conducted in the commercial satellite industry. We focus on the front end of the project, which ends when the satellite has successfully reached its operating orbit and entered service. From that point, there still are considerations of how satellites may need to be replaced due to normal or premature end of service (for more detail on this topic, see [4]). Our discussion is by no means exhaustive and should not be applied directly. Particular situations will differ widely, so a unique risk management strategy must be considered carefully and subsequently implemented. Help for doing that can be obtained from the relevant spacecraft manufacturer, insurance underwriters, and experienced industry consultants.

10.5.1 Launch Insurance

Commercial communications satellites typically are insured against loss due to launch failure. The launch is the part of the mission of highest risk and therefore poses the greatest potential for financial loss to the satellite operator. Of course, loss of the satellite at any point in its life is a serious concern, particularly when high-valued services already are being rendered from the space segment. For the established LVs and launch service providers, the likelihood of successfully being injected into the prescribed orbit is in the range of 90–95%. Stated another way, there is between a 1-in-10 and a 1-in-20 chance of launch failure. That is further affected by the remaining orbit change maneuvers, if any, and successful completion of initial on-orbit testing. These particular steps add to risk somewhat and cannot be ignored. Because of those considerations, launch insurance includes coverage of all those steps, plus some period of initial operation, usually 6 months.

Satellite projects that depend on commercial sources of funding generally must use launch insurance to satisfy the demands of investors and banking institutions. In the past, some large organizations have self-insured by taking account of the odds and holding sufficient reserves, either in terms of extra spacecraft or a financial pool of money. However, that may not be feasible for many, so launch insurance is the most logical way to go. That being the case, the question remains as to the amount of coverage to purchase and the likely premium cost. By practice, owners of GEO spacecraft typically insure the total value of the spacecraft and the launch, service, as well as the cost of the insurance itself. The value assumed should be based on the current replacement price, which can be called out in the original contract as an option. Insurance premiums are in the range of 12 to 15% during normal times but can increase to 20% or more after one or more launch failures.

Launch insurance is purchased through a broker or an agent, who actually obtains the coverage on the commercial high-risk insurance market. In the past, Lloyds of London was famous for being the house that places this type of coverage, but it largely has been replaced by a multiplicity of insurers in the United States, Europe, and Asia. It is the job of the broker to work with the insured to fix the coverage requirements and desired costs. From that point, the broker goes out into the market and assembles a syndicate of insurers to take on the risk. The typical insurance policy assumes risk of loss at the time of intentional ignition of the LV and terminates 6 to 12 months later. Readers should be aware that the precise definition of intentional ignition depends on the particular spacecraft and LV design and should be specified before going to the insurance market.

Many insurance brokers have technical experts on staff to review the reliability history and performance of the LV and the spacecraft. They may have come from the same manufacturers that provide the hardware and services, so they tend to know what questions to ask and what to look for. Insurance rates can depend on the particular design and manufacturer and fluctuate depending on recent experience. The insured party can place the policy well in advance but may not be able to lock the rate until some months ahead of launch. All those matters require careful consideration because of the large sums of money involved.

An approach taken by some operators is to contract with the spacecraft manufacturer for delivery in orbit (DIO). In that arrangement, the manufacturer is responsible for procuring the LV, managing all aspects of its use, and possibly purchasing the insurance as well. On the one hand, that simplifies things for the operator but, on the other hand, takes certain matters away from direct control. It is a tradeoff that needs to be considered; it can be advantageous for some (but not all) satellite operators. Some spacecraft manufacturers (and operators, as well) avail themselves of good deals for launch vehicles through the facility of the long-term agreement (LTA) with the launch services provider. That assures them of a supply at an agreed price per launch in exchange for guaranteeing a minimum buy quantity. For the DIO type of contract, the LTA can reduce the risk of having another launch available to get a replacement up.

The other, more common approach is to purchase the spacecraft for delivery on the ground (DOG), thereby placing responsibility for obtaining launch service and insurance on the operator's shoulders. The buyer must have adequate technical and administrative resources to manage the launch services provider and insurance broker. Operators that do this generally have a need for many launches over time and so can justify the expense and the complexity. There are, of course, hybrid approaches in which the DIO contract does not include the launch insurance.

Satellite operators may purchase another type of coverage, called life insurance, for the period of actual operation. The cost is more nominal, being in the range of 1–2% of the insured value per year. At the beginning of a satellite's life, the covered value is essentially the same as for the launch insurance. That coverage can be reduced as the satellite ages, to account for depreciation.

At the satellite's end of life, life insurance is theoretically unnecessary because a new satellite normally is made available through a new contract.

10.5.2 Backup and Replacement Satellites

Our last topic is perhaps one of the most complex and difficult for satellite operators to resolve. We have spoken about a satellite launch as a discrete event taken out of a general context. In reality, the satellite operator has to consider not just one but many satellites. A system that depends entirely on the successful launch of one spacecraft is a system that is taking a decided gamble. However, commercial business and government operations as well need to view their position as that of absorbing risk so that the ultimate users are not themselves having to gamble on a successful launch. (In spite of that, using organizations may find it prudent to arrange for alternative capacity from one or another satellite system.) The fundamental way to improve the probability of success is through diversity, which simply means the satellite operator needs to have multiple ways to deliver service besides the spacecraft involved in the specific launch of the moment. That is easier for established operators with many working satellites and more difficult (but not impossible) for newcomers to the business.

Diversity can represent a high barrier in financial and operational terms and is a discriminator in the marketplace. There obviously are many approaches and options for providing diversity, and there are price tags to go with them. The most basic package is to purchase, along with the spacecraft, either a second backup spacecraft or major portions of one. Then, in the event of launch failure, the second spacecraft can be readied (or completed, as the case may be) for a second launch. The LV also must be made available for the second try, which is something that can be accommodated in the LSA. If the first launch is successful, the backup can be placed in storage until it is needed to expand capacity or deal with an on-orbit failure.

The use of the second and subsequent spacecraft for expansion also ties in with providing on-orbit diversity. That is the most common approach to both developing a growing business and reducing risk for the operator and its customers. As a general practice, operators tend to utilize all their on-orbit resources to provide revenue-producing service. However, there could be excess reserved capacity for use for restoration of primary services in the event of launch or on-orbit failure. Then the operator can use the insurance proceeds to initiate construction of a replacement on a more methodical (and leisurely) basis. The objective is to better manage risk so that operations and finances are evened out over the long term.

This section has been only a brief introduction to the subject of risk management. No one can determine the particular risk environment and remedies better than the satellite operator itself. Individual situations differ greatly and depend on many factors, some of which are under the operator's control and others that are not. For example, the approach taken by an established operator with many functioning satellites already in service is much different from that of a new operator just starting out. Also, a different approach must be taken by a private company that has investors or public stockholders, who must be protected from financial downside losses, in contrast to a government organization that may be more concerned about the risk of not being able to service a vital need. Another important point is that all LVs are not the same because their reliability records differ, both in absolute terms and over time. Also, what seems to be a simple change in a rocket configuration (such as adding a different stage or changing some design feature) often has a significant impact on launch success. Those issues demand the greatest attention of all levels of management. It is well worth the time, effort, and resources to find and maintain the most effective risk management strategy.

References

[1] Isakowitz, S. J., *International Reference Guide to Space Launch Systems,* 2nd ed., updated by Jeff Samella, American Institute of Aeronautics and Astronautics, Washington, D.C., 1995.

[2] Miller, B., "SeaLaunch," *Launchspace,* Vol. 2.04, Oct./Nov. 1997.

[3] Cromer, D. L., "Risk Management—Program Opportunities and Challenges," keynote address, *Air Force Space and Missile Center Symp. on Aerospace Risk,* Manhattan Beach, CA, June 3, 1997, Hughes Electronics Corp., reprint.

[4] Elbert, B. R., *The Satellite Communication Applications Handbook,* Norwood, MA: Artech House, 1997.

11

Satellite Operations and Organization

There is a saying that a chain is no stronger than its weakest link. That is certainly true when it comes to ground control operations of a satellite communications system. This chapter is a discussion of the overall operation of the ground segment, utilizing both technical and human resources. Vital to the ground segment is a dependable satellite control network (discussed in Chapter 9) that provides appropriate and adequate contact with all working satellites, allowing a trained and experienced staff of professionals to perform the tasks necessary to achieve overall performance requirements. In our experience, that requires considerable investment in quality hardware, software, and human capital. What good satellite operation strives for is zero defects in terms of all satellite control functions. In practice, problems occur and mistakes, unavoidably, are made.

Satisfactory performance of the satellite control system depends on how well the operator deals with contingencies. We previously dealt with the physical hardware of the space and ground segments. Those elements must be designed and implemented to provide a working system. Keeping the system in operation and delivering services is the challenge of the soft side of the business—the assignment and management of the staff of professionals who attend to the immediate and long-term needs of the hardware. To that is added the users and the customers who access the space segment; they are partners in the important process of maintaining reliable and effective services. We provide this review to give the reader an appreciation of the requirements and issues that impinge on the quality and reliability of an operation. This chapter should not be taken as a comprehensive set of recommendations but rather as a brief introduction to the topic and a starting point for more thorough planning and development efforts.

11.1 The Satellite Control System

An overall satellite control system can take many forms, corresponding to the requirements of the satellite constellation, whether GEO or non-GEO. GEO satellite operation principles have been well developed over the years, having been initiated for the INTELSAT system and later employed in domestic satellite systems, first in the western hemisphere and later in the rest of the world. The approach was reviewed in Chapter 9 for typical TT&C ground facilities comprising one or more TT&C ground stations and an SCC. Figure 11.1 is an overview of that type of arrangement.

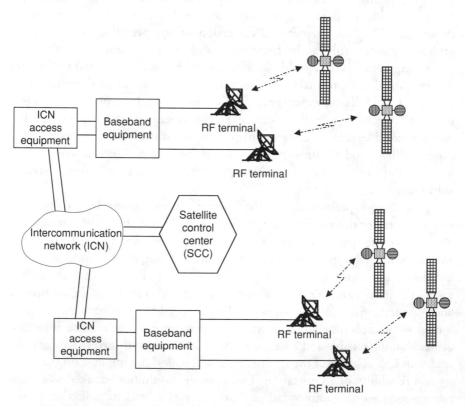

Figure 11.1   A typical TT&C ground segment, using two TT&C Earth stations and a centralized SCC.

The equipment installed in the TT&C stations and the SCC (illustrated in Figure 9.18) satisfies the basic requirements of satellite monitoring through the telemetry downlink, control through the command uplink, and orbit correction ranging and orbit determination functions. Those are the most basic technical requirements of satellite control and must be dependably performed over the satellite mission and lifetime of the system. Equipment of that type typically is purchased along with the spacecraft themselves and tested prior to and just preceding launch. Over time, the equipment must be properly maintained and upgraded when necessary to expand capacity or improve the quality of operation. The difficulty of making such changes should not be underestimated, because of the problems associated with working on a "live" system in a way that will not disrupt the ongoing operation and the ground control staff.

11.1.1 GEO Satellite Control

The first step in gaining an understanding of GEO satellite control is to identify the control requirements. They can be summarized as follows for principal aspects of satellite operation with the exception of the monitoring of payload services, which is discussed in Section 11.3.4.

- *Support the initial get-to-station activities following launch.* The spacecraft manufacturer usually is responsible for carrying out this phase of the mission but might need to use the TT&C ground stations and SCC of the satellite operator in either a primary or a supporting role. Operations personnel often benefit from being part of the transfer-orbit team, since they ultimately will take control and responsibility. Also, it is an excellent time to verify and validate the function and performance of new or modified hardware and the software elements of the ground system.

- *Conduct or actively monitor initial on-orbit testing.* This can follow the same line of thinking as for the get-to-station activities. Since there is a heavy emphasis on the long-life capability of the satellite following testing, the satellite operator takes on a stronger and possibly leading role in the conduct of these functions. Handover of the satellite from the spacecraft manufacturer to the satellite operator occurs when both parties agree that the system meets contractual requirements. If that is not satisfied, it still might be possible to commence service, with the manufacturer retaining responsibility to complete any open action-items (not unlike the "punch list" common in the purchase of a newly constructed house).

- *Maintenance of the satellite in its assigned orbit position with sufficient accuracy to support the ground network.* The common practice is to maintain orbit position to within ±0.1 degree north-south and east-west (or, stated another way, within a square of 0.2 degree on each side and centered on the assigned location). That generally will meet agreements to control adjacent satellite interference and is within the technical capability of most TT&C ground and spacecraft systems. If multiple satellites are to coexist at the same orbit slot (something that is common and can now be viewed as a requirement), accuracy must be improved by a factor of 2 at least. Individual colocated satellites can be precisely located from the ground and their position "choreographed" to preclude collision (there is only a small probability of collision in the absence of such active avoidance, but satellite operators prefer not to have to worry about that potential for loss). During all these activities and over the life of the satellite, fuel usage must be carefully controlled and budgeted so life expectancy is achieved or surpassed.

- *Monitoring of satellite health data.* The SCC is provided with facilities for receiving, displaying, recording, and analyzing spacecraft health data provided by the onboard telemetry system. The practice in commercial GEO operations is to perform those functions continuously on a 24-hour basis and without interruption. The reason is that GEO satellites are also expected to provided uninterrupted service to users, so their operators, of necessity, feel compelled to know what is going on with their in-orbit assets. Any difficulty with service can then be traced back to telemetry data and analysis performed by spacecraft technical experts. Also, the data permit a long-term evaluation of satellite performance and the identification of potential failure modes before they occur.

- *Control of spacecraft subsystems.* Modern GEO spacecraft contain sophisticated onboard computers, which make their operation nearly autonomous. While that is a desirable objective in the long term, there are many circumstances when human intervention is required to ensure a safe operation and proper service performance. Most station-keeping actions are initiated and conducted from the ground, using the SCC hardware and software systems. Response to anomalies and faults onboard the satellite almost always will be through human intervention via the command link. For those reasons, the SCC must be ready at any time of night or day to take control, either by plan or in the event of an unexpected contingency situation. Routine changes in the configuration of the payload or bus will also involve the SCC.

- *Operation and maintenance of the TT&C Earth station and SCC system.* The previous requirements relate to the proper operation of the satellites in GEO. In this paragraph, we are concerned with the like performance of the ground system itself. The ground facilities are just as essential to reliable service provision as any of the bus or payload elements. For example, if the TT&C antenna were to move off the satellite due to a failure of its tracking system, both the command and telemetry links would be cut off. The satellite would then have to function on its own, a situation that would have been considered in its design and that ordinarily would not produce a loss of service. However, the number of possible situations that the system could encounter cannot be anticipated in practice, so the best philosophy is take the steps to ensure as close to 100% availability of telemetry reception and command transmission as is humanly possible. An approach that improves on that situation is to have a fully functioning backup TT&C station (see Figure 11.1).

- *Ensure that satellites can be replaced or relocated as demanded by the service plan.* Satellite replacement is often an important responsibility of the satellite operator, since users prefer to view their on-orbit repeater as a permanent feature of the network. Just as with launch risk management, there are complexities in retiring old satellites and providing a smooth transition to the new on-orbit resources. For example, a new satellite may have to be configured and tested in an unoccupied orbit slot, then drifted into place before transferring traffic to it. In the case where capacity at a given slot is increased, the operation of the existing satellites may have to be modified to allow for the new inhabitant.

- *Develop and maintain contingency plans.* The satellite operator and user need to recognize the true meaning of Murphy's Law: "Anything that can possibly go wrong, will (go wrong)." (It also has been said that Murphy was an optimist.) Engineering and maintenance must identify any weak links and failure modes and create backup and replacement strategies. At a high level, the satellite operator needs a set of plans that identify how the organization would react to a partial or full satellite failure. Any major user of GEO satellite capacity, such as a TV network, should have alternate or diverse resources to maintain transmission in the event of a failure. Through attention and commitment to continuous service, organizations may stay ahead of Murphy.

The requirements and considerations described here are based on experience and should provide some guidelines for GEO operations in general. For a specific GEO satellite and system, operations managers should conduct their own evaluation and produce an operations plan appropriate to their particular needs. Many of the issues concerning human resources for the operation are covered Section 11.4.

11.1.2 Non-GEO Satellite Control

Non-GEO systems employ centralized satellite control with TT&C stations at critical points around the world that transfer the data to an SCC located at or near the general headquarters of the satellite operator. (INTELSAT has used this concept since it came into being.) The non-GEO approach has unique features, particularly that service availability is dependent on having enough satellites to afford adequate view by users around the globe. The following list summarizes some of the key ground support requirements for a non-GEO constellation, based on the requirements of a typical overall system [1].

- *Establishment of the constellation.* The satellite orbits are determined primarily by the launch vehicles, with limited ability of the spacecraft propulsion system to make the kinds of significant orbit adjustments common in GEO systems.

- *Maintenance of an adequate satellite complement.* A given satellite could be removed or replaced at any time without significant impact, provided coverage remains available from the overall constellation. Also, there is likely to be a need to maintain a continuous flow of replacement launches over the system lifetime, something that is much less frequent for the typical GEO system.

- *Assurance of continuous service through coverage.* The ability of the constellation to provide continuous coverage to users depends on the architecture of the system. An approach based on the bent-pipe type of repeater and without the benefit of intersatellite links must rely on simultaneous linkage between remote user and hub station (or between pairs of remote users for a mesh architecture). For such a system, sufficient hardware must be maintained in orbit to ensure that at least one working satellite has simultaneous view. Diversity operation, provided by Globalstar, is based on at least two satellites. The situation for a network like Iridium, with intersatellite links and without a requirement for gateways, is that each user be in view of one satellite (and not necessarily the same satellite). Such a system can tolerate the loss of a significant quantity of any satellites before there is an impact on availability (the system capacity may, on the other hand, not satisfy service demand because it would depend on the total satellite count). If too many satellites are lost, the resulting interruptions are particularly burdensome because they can affect traffic from distant regions. The MEO-type constellation of ICO reduces some of that difficulty because each satellite covers a much wider piece of the Earth's surface and remains in view for much longer periods than do the LEO satellites. Another important ICO feature is that gateways are interconnected by a terrestrial fiber-based routing ring which emulates Iridium's ISL grid.

- *Non-GEO satellite monitoring and control philosophy.* As introduced in Chapter 9, the operator of a non-GEO constellation might not have a continuous link from the SCC to every satellite. (As will be discussed in Section 11.2, the system using intersatellite links like Iridium may allow the SCC that 100% availability.) Without continuous availability, telemetry could be collected once per orbit using a few well-located TT&C stations [2]. Control of the satellite could take place as needed, either simultaneously with monitoring or during a subsequent period of contact. There is the obvious impact on spacecraft design that the system must operate autonomously and record relevant data between downloads to visible TT&C ground stations. That general approach of onboard data storage helps reduce the number of TT&C stations. Also, because the satellites presumably are of the same design, the system need only keep track of unique spacecraft signatures through a well-structured database.

- *Simplified on-orbit control.* Considering the preceding factors, the bus system of a non-GEO satellite might be simpler than its GEO counterpart. That is in keeping with a desire to reduce the parts count, cost, and complexities involved with satellite operation. One would hope that such measures do not hamper proper identification and correction of faults (detected or otherwise), but that has not been well established at the time of this writing. For ICO, the first MEO constellation to be introduced, the spacecraft is very similar to a GEO design because it draws from the subsystems of the Hughes GEO bus. Therefore, those satellites provide the facilities to which GEO satellite operators are accustomed.

- *Replenishment of satellites.* Operators of LEO constellations have built their strategies on expecting 5–8 years of satellite life in orbit. From public announcements, that estimate is based on the fuel usage to adjust for atmospheric drag and, in at least one case, a desire to upgrade and improve the design of the initial constellation (this, after all, is a new technology base and type of service operation). Production and launch schedules involve 1 to 2 years to achieve the full complement of satellites. Considering that there could be launch or on-orbit failures, a LEO operator must continue to construct and launch spacecraft throughout the operating lifetime of the system. Many view that to be, in and of itself, a substantial burden on the company, when coupled with maintenance of the ongoing constellation. A network that can tolerate the absence of several satellites without disruption of service (or perhaps only brief interruption, similar to when an automobile hits a dead spot in a local cellular network) is best able to withstand unexpected difficulties with incipient technical problems (or ones that appear after a few years of service) as well as short-term difficulties delivering new satellites to orbit.

The obvious and subtle differences among non-GEO constellations and networks make it extremely hard to generalize on the requirements for TT&C ground operations. Each system must develop its own unique operation approach. However, the guidelines presented here should give the reader a basic idea of the factors involved in this type of effort.

11.2 Intercommunication Networks

The majority of satellite operations depend on the reliable transfer of telemetry and command data between the actual TT&C Earth stations that are in direct contact with the orbiting satellites and the SCC, where the full range of technical and management functions are performed. Those links, which can employ both terrestrial and satellite networks, must satisfy the data throughput and quality-of-service (QoS) objectives of the overall satellite control system. Included in QoS objectives are such measures as service availability (in percentage of time), error rate performance, delay or latency (as well as the variation over time), lost packet ratio, and call setup delay (for connection-oriented services like ISDN). There has been renewed interest in QoS of commercially available communication services, since they affect how well the network meets (on a continuous basis) the overall needs of the system and the users. With the advent of modern fiber optic systems and digital satellite communication on a worldwide basis, it no longer is the challenging problem it once was. The difficulty now posed is how to interconnect the elements of the network into a reliable fabric that is manageable by the satellite operator under all expected conditions. To that end, we provide a brief introduction to the subject. Readers are encouraged to review our other work to gain further insight and appreciation for how to proceed on a particular communication problem [3].

11.2.1 Backbone Communications

The fundamental matter for the intercommunication network (ICN) is to connect the various sites with sufficient bandwidth to meet the needs of the SCC. That is the purpose of the backbone portion, suggested in Figure 11.2. Locations usually are determined by the basic design of the TT&C system, considering satellite locations and orbits, the requirements for maintaining continuity of service, and the needed support for the personnel who work throughout the system. All those factors have to be taken into account, which often requires several iterations during the development period as well as after the system begins service. That is a natural consequence of the learning process that goes on in the organization charged with managing the satellites and the services they provide. Also, the particular mix of telecommunications technologies

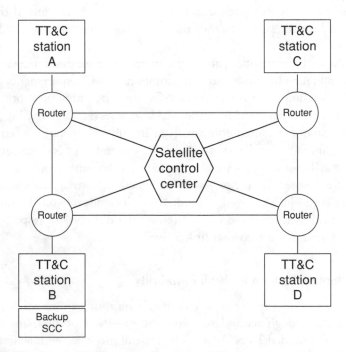

Figure 11.2 A fully redundant communication network for service to a TT&C system composed of four TT&C Earth stations and an SCC.

will change over time—what was once state of the art could, within even a few years, become obsolete and possibly taken off the market! Even when an appropriate technology is located and vetted for use, it may not be available at all locations or with a common feature set. That is, after all, one important reason why satellite communications itself is of benefit to telecommunications users.

The designer of the backbone, therefore, must determine which telecommunications systems and services are available for application in the ICN. The ideal solution for some is to select and contract with a single service provider, likely to be an international carrier like AT&T, MCI-WorldCom, British Telecom, Cable & Wireless, France Telecom, KDD, or the like. In reality, whichever carrier is selected must work to pull the resources of multiple foreign service providers together to provide the backbone solution. Satellite operators in many cases prefer to carry out that task themselves, using their own qualified telecommunication staff.

The technologies that are most attractive for the backbone are no different from the hottest products and services on the world market. They include user-owned equipment such as routers, intelligent multiplexers, and ATM switches for WANs. Interconnecting those elements are backbone links over terrestrial fiber links, undersea fiber cables, and satellite links from GEO satellites. (Current non-GEO satellite networks generally do not afford the bandwidth needed for the ICN, but they do offer a good alternative if needed in an emergency.)

Access to the backbone is another concern, because that often is the cause of service and reliability problems. The communication requirements generally can justify installing fiber between the SCC and the public network point of presence (POP). Where that is not available as a service, the satellite operator may have to pay for construction or even install and operate the equipment themselves. Similar considerations apply to the remote TT&C stations, wherever they may be located. As we have stated, that becomes troublesome where facilities are located in regions without good terrestrial communications. Satellite links can be employed, provided one does not rely on the same satellite being controlled (to do so could increase the difficulty of responding to a satellite problem that affects that link as well).

11.2.2 Alternate Routing for High Reliability

It is our general philosophy that a reliable telecommunications network can be achieved only through redundancy and the provision of alternate means of transferring the needed TT&C data. The general problems with achieving that in practice are cost and local availability. Many satellite operators find that the budget they can allocate to their ICN is fairly small, suffering from the problem that it derives no direct revenue. (Consider what would happen to revenue, however, if the network were to have a hard and lasting failure.) How does an organization justify the rather high expense and management challenge of developing the needed reliability?

In our experience, the telecommunication manager for a satellite operator must create a vision for the organization and develop a strategy for meeting the need. This person must become very familiar with the business and technical strategy of the satellite operator and be sure to achieve as close a match as possible. Through an interactive process of listening carefully to other satellite operations and engineering personnel, communicating the options, and working out the solutions, the telecommunications manager becomes the solutioneer for the whole organization. He or she learns to recognize the specific needs of each site and department, and responds proactively to calls for help. That methodology works whether we are talking about the design phase (when the ICN as well as the TT&C system itself are still on paper) or through implementation and during the operating life of the satellites.

The power of diverse routing has been demonstrated time and again. Primary links cannot be depended on for 100% availability on their own. An alternate path could follow a reverse direction around a ring, as illustrated in Figure 11.2, or might be connected on demand. The latter could be available in some regions, in which case it can be employed for a reasonable cost. Installing dedicated alternate paths, on the other hand, almost always is costly. A rule of thumb is that if an alternate path already exists or might be included at little or no direct cost, it should be maintained in service. It would be embarrassing, to say the least, if a desperately needed alternate path had been removed from service only because someone thought it had no current use. That may seem far fetched, but it is all too common in practice. Also, the best way to ensure that a redundant path is working is to pass a portion of the ICN data over it at all times.

11.2.3 Network Management

The ICN is a distributed facility, with potentially many remote locations. As a result, it, too, requires an effective system to monitor performance and respond to problems when they arise. Fortunately, there is a good selection of specialized hardware and software on the market to greatly simplify the task [4]. Underlying that is an open communication protocol, called the Simple Network Management Protocol (SNMP), developed for the Internet and already applied in a majority of large corporate WANs and intranets. SNMP provides the communications services that deliver network management data from remote communications devices (the routers, switches, and servers in the network) to a centralized network management station (NMS).

At the time of this writing, the leading NMS on the world market is HP OpenView, a workstation-based software product that runs under Windows or UNIX. Similar products are available from IBM (NetView) and Sun Microsystems (Sun Net Manager), but OpenView appears to have the widest overall support for networks such as the ICN. The benefit of that approach is that software modules can be acquired from either HP or the telecommunication equipment manufacturer and run on the common NMS. Alert messages, performance data, routing information, and critical alarms are collected, displayed, and analyzed using the same interface and database. It is even possible to write custom software modules for special equipment, as long as SNMP is utilized for the communication protocol. That usually is not a problem because of the popularity of SNMP and its close relationship to the Internet protocol suite (e.g., TCP/IP).

There are a number of specialized NMS solutions on the market that support a wide variety of proprietary telecommunications and computer networking products, including older designs no longer offered for sale. That may be appropriate when an existing system must be upgraded for centralized control without the benefit of SNMP. Also, NASA already uses its own system for satellite control and network management, which could impose a constraint on commercial users as well. Some specialist software companies offer NMS platforms that support SNMP as well as a number of proprietary network management approaches, particularly in the computer and telecommunications areas. As with OpenView, there are programming interfaces to create special software modules for unsupported monitor and control approaches still in use. That work can be performed by an experienced buyer, but often it is contracted out to the vendor of the NMS. Costs for doing that can be substantial, so it would be wise to do a thorough investigation of the real requirements for integrated network management and the available off-the-shelf products and software tools.

11.3 Network Operations

Sections 11.1 and 11.2 dealt with background activities that facilitate the reliable operation of the satellites in orbit. From the users' perspective, those functions should be invisible and, for all intents and purposes, perfect. They more likely will interact with the network operations element of the space segment, which is established to coordinate the actual communications services. Users make contact with network operations to establish connections, obtain assistance and information during routine activities, and seek assistance when difficulties arise. The last function is perhaps the most critical because network operations is the one place where users can have their problems addressed and is the first line of defense for the satellite operator when exercising control over the ground segment.

From a general perspective, the design and management of the network operations function needs the highest level of attention from the satellite operator. Every satellite system will, at one time or another, depend heavily on that activity to bring new services into being and address critical problems that are being experienced by current users. That cannot be done if network operations is treated as an additional duty by otherwise overworked satellite control personnel or inexperienced staff who lack the training and technical resources they need.

11.3.1 Standard GEO Transponder Services

Network operations for conventional GEO satellites with bent-pipe transponders are the most developed and, in many ways, the most straightforward with which to deal. That is because every transmission to each satellite can, in principle, be monitored at one or more locations on the ground. An area-coverage type of beam is the most convenient because, wherever the satellite operator is located, it is possible to monitor the full downlink spectrum from the GEO satellites under its control. We consider later the case of monitoring satellites that are outside the local reach of the operator, such as GEO satellites in another region or non-GEO satellites during the majority of time they are out of view.

Network operations in this environment can be routine but quite busy during periods of heavy customer traffic (e.g., prior to the news hour or during a major event such as the U.S. presidential campaign, the Olympic games, or the World Cup). The major functions that one would expect to address are reviewed in the following tabulation.

- *Development of requirements for the network operations facilities.* This is where the manager of network operators works with a technical and marketing/sales staff (who are involved with defining and coordinating the actual services) to produce the technical requirements for the facilities that support the users. Besides the basic monitoring equipment, another necessary facility is a computerized database that contains contact information for users as well as scheduling information for daily transponder activity. The best way to develop a service strategy is for these players to work as a team. The people who market and coordinate the services work with the network control personnel to resolve the specifics of how users will make contact and ultimately are served as quickly and as professionally as possible.

- *Qualification of user Earth stations for access to the space segment.* In historical terms, the procedures for technically qualifying Earth stations for their first access were created for the INTELSAT system. The requirements usually are attached to the satellite service contract with users and should be as effective as the force of law. User Earth stations must be willing to take direction from network operations, both during testing and later when services are being rendered. Embodied in the INTELSAT's Standard System Operating Guide (SSOG), the procedures covered every important aspect of Earth station performance and service capability. Over time, other commercial satellite operators saw that the SSOG style, while technically excellent, placed an undue burden on their customers, both current and potential. The procedures to qualify new Earth stations have been streamlined for that reason; however, the critical need is that any new Earth station meet the minimum technical requirements and therefore not introduce difficulties into the current operation of the satellite system. An example of a typical set of qualification requirements is provided in Table 11.1.

Table 11.1
Typical Qualification Test Requirements To Verify Earth Station's Technical Performance

Antenna pattern, sidelobe levels
Cross-polarization isolation
EIRP
G/T
Carrier frequency
Modulation bandwidth
Burst timing stability (if needed)
BER performance (if needed)
Channel amplitude response
Channel group delay response (if needed)
Threshold performance
Frequency stability
Operator control
Contact information

The extent of Earth station qualification validation depends on the degree of harm that a maladjusted station could cause. Hence, large Earth stations (such as those used for DTH uplinks and major hubs) are still tested according to the SSOG style with only minor procedural differences. The trend for small Earth stations is to try to validate their performance on a network basis. For a VSAT, the antenna pattern and HPA power likely would be measured on one unit (a process called type acceptance). But frequencies, timing and protocol, and control-associated parameters would need to be verified for each terminal individually over the satellites. For an Imarsat mobile laptop terminal, there is a type acceptance (which may presume certain tests at the manufacturer) so the user can take it out of the box and begin communicating in a half hour. The same can be said of hand-held MSS terminals.

At the extreme are RO terminals, as for DTH receive systems. For those, the validation process is simply verification of compatibility with conditional access systems and assurance of consumer standards.

- *Monitoring of all transmissions to and from the satellites.* This is accomplished with a combination of automated monitoring and test equipment along with manual systems that can be used by network operations personnel, who are available on a 24-hour basis. The automated equipment would be connected to sufficient downlink antennas to view all the operating transponders on the satellites and deliver the data to computers and video screens in the network operations center. A photograph of a typical facility is provided in Figure 11.3. Special software can measure and record carrier power levels, frequencies, and bandwidths and thus track the actual occupancy over time. Those data can be compared with what is recorded in the database of authorized services and users and can be an input for billing, if necessary. Operators can review that information and act on irregularities that indicate that either the Earth station(s) or the satellite transponder(s) are not working within prescribed limits. Ideally, this approach allows the satellite operator to identify and respond to problems before customers call in. In addition, monitoring software can tie in with the commercial management of services, in particular billing and customer account maintenance.

- *Telephone, fax, e-mail and other facilities to permit direct and effective communication with users who have access to the space segment.* There is no substitute for reliable direct communication between users and the network operations function of the satellite operator. With all the potential ways to do that, experience has shown that the telephone is

Figure 11.3 Automated test facilities for GEO satellite transponder management. (*Source:* Courtesy of Hughes Communication, Inc.)

usually the most effective communication tool. (Some systems require that all uplinks use a standard satellite telephone integrated into the operator's network.) The service agreement contract should require that all users provide a working fixed line or mobile telephone number where network operations can directly contact the uplink. That is more complicated in the case of VSAT networks, where individual terminals are usually unmanned. In this case, the VSAT network operator provides the control function and point of contact with network operations. Written or record communication is needed to provide specific instructions to users, particularly when RF interference or other unidentified problems arise. That often will amount to a broadcast message to all users and potentially other satellite operators as well.

- *Good and effective internal communication with satellite control and sales/marketing.* Network operations is in direct contact with users, who are actual customers or their Earth station operators. Therefore, they should be treated with care. That can be done most effectively through a good working relationship among the other key activities of the satellite operator: satellite control, where spacecraft problems often are identified first, and sales/marketing, where user services are put into place and customer relationships maintained. Teamwork and cooperation among those internal organizations are vital to the success of the business and then the provision of reliable communication services, as well. Frequent formal and informal contact is highly recommended. In instances where those functions are physically separated, there is a need for effective use of the ICN and face-to-face meetings, when appropriate. There is another opportunity to involve an important outside constituency: the users or customers. A technique that has worked is to conduct user conferences and seminars and to create a satellite user group (SUG), which is similar to the user groups that associate themselves with particular types of computers or networks.

This brief summary of the requirements and considerations for the network operation function is not exhaustive and should not be applied directly in specific situations. The best approach is for the managers of the particular satellite operator to develop their own network operators' organization and technical strategy and then to implement that strategy in the particular context at hand. The size and composition of network operations will change over time as communications traffic grows and services evolve.

11.3.2. Potential Non-GEO Services

We need to offer some thoughts on services to be applied for network operations services for non-GEO satellites. First and foremost, non-GEO satellites are intended for niche or targeted services to specific types of user terminals. Management of the user segment is covered in the next section. However, monitoring of satellite payload transmissions is not easily accomplished on its own. Certainly, bent-pipe satellites such as those used by Globalstar can be treated in this manner, but digital processing repeaters, common to ICO and Iridium, are too complex.

11.3.3 User Network Monitor and Control

The overall network operations function of the satellite operator may include the direct control of Earth stations in the user network. That, of course, is a very different kind of activity from qualifying Earth stations, monitoring transponder signals, and dealing with RFI problems (although each of those can and does affect the overall network). A user network of Earth stations requires the type of network management discussed in Section 11.2.3. The difference in this case is that, rather than dealing with the ICN (which is, by definition, internal to the activity of the satellite operator), the management of a user satellite network involves the customers and their respective activities. The range of possible needs and problems increases, sometimes exponentially, with the number of networks under management responsibility.

A classic example of such network management relates to a hub-based VSAT network serving one or more corporate customers. In the shared hub environment, the network operator monitors and controls a quantity of remote locations (sometimes numbering in the thousands), a backhaul circuit to the customer's host computer, and a set of end-to-end communications applications as required for the particular customer. If the satellite operator is fulfilling the network management role, there likely will have to be a separate function with the appropriate expertise made available. As the demands of customers increase, the size of the organization and associated resources must increase correspondingly. The point here is that what develops is a different kind of activity from that described in Section 11.3.1. On the other hand, there will need to be continuous coordination and cooperation among those activities so the Earth station network can be installed, activated, operated properly, and expanded as required.

Integrated network management for an advanced non-GEO network represents a challenge not previously undertaken during the development of GEO satellites [5]. The two different architectures, represented by the bent-pipe/ground hub approach of Globalstar and the routing/intersatellite relay approach of Iridium, have different requirements for user network management. As shown in Figure 11.4, the bent-pipe/ground hub approach allows user network management information to flow when the satellite is in view of both the user terminal and a gateway Earth station (or specialized control and monitoring station, if applicable). Information can then pass from user terminal over the satellite, through the gateway, and on to the network operating center (NOC) through an associated network such as the ICN. Using two-way communication, the NOC can send control messages in the other direction to remote users.

Normal hand-over procedures from satellite to satellite would be required to maintain continuity of management communication (which is how subscriber telephone and data connections are maintained as well). The overall network can therefore control user terminals and user services when gateways are available to provide the point of relay. If a satellite is in view but it cannot simultaneously connect to a gateway, then no management information can be delivered to the NOC (this may be academic since no calls can be established either). An independent terrestrial or satellite network is required between the gateway relay points and the NOC, since there are no intersatellite relays.

The routing/intersatellite relay, indicated in Figure 11.5, provides continuous paths for management data independent of gateways and terrestrial networks. As with subscriber calls, the management data may be exchanged directly between user and NOC on a unified basis. The messages can pass over any available links so as to leave maximum bandwidth available for revenue-

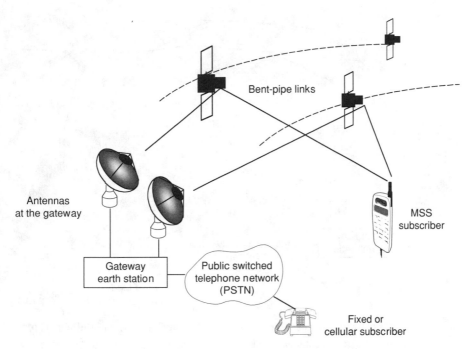

Figure 11.4 User network management for a non-GEO bent-pipe satellite constellation (based on GlobalStar).

bearing calls. Because of the routing nature of this architecture, the same basic procedures and protocols can be used for all classes of communication, whether associated with network management or the relaying of user traffic information. The network is robust in that, with an excess of satellites and links between them, a communication path can be established under nearly any circumstance. The one exception is when the user terminal is blocked from view of any satellite. That sort of thing is a normal characteristic of mobile communications using a non-GEO constellation but can be made acceptable by supplying an excess of satellites and link margin.

11.3.4 Payload Configuration Management

Requirements and procedures for management of a communications payload heavily depend on its particular design and complexity. The simplest is the basic bent-pipe repeater with service to a simple area-coverage footprint. Repeaters of this type are configured for service at beginning-of-life and may never require a configuration change. The exception is when an anomaly or failure is detected, requiring the replacement of a bad active component by an available spare.

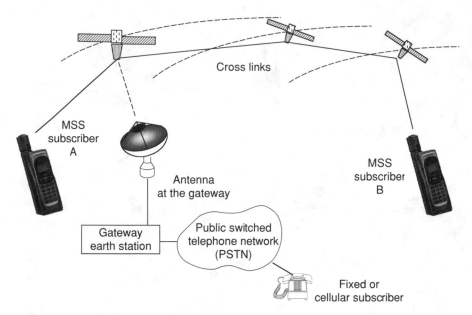

Figure 11.5   User network management for a routing/intersatellite relay non-GEO digital processing satellite constellation (based on Iridium).

Transponders that contain passive or active gain control also can require adjustment by ground command, in response to a request from users or network control personnel. Such changes possibly occur only once or twice a year and can be implemented by manual SCC commanding.

As payload complexity increases, the requirements for ground-commanded changes increase as well. On the extreme is the digital processing repeater with the capability to route narrowband channels or ATM cells. The particular routing pattern or channelization can be preloaded through the command uplink or as a separate processor-control link, checked for accuracy and then activated to provide a particular service arrangement or pattern. That can be changed as frequently as dictated by the network, in periods as short as one hour or as long as one month—it all depends on the processor design and the dynamics of the traffic.

Another type of payload is one that can establish connections on an individual call basis (similar to dialup telephone service). The request for service comes directly from the user and is relayed over the control link to the payload processor. There is no requirement for human intervention at the control center; rather, the system operates autonomously like a telephone exchange. The network control system on the ground should manage the end-to-end connection and perform such functions as collecting call records, tracking problems, and billing for services.

As payloads become more complex, particularly for non-GEO constellations with such features as intersatellite links, management of onboard communication equipment and the actual links themselves become the most complex task of satellite operation. Traffic in such a system is dynamic and can produce interesting and unexpected effects, depending on the particular distribution and timing of user demand in local areas and throughout the entire network. Examples of such networks have existed for many years on the ground, the U.S. long-distance telephone network operated by AT&T being a good example. For this case, AT&T maintains a major network management center in Bedminster, New Jersey, that can track telephone traffic demand at every node and on every link in the network.

INTELSAT has not had to go that far for its international satellite system because its primary role is to provide transponder bandwidth to users. Individual satellites are monitored through TT&C Earth stations that are strategically located to view every satellite. Traffic is monitored using automated test equipment that then transfers the information to the headquarters in Washington, D.C. Communications controllers can observe the overall downlink performance of each satellite and coordinate corrective action wherever it is needed.

The approaches to managing user terminals in non-GEO systems were reviewed in Section 11.3.3. The first global mobile satellite system to go into service, Iridium, demands substantial resources and capabilities to carry out the needed payload management function. The control center is located in Reston, Virginia (outside Washington, D.C.), at the company headquarters. Each payload acts as a channel router for information packets that contain digitized telephone traffic and control information. The integrated system of satellites, control stations, and network control places large demands on network management computers and software. Iridium gave itself well over one year, from the time of its first launch until any telephone calls could be completed, to implement and refine the overall system of traffic control and management. The difficulty of that task and the sophistication of making all the pieces work together correctly cannot be underestimated.

11.4 Human Resources for Satellite Operations

The human side of satellite system management is of equal importance to the technology. Additionally, it poses difficult challenges to the new operator yet offers opportunities for people and societies wanting to advance in the new millennium. While developing the required technologies for digital processing payloads and new launch systems is the domain of PhD-level experts, the operation of the resulting satellite systems and networks often can be delegated to qualified technicians with specialized training, both formally and on the job. In fact, the highly skilled technologists who develop the system often are not best suited to operate it on a long-term basis. That is because those experts find more challenge from creating the system than from keeping it in working order. Good operations people, on the other hand, are motivated by meeting the day-to-day operating needs of the system and seeing that users are satisfied. Of course, there is an ongoing need for technologists to support the maintenance and upgrade of the system, and operations people should have a strong input into the design of the system to enhance operational performance and efficiency after services are cut over.

In spite of the fact that the critical hardware and software elements are produced in highly industrialized countries, virtually every country and region will have candidates who can satisfy the educational and background requirements suitable for operations and management staffing purposes. The key to success on the human resources side of the equation is to select the proper staff and provide them with quality training and support during system development.

The initial months and years of service are a period when bugs are worked out of the systems and the people become proficient. In the words of a new satellite system manager in a developing country, "First, the contractor puts the system in and operates it for a trial period—after that, it's our turn." Experience has shown that the staff of new satellite operators are well up to the task, provided the right steps are taken along the way. The guidelines in this section are offered with that in mind.

The structure of the satellite operator organization must follow the technical architecture and resulting demands of the hardware and software. In addition, a satellite operator is usually a business organization as well, which has to be properly considered. We assume in this discussion that the business side has been adequately taken care of, so we concentrate on the technical side of the organization design.

Figure 11.6 is a typical organization chart of a medium-size GEO satellite operator that services one country or region. The following list summarizes the major organizational functions and how they relate to one another. We have identified a total of 10 different activities, which generally fall along lines of particular functions. That generally eases the problem of identifying experienced talent or obtaining appropriate training. In any particular organization,

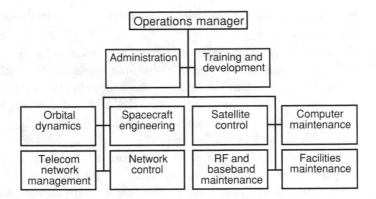

Figure 11.6 The technical organization of a typical GEO satellite operating company.

some of the functions are combined in a common department, based on the particular skill sets available from staff.

1. *Operations management.* Any technical operations structure demands experienced and qualified general management. In the case of a satellite operation organization, it becomes somewhat difficult to identify a senior leadership team with the requisite background. The operations manager must have experience with an organization of this type and be familiar with the local style of doing business. The leadership team will also have the responsibility of building the organization, typically from scratch. The operations manager should work closely with staff, conducting daily meetings to review routine occurrences from the night before and to discuss upcoming events of general importance. Special problems must be addressed quickly and efficiently to prevent their escalation into a system failure.

2. *Satellite control.* This function was outlined at the beginning of the chapter and represents 24-hour-per-day controller activity that ensures safe operation of all operating satellites. The controllers are usually organized into shifts, each under the direction of a shift supervisor. While on duty, a controller spends much of the time behind a computer graphics console that displays all vital facts for one or more satellites (depending on the control and management philosophy). The supervisor generates daily plans and shift assignments and monitors total performance to meet quality and reliability objectives. Unusual data and suspected anomalies must be reported immediately,

in accordance with well-crafted procedures for their evaluation and resolution.

3. *Spacecraft engineering.* Spacecraft engineers are to operating satellites what physicians are to their living human patients. Through basic technical education in electronics, spacecraft systems (as reviewed in Chapter 6, 7, and 8), ground TT&C facilities, and the principles of satellite control, spacecraft engineers are in a position to better learn the particular characteristics and needs of the satellites under their charge. They are called on to review routine spacecraft data obtained through telemetry processing and to identify long-term trends that might indicate incipient trouble. They are available to assist the controllers on duty who detect potential or actual problems with the operating satellites. To aid in that type of investigation, spacecraft engineers must have access to the spacecraft design information from the manufacturer or be able to make contact with technical specialists at the manufacturer or consultants, as appropriate.

4. *Orbital dynamics.* Also called flight dynamics or maneuver planning, this is the rather technical function that determines the precise orbit parameters for each satellite and produces the needed orbit correction maneuvers for GEO stationkeeping or LEO orbit maintenance. The staff members are skilled in the mathematics of orbital mechanics and are capable of using special satellite orbit maintenance software to create good maneuver plans. The object of those plans is to satisfy the position control requirements with satisfactory error and to extend fuel lifetime to the maximum extent feasible using the available technical and human resources.

5. *Network operations.* This function was reviewed in detail in Section 11.3. As with satellite control, network control is a 24-hour-a-day job, because user services generally are available at all times. In a startup situation, this function might be combined with one of the others; however, as activity builds, it becomes a requirement that a dedicated staff be selected, trained, and managed. That ensures that the needs of users, many of whom are paying customers, are addressed quickly by staff who can focus on the demands of achieving a high degree of customer satisfaction. (Complex networks, particularly those using non-GEO constellations, should start with a dedicated network operations staff.)

6. *RF and baseband electronics maintenance.* Since all satellite monitoring and control are exercised through the TT&C system, the ground equipment used for that function must be maintained in full working

condition at all times. The maintenance technicians who do that are familiar with the RF equipment for the uplink and downlink with the satellites and the baseband equipment that provides the modulation/demodulation and encoding/decoding functions. The type of training and experience base may be sufficiently different between the two areas to warrant separate teams. Also, in cases where the RF equipment is part of a remote TT&C Earth station or gateway, the separation of the specialties also will be required. Repair of particular failed subsystems usually is effected by replacing printed circuit boards, modules, or entire units, as appropriate for the specific piece of equipment. Often, the TT&C ground maintenance engineers work hand in hand with the other maintenance experts discussed in the following functions.

7. *Computer hardware and software maintenance.* Both the TT&C station and the SCC rely heavily of computer systems and software to carry out their many functions. There is often a variety of computing machines in each of those facilities, ranging from PCs on desktops, servers, and minicomputers in equipment rooms, and possibly even a mainframe computer to serve the operator's needs for centralized data processing. In some organizations, all the computing resources are put under one manager and his or her group. That promotes standardization and the possibility for efficiencies in buying and maintaining equipment. On the other hand, some organizations prefer to segregate computing and software according to the activities of the particular major group. If that is the case, multiple computer hardware and software maintenance activities will be spread around the satellite operator's organization. There has been a trend since the 1990s to outsource the generic aspects of computing; however, the specialized needs of satellite control often dictate that some software expertise be maintained in-house.

8. *Telecommunications network management.* The telecommunications requirements of the satellite operator are satisfied by the ICN, discussed in Section 11.2. The size and sophistication of the ICN reflect the complexity of the space segment and service networks being managed. In the case where the satellite operator is able to perform the mission using a single combined TT&C station and SCC, the ICN is reduced to providing good access to the public telephone network. That is still a minimum requirement because of the need to interface with customers and other outside entities. As more sites are added, particularly for a non-GEO constellation, the network and its

management become a considerable challenge. Systems of this type demand that a separate telecommunications network management function be established and supported throughout the operating life of the system. Personnel for this area would have received specialized training in telecommunications and computer networks. Technology has been changing rapidly in this field, so it is important that the manager and certain of the lead engineers and technicians continue their educational and research activities. A detailed discussion of network management can be found in Section 11.2.3.

9. *Facilities and mechanical equipment maintenance.* Just as a communications payload needs a properly working bus, an Earth station or SCC is just as dependent on support functions that allow the critical electronic equipment to continue to work properly. These systems include the prime electrical power, derived from the commercial power grid of the country where the facilities are located; backup power in the form of the UPS; heating, ventilation, and air conditioning (HVAC); internal telecommunications (PBX and LAN); water; gas (if required); and physical housing in buildings. Maintenance of those facilities must be performed on a timely basis and staffed to respond immediately to problems. This is one of those areas where, when things are going normally, no one notices or cares. But when an element of the facility systems fails to perform its function, everyone's patience is put to its limit. A simple thing like a heavy downpour of rain or a snow-clogged access road can put the system at risk, so the right human resources with the right tools need to be available, either on site or on call.

10. *Training and development.* The final organizational capability deals with what is the fundamental purpose of this book, that is, to develop and enhance the capability of professionals who are either new to the field of satellite communications or who need to update and extend their knowledge and skill level. Any well-run satellite operation needs to have a permanent training and development function that has the role of providing the training systems and materials and ensuring that they are used properly. Satellite control personnel, maintenance engineers, and other experts must be trained and retrained over time. That can be done in-house with trainers who are on staff, combined with training provided by the various vendors that supply the hardware and software. For a new system, it is common for the satellite or ground station provider to conduct training courses and on-the-job training

both prior to handover and for several months after the first satellite is launched. From that time forward, it would be an internal responsibility of the satellite operator to make sure training and development needs are satisfied.

Our final word on human resources is simply to say that the most important activity in bringing a new system into operation is the proper selection and training of staff. That is not as simple as it sounds because there are many considerations in finding the right people and making sure they are prepared to do their jobs. In a new organization, the structure is never perfect (the same can be said in an established organization), so staff will always find that there are overlaps and gaps. That is normal and to be expected. However, every organization must locate and retain people who are willing and able to pick up the slack and make sure the needs of the system are met. Choosing high-quality people is extremely important; providing them with the right training and tools is just as crucial.

References

[1] Howard, R. E., "Low Cost Little LEO Constellation," *AIAA 16th Internat. Communications Satellite Systems Conf.*, Washington, D.C., February 25–29, 1996, Collection of Technical Papers, American Institute of Aeronautics and Astronautics, New York, paper no. 96-1050, 1996.

[2] Smith, D., "Operations Innovations for the 48-Satellite Globalstar Constellation," *AIAA 16th Internat. Communications Satellite Systems Conf.*, Washington, D.C., February 25–29, 1996, Collection of Technical Papers, American Institute of Aeronautics and Astronautics, New York, paper no. 96-1051, 1996.

[3] Elbert, B. R., *International Telecommunication Management*, Norwood, MA: Artech House, 1990.

[4] Elbert, B. R., and B. Martyna, *Client/Server Computing: Architecture, Applications and Distributed Systems Management*, Norwood, MA: Artech House, 1994.

[5] Aidarous, S., and T. Plevyak, *Telecommunications Network Management—Technologies and Implementations*, IEEE Series on Network Management, New York: IEEE, 1997.

12

The Economics of Satellite Systems

The successful implementation and operation of a system or network depend as much on economics as on technology. Once the development risks have been identified and addressed properly, the economics of satellite communication can be defined using the basic principles of capital budgeting and service development. In the 1990s, geostationary satellite systems reached maturity similar to other industrial fields, including public telecommunications services. DTH and VSAT networks employing bent-pipe repeater satellites are well understood, so it is possible to make accurate predictions of what it will cost to implement and operate a system. The total of GEO satellite service revenues, amounting to $4.5 billion in 1996, has been projected to reach $31 billion in 2002 [1].

Unknown factors in system economics often have to do with new technology that must be developed and introduced for commercial service. The non-GEO constellations to service hand-held mobile subscribers and the coming generation of broadband Ka-band systems pose challenges for both the system architect and the business manager. Another set of unknowns involves the expected amount of traffic the system can carry and the prices users are willing to pay. Furthermore, competition, such as exists in the United States, is an extremely strong factor that can render even the best system implementation concept uneconomic.

A satellite system in its entirety is costly to implement, but the technology provides versatility and flexibility that are potentially greater than that of any other telecommunications technology. The cost of space segment implementation ranges from approximately $200 million for one conventional bent-pipe satellite to more than $10 billion for an advanced digital broadband LEO constellation. A medium-power GEO communications satellite delivers any combination of voice, data, and video services between a potentially large set of Earth stations located within the footprint. A comparable terrestrial network must be implemented on a wide-scale basis, and rapid relocation is virtually impossible. A non-GEO constellation, on the other hand, promotes the concept of ubiquitous global telecommunications service, independent of the terrestrial infrastructure. For many payload designs, communications satellites are attractive from an economic standpoint because the particular mix of services can be altered rapidly and the value of the investment is not lost due to changing demand. A satellite that has been launched will see its use increase over time even to the end of orbital life, when it must be retired from service. A classic example is the Marisat satellite series, which was the first to provide commercial MSS communications. Launched in 1976, Marisat continued to provide revenue-producing services for over 20 years. One of the satellites was still in service at the time of this writing.

The communications Earth stations employed to deliver services cover a wide range of applications and costs. Large Earth stations used as broadcast centers and telecommunications gateways have long economic lives, since it is possible to expand, modify, or even replace the specific equipment used to provide new capabilities. Low-cost end-user terminals, particularly DTH receivers, VSATs, and MSS mobile phones might have relatively short lives of two to five years. Part of the reason for that short economic life is the trend toward obsolescence as newer satellite services are introduced. The advent of microelectronics has greatly increased the versatility of ground equipment at the same time that unit costs have decreased. Also, higher power satellites permit a significant reduction in ground antenna size and RF electronics costs. Those mitigating factors encourage the replacement of functional but obsolete Earth station equipment.

The approach that follows is meant as a framework for understanding the underlying economics of a satellite system. It should be possible to use this framework to create a specific economic model of any system so the detailed costs can be determined. As a rule, the economics of the system go hand in hand with the functions it is required to perform. Typical studies of system economics consider the recovery of the original investment by including the time value of money and computing equivalent annual payments. The investment costs are embodied in the block diagrams of the elements and for the system as a whole. The annual costs usually are more difficult to assess, involving labor and other expenses for managing, marketing, and operating the system, and financial matters such as the cost of money and taxes. In large, complex networks, costs and therefore revenues must be split among the space segment operator, the Earth station operator, and the ultimate service provider or distributor. That can be exceedingly difficult to assess in advance of actually creating the required business structure. Economic studies are used to size the business and to attempt to convince investors (internal or external) that a satellite system is an attractive investment.

12.1 System Development Methodology

The basic approach for evaluating system economics is to view the system as an investment. When in operation, the system produces services that can be sold to customers. That generates revenues that, hopefully, will recover all costs and ultimately yield profit to reward the investors. Figure 12.1 presents a simplified flow chart that identifies the various stages of system implementation and operation. Financial inputs are required for each aspect of system evolution, that is, investment to create the system and service revenues from users (e.g., customers,

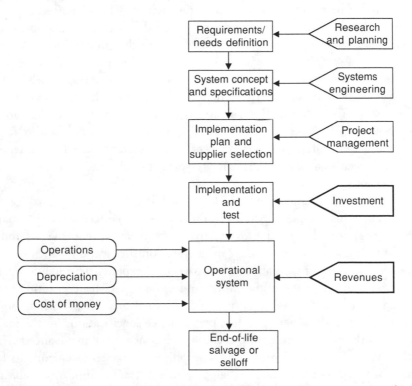

Figure 12.1 Flow chart illustrating the economics of implementation and operation of a commercial communications satellite system.

subscribers). The objective is to end up with an operational system that is self-sustaining in an economic sense. This same principle applies to other investments in the telecommunications fields, including cellular telephone, cable TV, and undersea fiber optic cables. The unique aspect of satellite communication is the implicit risk associated with launching and operating spacecraft in orbit. Risks to the ground segment exist as well, but there is a perception that the risks are addressed in the routine course of running any terrestrial business.

Beginning at the top of Figure 12.1, the first step in the planning process is to determine the service requirements or needs the system is to satisfy. Research into the needs of potential users is performed over some period of time either by investigating historical patterns of telecommunications usage (satellite and terrestrial) or by conducting surveys among the prospective users. Such a project represents a one-time cost, being the first fixed expense in system development. It can be considered part of the investment in the system or it can come under the normal operating budget of the private company or government agency doing the research. Developing requirements can be costly because data must be collected for each market segment and geographic region to be served. The most costly is primary research, in which one actually goes out into the field and gathers data directly from the targeted user population. Secondary research is considerably cheaper and quicker to conduct because it amounts to assembling already available data from published sources. It includes reports from consulting or market research companies, as well as data published by government and international bodies like the U.S. Commerce Department, the National Association of Broadcasters, the European Commision, the World Bank, or the ITU. Most market research studies begin with secondary research and move on to primary research only when initial feasibility has been established and there are some special (and extremely important) questions to be answered.

With the requirements more or less quantified, the next step is to use the process of systems engineering to determine the system concept. This is a major planning project and is best performed by an organization that is very familiar with the technology and economics of satellite communications. Such organizations include large manufacturers of spacecraft, experienced satellite operators, major technology consulting firms, and certain specialist consulting firms with experience on these types of projects. The final product of such an effort is a system design document that lists performance specifications for the major elements of the system. For example, the system concept for an advanced GEO satellite network would identify the geographic coverage requirements, its frequency band(s), repeater design (analog or digital), communication capacity, and other specific characteristics. Part of the design would be to determine, through appropriate tradeoff studies, the modulation and multiple access methods to be employed for each service. The locations and service capabilities of Earth stations also would be identified.

One of the most difficult aspects of determining the system concept is to know what technologies are available and appropriate. That information comes from the resources of the system designer or can be learned through exposure to the marketplace. It is common to issue a request for information to relevant developers and manufacturers so they can suggest design methodologies and solutions. Besides the technical aspects of that effort, the system designer needs to prepare estimates of the investment and operating costs. A thorough systems engineering study includes tradeoff analyses of various alternatives, yielding an approach that is nearly optimum for the particular set of user requirements. Another important aspect of the process is the consideration of how the system can be expanded or modified after it has gone into operation. That can have a decided effect on the approach taken for the initial system design.

The design is not finished, however, until the requirements/needs definition is revisited to verify the original assumptions. Done properly, the top two boxes in Figure 12.1 represent an iterative process, seeking the best technical and business solution. One result could be that one or more of the original services is discarded in favor of another more viable market opportunity. Likewise, a technology solution, which appeared attractive on first examination might have an excessive business risk associated with it.

Perhaps the most dynamic phase of the process is system implementation, during which most of the investment is made. Before construction begins, the functional requirements from the system concept are turned into detailed hardware specifications. With respect to major Earth stations, locations around the service area are firmly established with consideration given to the footprint of the satellite antenna. Contractors and manufacturers are selected and schedules set for construction of all the elements of the system. The system operator can perform the role of system implementation manager, but in many instances another experienced system integrator is selected. Depending on the types of facilities involved and whether a spacecraft must be constructed, system implementation can take anywhere from one to five years. Essentially all the expenses incurred during implementation become the investment in the system itself and are recoverable mainly through collecting revenues during its useful life. Alternatively, presale of satellite and Earth station capacity to one or more large users can provide a needed financial boost to offset some of the capital and initial operating costs.

After the system is implemented, the major space and ground elements are tested to verify performance and functionality. Then the operational phase begins and lasts as long as useful services can be provided. The economic input to the operating system usually is in the form of service revenues from users, as indicated at the right of Figure 12.1. Revenues can be collected monthly, yearly, or in a lump sum before service is even rendered. Examples are the monthly telephone bill that most of us pay and the up-front payment for a transponder on a condominium satellite. The system must be operated and maintained by ground personnel over its useful life, indicated at the left of the figure. Operations expenses cover the labor resources discussed in Chapter 11. Depreciation is a charge against revenue to allow for the fixed lifetime of the space segment and ground assets. For example, a 15-year GEO satellite depreciates at the rate of 1/15th its initial investment cost per year (in actuality, a much shorter economic lifetime is assumed to allow the operator to recover costs faster). The third charge, finance, is simply the cost of money, amounting to the effective interest that might be paid back to lenders. On top of that is the profit that would result when all costs are paid, delivering value for the owners/investors.

Additional investment would be made over the life of the system to add new users and to provide expanded services, in terms of both quantity and type. Any proposal for new facilities would go through the entire process described above, although on a smaller scale. Repair of existing facilities associated with current services comes under the expense of operating the system and would not be a new investment. When the system or an element has reached the end of its useful life, due to wearout or to obsolescence, it is disposed of. If it can be salvaged or sold, some additional recovery of investment money is possible. Otherwise, it is disposed of in a manner that cannot interfere with continued operations or future activity.

A business plan is a document that delineates all the information discussed in this chapter to this point. The following is a brief list of typical major sections of a business plan that might be used to describe a proposed satellite communication project:

- Executive summary;
- Basic service proposition (services to be offered);
- Satellite system description;
- User terminal equipment design, manufacture, and distribution;
- Spectrum, orbit assignment, and regulatory matters;
- Expected usage patterns and service take-up (market research);
- Competitive analysis;
- Marketing and sales plan;
- Financial model;
- Source of investment funds;

- Financial results and sensitivity to critical parameters;
- Risks and their management;
- Organization structure and implementation plan.

The breath of the list indicates the wide range of the type of information that a good business plan must contain. Of course, there would be additional descriptive information to help potential investors understand every aspect of the system and the business. Most important is how investors' money will be protected, increased, and returned. Good business plans are not easy to assemble. They consist of credible and consistent market research, development plans, financial analyses, and a clear business structure. Often, the people behind the business are the ones who are in a position to assure potential investors that the proposed services can be delivered in a profitable way. To sum it up, the definition of a good business plan is one that can be financed.

This has been a brief and generic discussion of the evolution and economics of a system used to provide telecommunications services. More information is provided in our other work [2]. The size of this type of planning effort should not be underestimated; it is the responsibility of the system developer to do adequate homework. In the rest of this chapter, the particular aspects of satellite communications are discussed using that framework.

12.2 Space Segment Economics

The economics of implementing and operating the space segment of a satellite system are dominated by the initial cost of manufacturing the spacecraft, providing the launch into orbit, and insuring the combined value against failure. To that is added a much smaller, but significant, amount to implement the required TT&C ground facilities. Figure 12.2 reviews the main elements of a space segment, including initial investment as well as operating costs. A spacecraft manufacturer typically enters into a fixed-price contract to design, manufacture, and deliver the required quantity of communications satellites. The exception is when the developer of the service is also the manufacturer of the system hardware, which has been the case for several non-GEO constellations. A separate contract usually is required with the launch agency, which in turn pays for the LV hardware and then provides the integration and launch direction services at the launch site. Some customers have included the LV services along with the spacecraft in what is called a delivery-in-orbit contract.

Following injection into the desired orbit, either the satellite operator or the spacecraft manufacturer takes over tracking and control of the satellite,

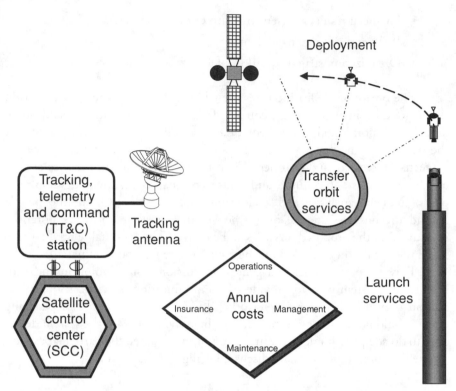

Figure 12.2 Major elements of the investment and operations expense of the space segment of a commercial satellite system.

performing the services indicated by the circle in Figure 12.2. Non-GEO spacecraft may be designed to configure themselves when released from the LV, thus reducing the demand for ground-based tracking. LV and tracking services normally are considered part of the investment cost of the system. Prior to the start of revenue-producing service, a GEO satellite is drifted into its final position, configured for operation and tested to ensure that the launch environment has not damaged any subsystems. An ever increasing element of space segment cost is the launch insurance, which provides coverage for a possible launch or injection failure. Premiums for such insurance have become extremely costly, running 12–20% of the cost of the spacecraft, LV, and insurance. That is roughly equivalent to holding a financial reserve based on one failure in five to one in eight attempts, which is conservative in comparison to actual experience. However, insurance has become nearly essential because of financial commitments to other parties, including public and private stockholders.

The TT&C ground facilities (shown at the left of Figure 12.2) are part of the space segment investment, to be continuously maintained during the life of the satellites. Due to the close tie to the particular satellite design, it is a common practice to procure most of the electronic equipment and software from the spacecraft manufacturer. More recently, some specialist technology companies have developed generic satellite control equipment and software to provide an element of competition into what has been a very closed market. In either case, the manufacturer must demonstrate prior to launch that all the ground and space components will work together properly.

The annual costs associated with operating the space segment, indicated in the diamond-shaped element in Figure 12.2, include expenses for the personnel who perform the technical and administrative functions at the satellite control center and the TT&C station. The cost of training the technical staff is high, and many personnel are required to be available around the clock. Fortunately, only a limited number of facilities are needed to operate a space segment consisting of several communications satellites. A global constellation also can be managed with limited TT&C ground assets, since continuous contact is neither feasible nor necessary. Because of centralization of the SCC, the efficiency of using such highly trained and specialized personnel is excellent and probably much greater than would be expected in most other capital-intensive activities.

The expenses of maintaining the ground facilities are usually fairly small due to their compact nature. The people who perform that task (introduced in Chapter 11) tend to be quite resourceful and versatile, which is probably the result of having to perform a wide variety of tasks on critically operating equipment. Nevertheless, an ongoing relationship with the spacecraft and ground equipment suppliers is essential to be able to maintain and expand the system over time. Perhaps the most difficult task in a commercial environment is the marketing of the services on the system. That can involve a large staff of sales engineers and service representatives, organized in teams to cover different user groups and applications.

Another annual expense is that of on-orbit insurance. This type of coverage, introduced in the late 1970s, allowed satellite operators to address part of the risk of loss of a working satellite. Later, purchasers of transponders needed this insurance coverage because it was demanded by the banks that provided the funds necessary to purchase the transponders in the first place. It could be argued that on-orbit insurance (also called life insurance) provides little benefit because there have been few significant commercial satellite losses once successful placement into service has been achieved. On the other hand, the financial loss is great in the event of a total failure of a satellite, which could justify the coverage. On-orbit insurance works like depreciation, paying an amount that approximately equals the remaining value of the satellite. A failure that occurs in the first year (but after the launch insurance coverage has run out) would result in payment of nearly the full replacement cost of the satellite (less inflation, unless that had been considered). The payment for failure at the end of life is, by definition, zero. The insured value in the intervening years is found on the line connecting those end points.

The relative magnitude of each major cost element for a two-satellite system is shown in the pie chart in Figure 12.3. Annual costs have been considered by taking the simple sum of such costs over an eight-year period, which is one-half the actual life of an assumed space segment. That takes account of the time value of money (money spent several years in the future is worth less than money spent in the current year). The most costly single element is for spacecraft, representing approximately one-third the total. The sum of the cost of launch and launch insurance is comparable. Investment in ground facilities for the TT&C station and SCC is relatively small (2%). Operating expense over the eight-year period takes another relatively large chunk, approximately one-third the total expense.

12.2.1 Space Segment Investment Cost Elements

As discussed at the beginning of this chapter, the investment cost elements of the space segment are purchased prior to the initiation of operation and are

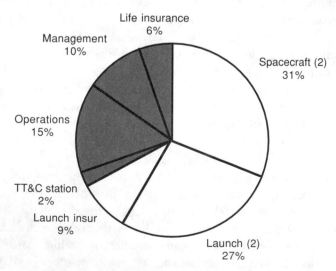

Figure 12.3 Allocation of the cost of a space segment composed of two satellites, including the initial investment and eight years of operations support.

intended to last through the end of life of the satellites. In larger GEO systems and constellations of non-GEO satellites, there is a need for methodical replenishment of spacecraft when they reach end of life. That places on the satellite operator the burden of maintaining a new flow of investment in spacecraft and LV services for as long as the business can be sustained. Replacement GEO spacecraft usually are quite different from their predecessors due to enhancements in coverage, power level, and even frequency bands. The elements to be included in a basic GEO space segment are indicated in Figure 12.3 by the shaded areas and are discussed next.

12.2.1.1 Spacecraft

A communications spacecraft is designed and built as an integrated system, as presented in Chapters 6 through 8. On delivery to the launch site, it is capable of being placed into orbit and operated for its useful life without direct servicing. Any repairs or adjustments must be made by ground command and with test and redundancy capabilities that are built in. Once a satellite is placed into orbit, it generally cannot be removed from service or stored for later use. There is an exception for a GEO satellite that has been placed into an inclined orbit. If the inclination is positive, meaning it will decrease to zero at a rate of 0.8 degree per year, the operating life can be delayed. Once a satellite is actually placed into service, its cost is completely "sunk," meaning it cannot be recovered except by selling services to users or by selling the entire satellite to another operator. The latter applies only to GEO satellites because non-GEO designs are highly customized for the particular constellation. Provided that the footprint and frequency plans are suitable, a GEO satellite can be repointed or moved to a new longitude to provide service in a different manner or to a different region.

It should be evident from these discussions that a communications spacecraft is a complex and therefore costly item. As reviewed in Chapter 2, the historical trend has been from small spacecraft with few transponders to larger spacecraft with many transponders, multiple frequency bands of operation, and digital processing. The power levels of the individual transponders have increased over the years. Even ignoring inflation, the cost of a spacecraft has continued to increase in line with spacecraft weight and complexity. Communications payload weight increases because the quantity of repeater electronics and the size of the antennas (as well as their number) have increased. Providing more payload power expands the quantity of solar cells and batteries; that, along with the enlarged propulsion system and fuel supply, increases the mass of the spacecraft bus. The weight of the structure also has to increase to provide the volume and physical strength to support the expanded payload and bus.

As a rule, it is possible to view the cost of a spacecraft as being directly related to its mass; the heavier the spacecraft, the more it will cost. Other factors have a direct bearing on cost as well, including the number of components, the complexity of the payload, and the maturity of the design. The relationship is perhaps not linear so that there is an economy of scale (i.e., a decrease in the relative cost in terms of some monetary unit such as dollars per kilogram) as one moves toward larger spacecraft with more communications capability. Components of a spacecraft such as battery cells, solar cells, and electronic amplifiers can be manufactured with modern automation techniques, which tends to reduce cost. However, larger elements of the spacecraft typically are custom made by highly skilled technicians, whose time expended tends to increase as complexity increases. That aspect is difficult to change because all work must be of the highest quality to ensure proper operation during orbital lifetime.

An area in which automation has played an important role is in the testing of subsystems and the entire spacecraft. Digital computers programmed to control modern test equipment can perform the testing during each phase of manufacture and at the launch site. That increases the accuracy and the amount of data and can decrease the number of personnel on test teams. The primary benefit of using this form of automation, however, is the improvement in reliability that comes from being able to spot potential problems before they occur. Once on orbit, this database becomes very useful during initial testing to see if there has been any detrimental change after launch and to aid in troubleshooting during the life of the satellite.

12.2.1.2 Launch and Transfer Stage

As evidenced by Figure 12.3, the fixed cost of placing a GEO spacecraft in orbit is comparable to the cost of the spacecraft itself. Non-GEO systems have the advantage that multiple spacecraft can be deployed in a given launch event. That reduces the launch cost per spacecraft by the same factor (although there is a requirement to launch correspondingly more satellites to complete a particular constellation). A satellite operator does not purchase LV hardware but rather purchases the services of a launching agency that is then responsible for the manufacture and performance of the rocket system. Whether the transfer stage is part of the launch service depends on the particular LV and mission, as discussed in Chapter 10. An LV such as Ariane 5 or Atlas 2AS injects the spacecraft into GTO, from which a normal AKM firing is all that is required to reach geosynchronous orbit. Direct injection into geosynchronous orbit is available with the Proton rocket, eliminating both the transfer stage and the apogee kick motor.

Some additional expenses associated with the launch mission involve the support services at the launch site and, if appropriate, during the transfer and drift orbit phases. The spacecraft contractor can bundle the launch site (launch prep) expenses into the price of the spacecraft as may the launch agency. Transfer orbit services are fixed and independent of the particular satellite design and can be purchased from another satellite operator. The spacecraft contractor normally directs the timing of apogee boost, drift orbit, and initial checkout activities and turns the tested satellite over to the satellite operator. However, the communications payload is tested by the satellite operator to determine the fitness for service. The costs of all those services, while essential, are relatively low compared to the investment for the spacecraft and the LV.

12.2.1.3 Launch Insurance

Launch insurance has been an important element of satellite system implementation, particularly when the system is operated as a commercial business and requires outside financing. The general trend toward higher insurance premiums for high-risk kinds of coverage (not just for satellites) also has contributed to the cost. The risks associated with launch have always been very high, but a good commercial launch record between 1970 and 1980 made the risks appear to be diminishing. Consequently, launch insurance premiums during that decade were low enough to encourage satellite operators to insure not only their investment but expected revenues as well. That removed most of the business risk from the satellite operator and placed it in the hands of the major insurance underwriters, notably Lloyds of London. With Lloyds now largely out of the market, satellite operators have discovered a new breed of launch insurance underwriters who better understand the nature of the game.

A recent trend is for satellite operators to make partial claims when a satellite can provide only a useful fraction of its design capability.

12.2.1.4 TT&C Ground Facilities

As discussed at the beginning of this chapter, the investment cost of TT&C ground facilities is tied primarily to the equipment block diagram and facilities layout. To that must be added the fixed cost of providing the needed telecommunications equipment associated with the ICN. It is a fairly straightforward matter to identify all the equipment necessary to perform the TT&C functions as well as those associated with system management. The buildings are designed so that the equipment is properly housed and environmentally controlled. It is also important that human factors be taken into account, since operations are conducted around the clock throughout the life of the satellites.

The relative cost of implementing the ground facilities (indicated in Figure 12.3) is small in comparison with that of any other element. Designing and implementing a proper ground environment, therefore, is a wise measure, even considering the extra cost in doing the job right. Placement of equipment for convenient access and the provision of well-designed operator consoles and support software tend to reduce personnel costs. It also results in more reliable operation of the satellites themselves—the proper resolution of problems depends on the capability of the people who staff the SCC and TT&C station.

Particular attention should be focused on the computer system used to monitor and control the satellites, perform orbital and engineering analyses, and operate other parts of the TT&C facilities. Having excess computing power and backup computer equipment almost always will prove to be of significant value during the life of the system.

The personnel and the equipment within the facility require an internal telecommunications network to allow the technical staff to speak to each other and to exchange data over a LAN. In modern facilities, those are part of a cabling system (preferably fiber based) and an IFL between buildings and antennas. The reasons for using fiber rather than coaxial cable or twisted-pair cable are (1) it virtually eliminates the possibility of conducted mutual interference; (2) it provides ample bandwidth for expansion of services and modifications for additions; and (3) it is more secure because of the greater difficulty to tap the cable. Another area is in the monitoring of the transponders and signals within them. That can be important in providing services to users, who rely on the SCC to control access to the communications payload and to resolve interference problems as quickly as possible. Costs of transponder monitoring systems vary widely due to the disparity in capabilities, including whether the receiving antennas (one per satellite to be monitored) are local or remote. Another cost factor is the degree of automation of the testing functions, something that can improve the efficiency of aligning new Earth stations.

The location cost of TT&C facilities can be substantial, depending on the particular country and closeness to urban areas. Real estate costs can be very high in attractive, built-up areas. For that reason, some satellite operators tend to place their ground operations in remote areas. That also reduces the problems associated with RF radiation and interference. On the other hand, the people who work at the site may or may not find rural living acceptable. There are issues concerning adequate utility services, access roads, and conveniently located communities where employees can live or, at least, find a place where they can eat. In extreme cases, the satellite operator provides bus or van transportation to help the employees deal with remoteness.

12.2.2 Annual Space Segment Costs

This section reviews the main annual costs of operating the space segment. In addition, the financing of the investment in the space segment usually results in an annual expense to the satellite operator for interest, depreciation, and return on investment. Because of the wide variety of ways for financing a system and the different circumstances that satellite operators are under, specific financing strategies and costs are not covered.

12.2.2.1 Satellite Operations

The performance of satellite control and network management is by the team at the TT&C station and the SCC. As discussed in Chapter 11, those personnel include engineers, computer scientists, and technicians, many of whom hold advanced degrees. Specific training in specialties such as spacecraft engineering, orbital dynamics, and communications monitoring often is provided by the spacecraft manufacture because the knowledge relates to the particular satellite and TT&C system design. The operations personnel who perform the ongoing tasks at the spacecraft and communications consoles need to be on duty 24 hours a day, seven days a week. As shown in Figure 12.3, the expense for that labor is substantial, amounting to approximately 15% of the total system cost, assuming an eight-year period. Because the quality and experience of the staff are vital to the successful operation of the space segment and the quality of services rendered, it would not be wise to economize in that area.

12.2.2.2 Equipment Operations and Maintenance

The ground equipment and facilities, which allow the operations personnel to perform their tasks, must be maintained properly during the useful lifetime. Those expenses, while significant, are relatively small in comparison to the annual operations costs reviewed in Section 12.2.1. Allowance is made for the utilities and maintenance on the buildings housing the SCC and TT&C stations. The electronic and computer equipment usually is fairly costly to maintain due to its complexity. There is a tradeoff between hiring a full-time staff to perform electronic maintenance versus contracting for such services with the original suppliers of the equipment or other maintenance specialists. Even if outside services are used, a skeleton crew of electronic technicians is still needed to respond to emergency situations at the SCC and TT&C station.

12.2.2.3 In-Orbit Insurance

The need for in-orbit insurance coverage depends very much on the circumstances of the particular satellite operator. For example, life insurance proceeds would compensate the satellite operator for lost revenues in the event of a loss of satellite capability. In the absence of such a need and if the satellite operator has sufficient on-orbit capacity to deal with any contingency, then life insurance might not be required. The significant amount shown in Figure 12.3 is based on the assumption that the insurance is to cover the cost of replacing the satellites in orbit (i.e., including the cost of the LVs).

12.2.2.4 Management and Marketing

The nonoperations functions of management and marketing might be viewed as unnecessary overhead. From an economic standpoint, however, a space segment cannot become self-supporting unless those critical functions are provided. Management organizes the various personnel resources of the system and arranges for hiring, training, and administrative support. In the case of a commercial system, management is responsible for operating the system at a profit in the face of internal technical and administrative problems and external difficulties with suppliers, customers, and competitors.

The role of marketing is to provide a continuing flow of new customers and revenues, without which the transponders onboard the satellites probably would remain empty. In that context, the sales function (i.e., locating prospective customers and executing service agreements with them) is part of marketing. GEO satellite systems have, over the years, built their businesses on meeting the demand of a limited market. It is not uncommon for a satellite operator to obtain more than half its annual revenue from just two or three customers. Building a business further involves introducing new and often innovative ways to divide the space segment, making it more obtainable by a larger customer pool. Going from three customers to 50 to 100 means accepting lesser amounts of revenue per customer. In MSS and DTH networks, the customer quantities are measured in millions, so the means for selling to and servicing such a large customer community represent a completely different paradigm. Companies usually address that situation by forming a new company and populating it with experts in the new class of service and consumer mass marketing. That requires a completely new look at the economic model of the project, along the lines of the business plan.

New applications for the space segment should be examined continually by the marketing staff so services can be adapted to changing user needs. Marketing usually is responsible for estimating the future loading of the current space segment and for the preparation of plans for replacement satellites. Another related task is preparing the applications for new orbit slots and working with the regulatory authority to protect existing slots, through the process of frequency coordination. That could involve a proactive role in domestic regulatory affairs and participation in ITU WRCs. For a regional or international system, staff become involved with the challenge of gaining access to different national markets and building relationships that lead to new business.

12.3 Earth Station Economics

The ideal way to evaluate the economics of an Earth station is to identify each major element and relate its cost to technical performance. Such a detailed approach gives assurance of realism in determining network cost even if the implementing organization purchases complete stations from a contractor. Purchase of stations on a turn-key basis means that the contractor designs, installs, and tests the complete station and simply turns the door keys over to the buyer. That approach can be taken on an entire network, a single station, or one or more major elements of a station. An alternative approach that puts more control in the hands of the system operator is to build up an internal staff of engineers and installation people who can procure the individual elements and then perform the integration themselves. There is a cross-over, above which the fixed cost of maintaining such a group can be justified because of the more stable workload placed on them. Below that cross-over, there simply is not enough work to keep the group sufficiently busy with useful technical assignments and project work.

A proper network design analysis and optimization must consider the relationship between the performance of the Earth stations and that of the space segment. This brings into play the requirements of the microwave links to and from the satellite and the technical parameters of modulation and multiple access, reviewed in Chapters 4 and 5, respectively. For an existing GEO satellite, the price of using the space segment could be determined from publicly filed tariffs or by contacting the marketing staff of the satellite operator. The analysis, therefore, is simplified because the satellite technical parameters and associated costs are givens. The case in which a new satellite is to be designed as part of the exercise is beneficial because the characteristics of the satellite can vary along with those of the ground stations. As discussed at the end of this chapter, there are useful tradeoffs that can help define an optimum space segment design that provides the desired capability for the least cost.

12.3.1 Earth Station Investment Cost Elements

The major elements of a communications station were reviewed from a technical standpoint in Chapter 9 and are illustrated in block diagram form in Figure 12.4. The RF terminal provides the actual uplink and downlink paths to the satellite, while the baseband equipment arranges the information (video, voice, and data channels) for efficient access. Interfacing of services with users is by way of the terrestrial interface and tail (if required). Indicated at the bottom of the figure are categories of expenses for monitor and control (e.g., network management), power and utilities, as well as annual expenses for operations and maintenance. Economic aspects of each of those sections are reviewed next.

12.3.1.1 RF Terminal

The two essential characteristics of the RF terminal are the transmit EIRP and the receive figure of merit (G/T). Besides being the leading parameters in the

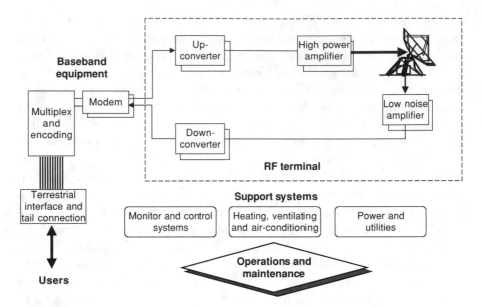

Figure 12.4   Typical Earth station configuration that identifies the major elements of investment and operations expense.

uplink and downlink, respectively, EIRP and G/T have a direct impact on the investment cost of the Earth station. In Figure 9.9, it is shown that the power output of the HPA and the gain of the transmitting antenna combine with each other to produce the specified value of EIRP. Figure 9.4 presents receive G/T as a function of the antenna size and LNA noise temperature.

The relationship of performance to cost of those elements is highly non-linear and should be considered on a case-by-case basis using the most recent information from equipment suppliers. For example, the investment cost of an antenna increases at an increasing rate with diameter. Doubling the diameter from 3m to 6m increases cost not by a factor of 2 but by more like a factor of 6. A Ku-band power increase from 50W to 100W (within the typical range for SSPAs) may increase cost only by a factor of 2. But increasing from 100W to 200W (requiring a TWTA instead of an SSPA) could increase cost by a factor of 4. The use of a small antenna (with a relatively low value of gain) could demand a relatively high HPA power output. On the other hand, increasing the size of the antenna permits the use of a lower powered HPA at reduced cost. There typically is an optimum point where the sum of HPA cost and antenna cost is minimum. If that point occurs under the condition of a relatively large antenna, there will be the benefit of improved receive gain and G/T as well. A nonoptimum combination with a small antenna may be forced by the need to minimize the physical size of the installation due to local zoning restrictions.

The upconverter and downconverter are essentially fixed cost items not dependent on the RF performance of the station. As discussed in Chapter 9, some of the investment expense can be reduced by obtaining integrated electronic units called exciters and receivers, which combine the functions of modulation with upconversion and downconversion with demodulation, respectively. The availability of such multifunctional elements depends on the service or specific user application. While a reduction in investment cost is possible, the combining of elements makes it more difficult to reconfigure those parts of the Earth station in response to changing requirements.

A major factor in the cost of the RF terminal is the quantity of redundant electronic equipment and the degree of automation with which it can be controlled. In terms of overall network economics, the ultimate cost of outage of a critical service could far outweigh the initial cost of providing sufficient redundancy. Ignoring the antenna, a full set of redundancy (i.e., 100% or two-for-one backup) more than doubles the investment cost in electronics. That usually is justified in a major Earth station serving as a video uplink or hub in a network. A VSAT or TVRO would have a single string of RF components (without redundancy) under the assumption that failures are less likely with such low-power solid state electronics, coupled with the likelihood that alternative means of communication would be available in case of electronic failure.

12.3.1.2 Baseband Equipment

The modems, codecs, and compression and multiplex equipment that make up the baseband section of an Earth station often represent a significant portion of the investment and operation expense. Modern stations employ digital processing, notably TDMA and packet data, providing significant flexibility in the manner in which the station is used. On the other hand, digital technology is constantly evolving, and few systems have been standardized (a notable exception is the DVB standard, used for DTH television broadcasting). The result is that annual expenses for baseband equipment are second only to RF. Trained maintenance personnel must be close by to address technical problems when they occur. An Earth station may be designed in such a way that the RF terminal can be retained even if the baseband equipment is changed out. Also, baseband equipment, like computing servers and other digital machines, tends to be sensitive to environmental conditions like high temperature and humidity. In VSATs or TVROs, on the other hand, part of the baseband portion may even be integrated into the RF terminal.

12.3.1.3 Terrestrial Interface and Tail

The cost of the tail is driven by the bandwidth characteristics and the distance covered. Other important factors include the type of technology (e.g., fiber optic or microwave radio), accessibility of installation sites, and cost of land or right-of-way. Minimum cost goes along with tying the station directly to the user, which is the concept behind the use of VSATs for private networks and TVROs for DTH service. On the other hand, a major Earth station can serve several users or user locations, requiring that local access or tail facilities be included. If an extended distance is involved, the choice must be made between renting the tail link from the local telephone company or installing a private link or bypass. An example of this type of trade off is presented in Figure 9.14.

12.3.1.4 Monitor and Control

The traditional method of monitoring network performance is to rely on users to complain when they experience problems. While that simple approach may have worked in the early days of satellite communications, the expectations of today's users cannot be met without continuous monitoring and extensive maintenance capabilities in the hands of the satellite operator or network service organization.

Having an automated system of monitor and control reduces or eliminates the need for on-site maintenance or operations personnel. That clearly is preferred in a large network where labor costs could be extremely high. Such a specialized network management capability raises the initial investment in the network. If not incorporated in the beginning, an automated capability is extremely costly to add because the electronic equipment at remote sites might not have been designed for remote control and monitoring. Personnel at the centralized control site need to be trained in network management. They should be able to dispatch resources to expedite corrective actions once problems have been uncovered and isolated to a particular location or piece of equipment.

12.3.2 Annual Costs for Earth Station Networks

Two areas of annual expenses for Earth stations are shown in Figure 12.4. *Power and utilities* refers to the cost of operating the building that houses the Earth station electronics and personnel (if appropriate). That particular expense is significant in the case of a major Earth station. The other annual cost activity is *operations and maintenance* (O&M). Reliability of solid state electronics continues to improve even with the trend toward greater complexity. Incorporating the M&C capabilities often provides the means to detect and deal with problems before they affect the network services.

While the cost of O&M for a single station may be manageable, the issue of supporting a nationwide network of hundreds or even thousands of remote sites, major uplinks, hubs, and so on, can be particularly challenging. For that reason, the services of an existing maintenance organization are attractive, provided the service personnel can be trained to maintain the type of electronic equipment employed. Providing sufficient redundant units at the remote sites could eliminate the need for a widespread service capability, relying instead on returning failed units to the manufacturer or a centralized depot.

12.3.3 Teleport Earth Stations

Earth stations to uplink video or transmit point-to-point digital wideband links can be rented on a full-time or part-time basis. That basically is the concept of a teleport, which is a telecommunications facility intended to serve multiple occupants with Earth station antennas and associated indoor electronics. According to the World Teleport Association (New York), there are more than 100 operating teleports in the world. Many of them are owned and operated by existing satellite communications service providers and international telephone carriers, often designed to provide access to the INTELSAT system. A teleport can be managed as an independent business, offering a variety of transmission services. Alternatively, the teleport is an industrial development that sells or rents floor space in a building to operators that find the arrangement attractive. As many readers can tell, there is no precise definition for a teleport, yet the term and the business arrangement have persisted for many years.

Perhaps the biggest teleport in the world is that operated by Hong Kong Telecom at Stanley on Hong Kong Island (Figure 12.5). The primary purpose of Stanley is to connect Hong Kong to the rest of the world through INTELSAT and Inmarsat links. The teleport serves many types of users who contract with Hong Kong Telecom for one-time events and long-term services as well. One particular tenant is AsiaSat, which uses Stanley for its primary TT&C Earth station site. Another type of teleport is PanAmSat's Fillmore Earth Station, located northwest of Los Angeles. Fillmore was originally constructed in 1982 to be the backup C-band TT&C station for the Galaxy Satellite System, formerly owned and operated by Hughes Communications. Over the years, dedicated communications antennas were added to the site to provide a variety of video and data transmission services in the United States and across the Pacific.

A teleport service represents the "buy" alternative of the make-or-buy decision. Basically, a user of satellite capacity has the choice to construct and managing a dedicated Earth station or to arrange for capacity from a teleport

Figure 12.5   Hong Kong Telecom's teleport facilities. (*Source:* Courtesy of Hong Kong Telecom.)

owner/operator. The option of using a teleport may or may not exist—it depends on whether there is an available teleport that is suitable to the purpose. A user, such as a cable TV network, may decide to locate its studio and uplink in a particular city with an excellent teleport. Evaluating the economics of the choices is a complex matter because the user must consider every significant cost, both initial and annual. The ideal situation is one in which the teleport operator has most or all of the necessary equipment and is willing to make an attractive deal at or near of the marginal cost of serving the particular user.

12.4   Analysis of Network Economics

The overall economics of a satellite network can be analyzed by building as complete a financial model as possible and evaluating the costs as a function of various performance parameters. That can use the techniques of system analysis and financial modeling, aided by computerized tools. It is particularly convenient to develop the model on a personal computer using conventional spreadsheet software. The model can be as simple or as complicated as desired. Once the model has been entered into the computer, it is relatively simple to change some aspect of the network and observe the results. Sensitivity studies are useful to identify the most critical parameters. More sophisticated models can include technical parameters, such as EIRP or G/T of the Earth stations, transforming them into element costs.

A basic financial model uses simple arithmetic to estimate system investment cost and annual expense for a specific configuration. More complex modeling incorporates mathematical search and optimization algorithms that look at many possible network arrangements for satisfying a variety of traffic requirements. A particular network design then is selected for detailed analysis and specification, as outlined in Table 12.1. The steps involve gathering traffic data, formulating network models, and evaluating alternatives for network implementation.

Table 12.1
Outline of an Analysis Procedure for Determining the Economic
Performance of a Satellite Network

I. Determine traffic requirements

  A. Collect current usage data

  B. Map the network

  C. Determine channel requirements

    a. Video

    b. Voice

    c. Data

II. Model Earth station elements

  A. Determine station categories

    a. Major Earth station

    b. VSAT

    c. TVRO

  B. Formulate Earth station cost models

III. Evaluate the total network

  A. Configure Earth stations

    a. Install traffic capabilities

    b. Determine factors such as redundancy

  B. Perform tradeoffs for network

    a. Sum total costs

    b. Test sensitivity

    c. Consider space segment

12.4.1 Determining Traffic Requirements

One of the most difficult tasks facing the system designer is determining the expected usage (and ultimately the revenue) of the project before it goes into operation, which is the first step in Figure 12.1. The term *traffic* actually refers to the quantity of information flow and its distribution over time. For example, telephone traffic is expressed in call attempts and call durations of telephone conversations, measured in minutes. Existing telephone switching equipment can measure telephone traffic in a terrestrial network currently providing service, yielding the accurate traffic data needed to design a new satellite network. A totally new telephone network via satellite may not have such benefit, so estimates must be made through other research techniques (not, one would hope, including guess work). In years past, telephone traffic followed repeatable patterns and could be estimated using a set of statistical measures developed in the early part of the twentieth century. More recently, telephone networks deliver fax messages and are used to connect to the Internet on a dialup basis. That has greatly complicated the problem of designing a telephone network with precision. What usually is done is to design using conventional traffic engineering principles and then to provide significant extra capacity to deal with peak loading situations.

For video distribution, the usage usually is predetermined by the TV programmer and can be specified as a requirement that is fixed for a period of time. A cable TV programming network operator would specify a quantity of full-time video distribution channels that provide a group of related programming formats. An example is a common network that must be delayed for different time zones and adapted to deliver some local advertising. Other types of video traffic are variable in nature, such as backhaul for news and sports events (discussed in Chapter 3). Those requirements could have a seasonal element (e.g., baseball and football) as well as a component that is entirely random (e.g., news events and visits by heads of state). Another significant variable has been introduced as a result of digital compression and multiplexing of several video channels on one carrier. Compression standards, particularly MPEG 2, allow for variable bit rate encoding; therefore, more video channels can be multiplexed on to one carrier.

Data communications cannot be treated like telephone traffic because there is wide variability of source information rates and their distribution over time (which is all too common in modern computer networking). Standard protocols, notably TCP/IP and X.25, have introduced some uniformity in network data traffic, but patterns are still difficult to characterize in real-world situations. The move to ATM will allow networks to better adapt to variation in data traffic loading and types. Multimedia satellite systems to be introduced after 2000 can employ ATM to more effectively deliver bandwidth on demand. The best method for evaluating ATM or other data traffic loading is to use computer simulation tools currently popular in terrestrial network design.

An example of the expected flow of traffic between the geographic nodes of a network is plotted in Figure 12.6. Primary locations such as corporate and division headquarters have large traffic requirements, justifying the use of a mesh of major trunking paths (shown as diamonds connected by heavy lines). Regional centers, indicated by squares, have significant requirements for bandwidth connections to the primary locations. The smallest predictable traffic requirements are between remote sites such as branch offices and customer locations. Those thin routes, which could number in the thousands, pass data and voice traffic to the regional centers and possibly the primary locations. Using terrestrial network technology, each location on the map is a node that can independently generate and accept traffic. For reasons of efficiency, it may be desirable to aggregate closely spaced nodes through a process called bundling. Many alternative arrangements of bundled nodes could satisfy the traffic requirements, but that must be analyzed with the overall economic model to yield the network configuration that is truly the most economical to implement. The disadvantage of bundling is that it increases the length and the quantity of local access lines and, in the case of a satellite network, the tails, since many user locations are some distance from the closest node.

The last step in determining the traffic requirements is to convert traffic flows into defined communications channels of the satellite system. That is necessary because an Earth station generates one or more modulated carriers that access the satellite repeater. Flexible multiple access methods, like TDMA and

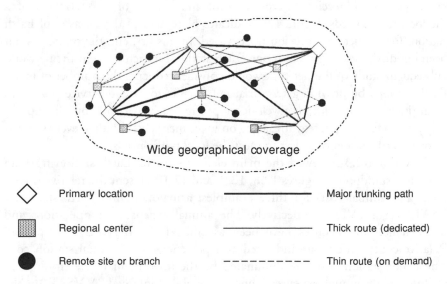

Figure 12.6 Interconnection of user locations in a private wide area communications network based on a terrestrial implementation.

ALOHA, are adaptive in nature. In any case, there will be a specific quantity of interface ports to deliver the required video, voice, and/or data to the terrestrial interface. Devices that respond to the time-varying traffic flow can be installed but there still is an upper limit to port capabilities. The model should allow the temporary selection of channel capacities and the testing of those assumptions to see if the projected traffic can be supported. Insufficient channel capacity between two nodes causes blockage, meaning that subsequent attempts to use the link will receive a busy signal or be denied access through some other mechanism. In a terrestrial version of the network, traffic from a remote site can reach a primary location by passing through a regional center node. That can pose a problem for a satellite network since such connections might involve a double hop. (The exception is for a LEO satellite system, in which the delay of each satellite hop is considerably less than the GEO delay of 260 ms.) Direct point-to-point connections are possible over the satellite if the demand assignment feature is included in the network architecture. Otherwise, a quantity of terrestrial circuits should be retained as part of the backbone to prevent a double hop when a circuit connection must pass through another node.

12.4.2 Laying Out the Network

A priority in adapting satellite communications to an overall network design is the selection of various categories of stations and their capabilities. As suggested in Table 12.1, there might be three different classes of Earth stations in the network. A site would receive the type of station and quantity of channels that match the local traffic needs. Figure 12.7 indicates placement of three classes of Earth stations for the network previously discussed. In this first cut, the traffic has not been bundled, meaning there is one Earth station for each node in the network. Subsequent study of the network would consider reducing the number of major Earth stations by providing terrestrial tail circuits to extend service points. In countries or regions without a well-developed terrestrial network, bundling probably is not a feasible strategy for saving on investment costs. Rather, every site gets its own Earth station of a class to meet the local traffic needs.

Section 12.3 reviewed the main elements in the Earth station from an economic standpoint. Figures 12.8, 12.9, and 12.10 present the relative investment and annual costs for three examples: a major or hub Earth station, a VSAT, and a TVRO, respectively. The annual expenses for operations and maintenance and utilities have been assumed for a five-year time period. Relative contributions are indicated as a percentage of the total station cost. The major station costs are dominated by the investment in electronics and facilities, with annual expenses being reasonably contained. A VSAT, on the other hand, requires more attention to its annual expenses, mainly because the

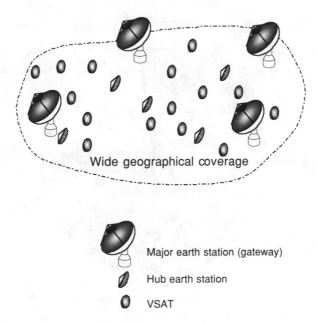

Major earth station (gateway)

Hub earth station

VSAT

Figure 12.7 Configuration of a private satellite network to satisfy requirements for interconnectivity between points in a wide geographical region.

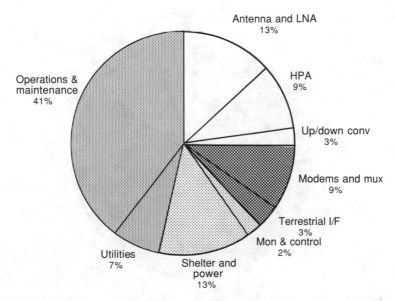

Figure 12.8 Allocation of investment and operating expenses among elements in a major or hub Earth station.

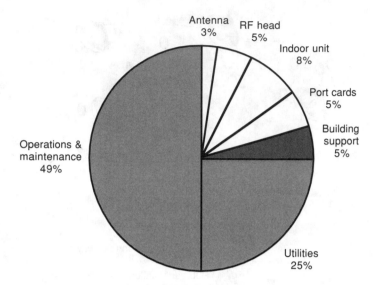

Figure 12.9 Allocation of investment and operating expenses among elements in a VSAT for interactive two-way communication.

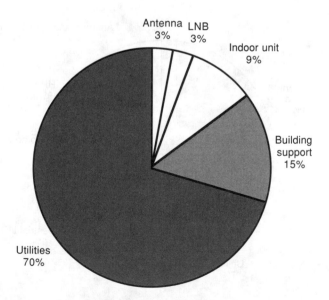

Figure 12.10 Allocation of investment and operating expenses among elements in a TVRO terminal for DTH service.

electronics are not particularly costly and placement in an existing building is assumed. Many organizations convert VSAT investment cost to annual cost by leasing or renting the equipment from the manufacturer. That is a good strategy when a using organization does not want to invest in equipment that they feel might become obsolete in a few years.

A TVRO station appears to represent an extreme case, in which the cost of power for the electronics over a five-year period is comparable in cost to the initial expense for the antenna and LNA. The type of equipment used in digital DTH networks is as simple to operate and maintain as a VCR and a TV set. Expenses for operations and maintenance are small because of the simplicity of the design of the TVRO and because repairs typically are accomplished by sending the failed unit back to the factory (or, more likely, discarding the unit and buying a new model). To compare the three stations with each other, the relative magnitude of total cost (i.e., the size of the individual pies) approximates 10,000 to 10 to 1 for the major station, VSAT, and TVRO, respectively.

12.4.3 Total Network Evaluation

In the last phase of the study, the constituents of the network traffic arrangement, Earth station deployment, and element cost are brought together for an integrated analysis. The stations are configured as nodes in the network, incorporating the channel capacities determined in the first phase. As indicated in Table 12.1, it is important that redundancy be included where appropriate. That can be accomplished by providing backup equipment in the individual Earth station or by adding alternative communication paths in the network.

The PC spreadsheet economic model can automatically sum up costs for the network and compute useful indices such as cost per channel or per minute of service. In a commercial system, it is important to determine the projected profit (or loss) that would result under various assumptions for system traffic loading and pricing to users. Various tests of sensitivity to key parameters can be made after the accuracy of the model has been established. It is common in investment analysis to evaluate the internal rate of return (IRR) for a nominal case and then to examine the change in IRR for different values of important parameters. Examples of such parameters include:

- Quantity of subscribers;
- Service usage per subscriber (minutes per month);
- Take-up rate (rate of increase in usage of the service) and churn (rate of customer turnover);
- Charge per month and per increment of usage;

- Market segments and penetration rates in each segment;
- Increase of 10% in the total investment in space or ground;
- Debt-to-equity ratio;
- Interest rate applied to debt.

12.4.4 Optimizing the Space and Ground Segments

As mentioned previously, it is desirable to bring the technical and economic aspects of the space segment into the analysis. One of the more interesting tradeoffs is the determination of the optimum Earth station G/T. The results of such a tradeoff for TVRO service using the Ku-band BSS are presented in Figure 12.11, where the underlying assumptions are indicated at the lower right. The cost of a certain quantity of satellite bandwidth usually depends on the amount of RF power (EIRP) delivered in that bandwidth. In this example, it is assumed that the total weight and cost of the satellite are fixed but that the power per transponder and number of transponders are varied (i.e., doubling of TWT output power halves the total number of transponders on the satellite). The benefit of more power is that the diameter of the receiving Earth station antenna can be reduced, saving ground total investment cost. That cost could be allocated either to the users (who must purchase their own DTH receivers)

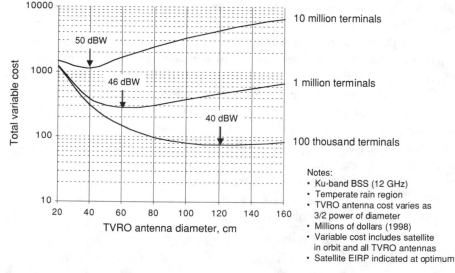

Figure 12.11 Determination of the optimum antenna size for a DTH TVRO terminal in the Ku-band BSS.

or retained by the network operator (who provides the DTH receiver free with the service). For a fixed system noise temperature such as 100K, the Earth station G/T is determined entirely by antenna gain. Consequently, the tradeoff is presented in terms of receiving dish size.

There will be a combination of satellite power and antenna diameter that minimizes the sum of the costs of those two elements for all stations in the network. It usually turns out that the more Earth stations there are in the network, the lower is the optimum antenna size (corresponding to a higher value of satellite EIRP). In terms of a nominal design point for the satellite, the number of transponders is 16 and their individual power is 200W, corresponding to a typical BSS satellite design with a per-transponder EIRP of 50 dBW. Investment costs are estimated for the satellite in orbit and for the receiving dish without the indoor settop unit because only the RF element costs are affected by the tradeoff of dish size and satellite EIRP.

The curves in Figure 12.11 indicate that in the broadcasting application using a very high powered satellite transponder, the minimum cost occurs at approximately 1m for a system of 100,000 terminals, while the corresponding minimum for 1,000,000 antennas is approximately half that diameter. To allow the diameter to be decreased to around 40 cm, the satellite EIRP should increase from approximately 50 dBW, as indicated along the X-axis of Figure 12.11. The BSS approach with relatively wide spacing between satellites is favored in those results because of the small dish size used to achieve optimum. Minimum cost conditions in the figure are within flat bottoms of the respective curves, demonstrating that the sensitivity of the optimum to the selected antenna size is not particularly strong.

This particular tradeoff and optimization study has been simplified to some extent, and the results should not be taken as conclusive. However, the power of the technique is evident. To be able to perform studies of this type, it is essential that accurate performance and cost data be used and that the model be properly constructed. Planning of this type requires a considerable investment in time and talent, beginning with a basic model and extending and expanding it as new data are gathered. Furthermore, the results should be checked against other hypotheses and available real-world data. Review by systems engineers and operations experts will provide greater confidence in the models and what they are telling their developers.

12.5 Satellite Communications: Instant Infrastructure

The systems and services to contribute to the business side of satellite communications have grabbed the attention of the financial community [3]. Satellites at any altitude have the power to deliver applications over the widest area, creating an instant infrastructure with a predetermined economic life. The established FSS, BSS, and MSS applications derive excellent revenues for established operators. The challenge for a new entrant is to develop a new niche, gain the spectrum and orbit resources, provide for market entry, and create and manage the instant infrastructure for maximum advantage. Companies like Iridium, ICO, and Globalstar are building a new future for the satellite industry and will create wealth for investors and affiliated service providers.

According to Bear Stearns, the companies that will be successful are those that:

- Properly play the space game. They obtain the right orbit slots (GEO) or orbit resources (non-GEO) and frequency assignments. From there, the satellites must have high capability and utilize proven technology to reduce risk.
- Lead the ground game. They get the right partnerships with equipment suppliers, service providers, and distributors. To that is added the key aspect of market access and local licensing.

Success is not a given—the project must have adequate funding to implement the system and deliver services during the initial lean years. Furthermore, having satellites in orbit is not enough. The ground partnerships need money, too. Making all that happen in a coordinated way makes the industry unique.

References

[1] "Activate the Money Star," *The Economist*, May 3, 1997, p. 56.

[2] Elbert, B. R., *The Satellite Communication Applications Handbook*, Norwood, MA: Artech House, 1997.

[3] Watts, T. W., and D. A. Freedman, *Satellite Communications—Instant Infrastructure*, Bear Stearns Equity Research, November 5, 1996.

13

Future Directions for Satellite Communications

It has been 12 years since the writing of the first edition, and the future of satellite communication is more promising than ever. Until 1990, advances in satellite communications systems and applications were evolutionary in nature. The beginning of the industry was technology driven, with the development of the first GEO spacecraft design and the proving of its feasibility in the early 1960s. Favorable economics—compared to terrestrial alternatives—gave satellites a needed boost to literally get them off the ground. The rapid introduction of fiber optic transmission is seen in some ways as a response to the advances made by satellite communications in reducing transmission cost. Today, networks of wideband fiber spread across continents and oceans, delivering digital communication service of the highest quality and at reasonable prices. However, a near revolution in our thinking about satellite systems occurred in the 1990s, with the development of non-GEO constellations like Iridium, ICO, and Globalstar. That approach, illustrated in Figure 13.1, promises to double the size of our industry.

The versatility of satellites should allow them to fill important and vital needs even as fiber optic networks extend to the home. At the same time that satellite applications are evolving in response to competition from terrestrial systems, the technology base also is undergoing evolutionary change. There is every reason to expect that the capability and versatility of satellites and Earth stations will continue to improve significantly over the coming years. That provides a technology push all over again, allowing system designers and operators to approach new applications and markets with powerful hardware and software

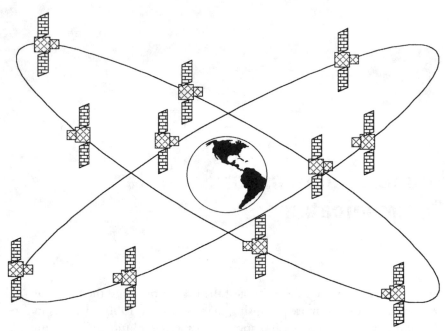

Figure 13.1 Satellite communications continue to expand in the coming millennium as new constellations, like ICO, come into being.

capabilities. Lower costs from microminiaturization in user equipment and advanced satellite constellations will make satellite communications as commonplace as the laptop computer and cell phone.

In this section, an attempt is made to project into the future using the current shape of the industry as a starting point. Much of this discussion should be viewed as conceptual and about possibilities because the precise makeup of the applications and technology is impossible to predict. Even the most conservative projection would show satellites as having a significant role in the future telecommunications picture. However as unpredictable advances will occur, the future for satellites probably will be even brighter than predicted here.

13.1 New Satellite Applications

The changing nature of the uses to which communication satellites economically have been put is reviewed in Table 13.1. The table is a compilation of the version in our original work plus an assessment and projection from the *Wall*

Table 13.1
Evolutionary and Revolutionary Trends in the Applications of
Satellite Communications Systems

1970s	First domestic GEO satellites
	Long-distance telephone: international links
	Video transmission: point-to-point
	Experimental use for defense applications
1980s	GEO satellites become commercialized
	Video and audio distribution: point-to-multipoint
	Data communication: point-to-point and point-to-multipoint
	Private networks: introduction of VSATs
	Maritime mobile satellite communication: ship-to-shore
	Defense-related satellite communications
1990s	Rapid expansion of GEO satellite business: first introduction of non-GEO satellites
	Video broadcast: digital DTH point-to-multipoint
	Mobile satellite communication: service to mobile users
	Extensive use of VSAT networks
	Extensive defense application using government and commercial satellites
	Introduction of satellite-based Internet service: point-to-multipoint
2000s	GEO and non-GEO satellite investment become comparable
	Growth of global mobile personal communication service (GMPCS) from non-GEO systems
	Satellites deliver Internet and multimedia services across the globe
	HDTV added to DTH systems

Street Journal [1]. We have seen the rate of new introductions of technologies, systems, and applications increase to the point that we might call it a revolution rather than an evolution. Prior to and during the 1970s, the predominant applications were in telephone communications, a use that employed 80–90% of the available capacity on domestic and international satellites. The INTELSAT GEO satellites formed the backbone of the international telecommunications system. Anik A, the first domestic satellite, was launched in 1972 to serve Canada and was soon followed by several other domestic systems. Video services provided at that time were primarily point-to-point in nature, while TV networks continued their reliance on terrestrial microwave to distribute programming to their affiliates. Toward the end of the decade, point-to-multipoint distribution of cable TV and radio programming grew rapidly. On the government side, GEO satellites were adopted for strategic and tactical communications, demonstrating the ability of the technology to serve large military units and ships. Those early experiments proved to the governments of the western hemisphere that satellites could be depended on to fill a vital need.

During the 1980s, the number of GEO satellites increased rapidly, and satellite operation became a commercial business as opposed to a government activity. The combined mass of all video transmission consumed approximately two-thirds of total domestic satellite capacity serving the United States. That occurred because cable TV programmers expanded their channel delivery and were joined by TV networks to take advantage of low cost and reliable satellite signal distribution. That was due to the growth of existing U.S. satellite operators adding four new players: SBS, GSat, Spacenet, and ASC. Their objective was to address the unsatisfied demand for business networks and telephony. The quality and reliability of the satellite link made data transmission attractive for both point-to-point and point-to-multipoint (broadcast) transmission, particularly using the new VSAT technology. Maritime mobile satellite services proved themselves as Inmarsat established itself and commercial HF radio links were nearly abolished.

By the 1990s, video-based media became an enormous industry, expanding on a global scale. That would not have been possible without the quality and economy of using domestic and international GEO satellites. The advent of high-powered DBS satellites and highly effective digital video technology greatly increased the quantity of TVRO dishes found at homes, principally because of diminished size and cost. As indicated in Table 13.1, small dishes also proved valuable for interactive voice and data services in the environment of Ku-band VSATs. Land mobile satellite services were introduced by Inmarsat, Optus, AMSC, and others, wherein users in vehicles and remote locations could access the public network for voice and data services. Finally, Internet access service was introduced in the United States by Hughes Network Systems, proving that satellite delay can be accommodated provided the TCP/IP protocol is managed properly.

The decade following 2000 brings both a new millennium and the first operation of the new class of non-GEO satellite constellations. With Iridium and Globalstar on line, satellite users can carry and use Earth stations in their hands. GEO systems are not to be left behind, as greater power, new frequencies (Ka band), and digital processing usher in an era of more bandwidth and improved interactivity. Terminals become a commodity, as users find that GEO and non-GEO satellites can deliver the services they need around the world. The wired networks advance as well, with improved access speeds and network protocols like ATM. However, the benefits of satellite communication will maintain this medium as important and cost effective for decades to come. Not least of those is HDTV, which the terrestrial networks strain to address but which becomes a normal facet of modern digital DTH systems.

This picture provided an introduction to the following discussion of the lineup of valuable services that satellites will provide in decades ahead. We see that the near term is dominated by the technologies, systems, and applications that were under development just prior to 2000. At the same time that satellites are addressing new markets and services, the terrestrial networks are evolving and improving. Keep in mind that applications often appear out of nowhere and some of the more important ones to come have not even been thought of yet.

13.1.1 Emphasis on the Broadcast Feature

The point frequently made in previous chapters is that one of the satellite's principal advantages is its wide-area coverage capability. In broadcasting, the downlink signal is available everywhere within the footprint. That capability will continue to be attractive for video, audio, and data delivery purposes. The decade of the 1990s has, in fact, seen the introduction of true DBS service, building on the foundation of medium- and high-power bent-pipe GEO satellites and digital compressing and multiplexing technologies. C-band DTH service offerings from cable TV programmers have given way to more attractive Ku-band approaches, which permit smaller antennas on the ground. The ability to deliver over 200 television channels from one orbit position is in place in many countries. Yet, DTH will become more competitive vis-a-vis terrestrial systems as DBS operators experiment with new media formats.

One exceptional aspect of DBS is its ability to reach a narrow segment of users dispersed throughout a nation. Programs in a particular foreign language such as Armenian or Mandarin Chinese can be sent directly to the home for convenient viewing by families. Likewise, custom training and specialized cultural events can be targeted to the individual rather than the community. That greatly simplifies the logistics of delivering programming that otherwise would have to come by way of the local cable TV system or UHF station, both of which may be unavailable due to limited channel capacity. Another possibility, currently allowed in Mexico but not in the United States, is the use of DBS to distribute local programming to metropolitan areas or provinces. That could employ a nationwide transmission using addressable receivers or through multiple spot beams that also afford frequency reuse.

Another strong potential area for DBS is in the delivery of HDTV programming. The digital standards for HDTV broadcast have been adopted in the United States and Europe, yet the technology still appears to be bound by the reluctance of broadcasters to make the required investments. Also, while receivers are available, their high cost and relatively low brightness have impeded initial acceptance. The DBS satellite can easily transmit such a signal, which requires about four times the compressed data as a conventional signal. While that is an attractive idea, it still rests on the assumption that enough viewers will purchase the more expensive HDTV receivers when they become widely available. The chicken-and-the-egg problem exists: Which comes first, the digital network or the millions of HDTV receivers? Add to that the challenge of making programming available in this format. Perhaps the resistance is due to the reluctance of manufacturers to produce the right receivers before the transmissions appear on the popular satellites.

The DBS platform, with its ability to deliver a high-speed digital multiplex, also provides an important opportunity to provide a variety of information services. Figure 13.2 indicates how data broadcasting can be introduced in the digital DTH systems now prevalent. As discussed in Section 13.1.2, broadcast data is an underexploited capability of bent-pipe satellites. The same dish that provides over 200 channels of compressed full-motion video is also an effective way to distribute multimedia accessible with a Web browser. An important extension of this medium comes from adding the return channel either through the PSTN or, ultimately, over the satellite. The latter can be supplied at either Ku or Ka band, using a low-power transmitter from the user terminal. Packet data are collected at the broadcast center, processed, and used as

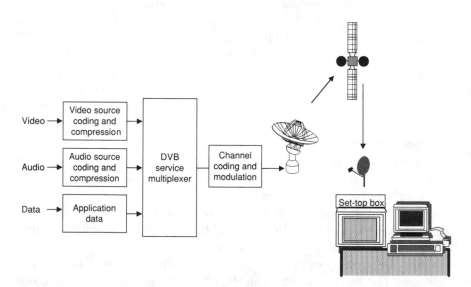

Figure 13.2 Adding broadcast data to the MPEG II digital video multiplex.

the means to select a new data flow and even for video on demand (VOD). The basic technology to do that has been available to the industry for many years; what remains is making the technology available to consumers at low incremental cost (or perhaps with no increase in subscription fee). A precursor to that is a service called Wavecast, which follows the PBS TV signal and resides in the vertical blanking interval (VBI).

13.1.2 Paralleling the Terrestrial Networks

The evolution of satellite applications illustrated in Table 13.1 clearly shows that while satellite transmission was at first attractive for long-distance links, the situation has shifted significantly since the advent of fiber optics and digital wireless services like PCS. However, satellite links will continue to be useful as diverse and alternative routes. That is demonstrated by large private VSAT networks operated by companies like Wal*Mart and Chevron, and agencies like the civil aviation ministry of the People's Republic of China. Private networks employ satellites because of the flexibility of the services offered and the ability of the user to own the ground equipment and exercise total control of the network.

While it often is said that satellites and terrestrial services compete with each other in the global markets, it is better to recognize the complementary nature of the two media. At times, one will dominate a given application segment, such as terrestrial cellular for mobile telephone and satellites for digital video broadcasting, both in the 1990s. The nature of that dominance and competition between terrestrial communications and satellite communications will change and shift from time to time. The first innovations in high-capacity long-distance transmission were in the terrestrial area. Those developments included microwave transmission and analog FDM. Satellites took off next, relying on the same technologies but greatly reducing the cost of transmission by eliminating the unnecessary microwave repeater sites. Satellites became well established and pioneered new technologies in digital processing and transmission, notably TDMA, digital echo cancellation, and video compression. That began to bleed over to the terrestrial side, providing the basis for advances in terrestrial communications such as digital switching and high-speed digital transmission. More recently, the obstacles to HDTV and digital TV broadcasting are being overcome through the innovation of digital DTH service, pioneered by DIRECTV. We are poised for greater expansion of satellite services by virtue of GMPCS projects like Iridium and Ka-band broadband networks like Teledesic. Quite logically, forthcoming innovations in both satellite and terrestrial technologies will aid in the redefinition of telecommunications services that can be provided by the next generation of both approaches.

13.1.2.1 Satellites Versus Fiber Optics

While fiber optic networks are well established in the developed world, satellite communications will continue to play a complementary and therefore important role. Conceptually, a fiber optic network in a given country provides the backbone of transmission between major cities and user locations. The economics of fiber are very favorable as long as the fibers and digital transmission groups within them can be adequately loaded with paying traffic. An unloaded fiber optic cable is not attractive economically, compounded by the fact that once the link is installed between two points it cannot be deployed elsewhere. Interestingly, excess fiber optic capacity in the United States lead to the development of the Vyvyx video interconnection network, currently popular with the major TV broadcasters.

The obvious role of satellite transmission is to provide the flexibility that point-to-point cables cannot. Thin route applications will always be attractive by satellite. A hybrid network of heavily loaded fiber and diverse satellite links can produce the lowest cost per call where service is to be provided on a widespread basis. The type of satellite that is called for would operate in the FSS portion of the Ku band, taking advantage of the ability to locate uplinks anywhere in the satellite footprint. A high-capacity Ku-band satellite can achieve cost per call for thin route service that is competitive with conventional service over the public network, even with low-cost fiber optic transmission between major nodes. Another approach to applying satellite technology is through fixed services from MSS systems. The low cost of the MSS user terminal in these systems makes the service attractive on an investment cost basis, if not in terms of per-minute charges.

Fiber optic bandwidth is not directly available to most residential users and to only a fraction of businesses. The real excitement in terrestrial networks is the promise of greater bidirectional bandwidth on the local loop. Technologies like digital subscriber line (DSL) and cable modems could open wider doors for a variety of interactive applications. Users frustrated with the underwhelming capability of even the best analog modems (claiming to offer 56 kbps) and narrowband ISDN all look forward to megabit-per-second access speeds. They would like to enjoy better Web surfing, with real-time video and audio and possibly video conferencing. Also of interest to both the telephone companies and cable operators is VOD, first introduced in Hong Kong by Hong Kong Telecom.

The reality at the time of this writing is that both DSL and cable modems are past the early development and standards-setting phases and are ready for prime time. However, a radio advertisement for DSL from a local telephone operator mentions that subscribers could obtain the service "where available" and likely at an increased cost. That must reflect the difficulty that local operators have in making a major (expensive) commitment to DSL technology on a network basis. In comparison, the satellite approach requires that the space segment be placed in service with minimal subsequent cost to extend service to any particular new subscriber. The issue for satellite operators is whether to launch a single GEO satellite or a constellation of non-GEO satellites.

13.1.2.2 International Communications

Satellites will continue to provide a cost-effective means of spanning oceans and continental distances between countries. A high-capacity satellite operated by INTELSAT, PanAmSat, or another operator provides a valuable point of central interconnectivity for the widest variety of traffic between dozens of international gateway stations within a hemisphere. In the non-GEO domain, a web of broadband LEO satellites can offer end-to-end connectivity, far better than can undersea cables. On the other hand, it is possible that a specifically designed heavy-trunking GEO satellite can effectively parallel fiber optic cables in the heavy route marketplace.

The use of international satellites to provide telecommunications service within domestic borders should continue to be viable into the future. A developing country that wishes to upgrade the quality and reliability of domestic telephone and television service by installing a satellite network would find that transponder leasing is an economical way of getting started. This type of business is attracting organizations other than INTELSAT that intend to aggregate several such customers and operate the satellite on a condominium basis. A dedicated domestic satellite system is another viable approach, in which demand can justify the investment.

13.1.2.3 Expansion of Business Video

The development of video communications for private business use has largely been made possible by advances in satellite transmission. Whether the signal is transmitted in analog or digital form, the application represents an innovative approach for business communications. Point-to-multipoint broadcast as well as point-to-point teleconferencing are both growing quickly in use in the developed world. The service is easily added to VSAT networks operating at Ku band because the relatively high power of the satellite permits the direct reception of the video signal. An interactive service is possible with the transmit capability of the VSAT. The advent of Ka-band satellite systems, with broadband access to individual rooftops, makes interactive business video services even more affordable and, more important, more available.

Business video is beneficial in both domestic and international satellite systems, in which a significant reduction in travel time and expense is possible.

Perhaps even more important is the increase in person-to-person communication with the added visual dimension. Business travel allows a few people to meet, but business video increases the number of participants since those who have not traveled can now participate. The key is to make the room or desktop equipment more available and prevalent and to offer low-cost long-haul transmission service that is available when needed (on demand, without significant setup delay). The technology of video transmission for business applications is expected to reduce the cost of providing the service, leading to even greater use in the future. Bandwidth compression will make terrestrial transmission attractive for point-to-point teleconferencing, but broadcast usage will rely heavily on satellites well into the future.

13.1.3 Mobile Communication Systems

Communications with and between vehicles, ships, and aircraft have been an important facet of radio. More recently, projects like Iridium and ICO are demonstrating how satellite networks can offer true global personal communications to hand-held user terminals, similar to cell phones. The use of satellites for such purpose accelerated in the 1990s after many years of slow and patient introduction. However, if it had not been for COMSAT and Hughes with their Marisat GEO satellites, mobile satellite services might never have appeared in the first place. Satellite MSS has replaced various forms of shortwave and VHF radio for mobility, and the future is very bright indeed.

Future mobile satellite networks will draw from the successes of the past, with the benefits of global connectivity being an important attribute. This can perhaps be foreseen through the concept of critical mass in networks. Even with all the space and ground factors properly considered, a satellite communications network still can fail as a business. That, of course, is a complex subject and cannot be answered in general. There is, however, a principle that helps explain the timing of when (and if) a new telecommunications service will achieve success in the marketplace. Figure 13.3 presents a conceptual curve of how a network service is developed and taken up by subscribers. We see that in the initial years, only a small percentage of the general population (early adopters) subscribe to the service. Subsequently, more users come on line to experiment and learn two things: what information is available and who else is connected.

Experience has shown that as the subscriber base increases, the network becomes attractive to yet more users. If that works as expected, the network is its own self-fulfilling prophesy. User adoption increases rapidly at the point of inflection in the curve, which is where critical mass is achieved. Network subscriptions accelerate as more potential users sign on and communicate with

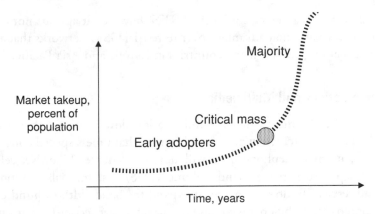

Figure 13.3   Conceptual service adoption curve, indicating the point where critical mass is achieved.

others. In some networks, the attraction to the network is irresistible, as people find that they must connect either to carry out their business or to enjoy what has become a widely adopted source of information, communication, or entertainment (whatever the case may be).

We can certainly see how critical mass worked to the benefit of fax machines, cellular telephones, and now the Internet. On the other hand, those areas are subject to intense competition, and it is hard to find a lucrative niche. In some countries, cellular service providers are protected by monopolies and import restrictions. In a few cases, companies have found a means of protecting their position in a market that has reached critical mass. Examples include Microsoft in PC operating systems and personal productivity software (which are not strictly telecommunications services but act like them in some ways) and Qualcomm in CDMA technology, due to their early lead and patent protection.

The non-GEO technical strategy has remained dormant since Telstar and Relay, until Motorola identified it as a way to better address pent-up demand for two-way wireless voice communications to remote regions of the world. The highly integrated nature of the non-GEO entries makes them attractive as platforms for exploitation of the principle of critical mass. Iridium is interesting in that sense because its network operates on a stand-alone basis, with only limited interconnection to ground-based networks. Iridium stands to capture all the benefits of service expansion (including financial rewards and protection of market) if a degree of critical mass is achieved. ICO can be seen to have many of the same supportive elements, and Teledesic delivers integrated service options that do not depend on other networks. In contrast, Globalstar is an extension of the terrestrial telephone network since all calls must pass through gateway Earth stations to reach other parties. That yields instant critical mass in the sense that it is an extension of the PSTN; however, it appears not to give the operator any particular advantage over terrestrial GSM networks that afford convenient roaming among many countries in Europe and Asia Pacific.

13.1.4 Interactivity and Multimedia

Broadband generally means that the information flow is greater than what a standard telephone circuit can pass, which means that unless special conditioning is used, normal telephone lines will not be adequate. Likewise, cellular, PCS, and Japanese personal handy phone (PHS) systems will not provide broadband service. To do that, we must expand the bandwidth to hundreds or even thousands of kilobits per second (e.g., megabits per second). That implies fast response time between a user and the ultimate source of the information, but not necessarily so. We can always broadcast a lot of information and yet provide only a very limited capability for direct interaction between the two ends. That is the principle behind the DirecPC™ service currently offered offer GEO bent-pipe satellites and is part of the new service options from the IMT-2000 terrestrial wireless standards development effort.

By convention, multimedia include all forms of information, such as text, image, audio, and moving picture. The World Wide Web provides many of those elements but lacks the broadband feature. Multimedia implies that we have an interactive link, but that can be simulated in a variety of ways. In Figure 3.20, we see a user in front of a PC connected by a network to an information server. With a Pentium or Power PC class of personal computer, the user has a high degree of interactivity on the desktop. However, when we connect that same PC to a network, the quality of interactivity degrades quickly. The performance drop happens at each point of information transfer within the network and finally at the information server. The degree and quality of interactivity for the user can be improved by making the network much simpler, with fewer intermediate nodes and points of congestion.

Advanced multimedia satellites can employ GEO and non-GEO systems, provided that adequate radio spectrum is made available. Figure 13.4 shows a view of the microwave spectrum, starting with L band at around 1.5 GHz and ranging up to the top of Ka band at 30 GHz. The major lettered segments are indicated from left to right, highlighting the available bandwidth. Ku-band satellites have 500 MHz of spectrum available to the uplink and downlink. A single satellite of that type can deliver up to about 32 analog TV channels or about 200 digital TV channels. That provides a form of broadband communications, lacking the interactive end-to-end feature. The real opportunity for new multimedia services is at Ka band, where up to 2,000 MHz of bandwidth could be available.

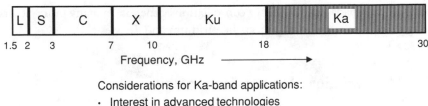

Considerations for Ka-band applications:
- Interest in advanced technologies
- More bandwidth
- More orbit slots
- New networking (and business) potential
- Technical and economic viability to be determined

Figure 13.4 Ka-band entry into the millimeter wave spectrum and promising new applications in multimedia and broadband communications.

Ka band is not a new idea because several governments and the satellite industry have been developing this technical capability for over 20 years. The difference is that for the first time there is a good business opportunity to support the required investment. When properly engineered, Ka-band satellite networks offer high capacity to relatively small fixed Earth stations that could be cheaper than the VSATs currently popular with many corporations around the world.

Ka band offers ample bandwidth for networks of small Earth stations, but there are a number of technical challenges. Based on experience in Japan, the United States, and Europe, we know that rain severely affects the satellite link at Ka band. In comparision to a Ku band, about three times the amount of margin is needed for Ka band. That translates into a requirement for much greater power on the link. Fortunately, the natural characteristic of Ka-band satellite antennas compared to Ku band is to provide a much higher gain spot beam for the same size of physical reflector in space. That is nearly enough to overcome the problem. The complexity, then, is in using a satellite that has 100 or more spot beams to cover a country the size of China or the United States. For that reason, many of the coming generation of Ka-band satellites probably will also include onboard digital processing.

These observations and projections of satellite communications suggest that there are many possibilities for new services and added competition for digital terrestrial networks. It will be up to the developers of satellite networks to make the decisions at the right time. That is not a simple matter because while it is easy to suggest options, it is much harder to construct a system with long-term economic viability. GEO satellites with bent-pipe repeaters are a known quantity and are capable of achieving very high rates of return. Accomplishing the same result for processing payloads and non-GEO constellations depends on many more factors, as suggested by the scope of projects like Iridium and Teledesic.

13.2 The Evolution of Space Technology

While satellite communications began through the technology push of the space program, it is the commercial application that produces a $50 billion industry. The decade commencing in the year 2000 should witness the introduction of new technologies that can play a leading role. This section reviews a number of these evolving technologies, many of which have already been applied in government and commercial programs. Readers should keep in mind that we do not pursue new technology for its own sake (even though that was perhaps the basis for many of the original innovations). Today, we operate in a business world, where performance is often measured in financial terms. That can simplify our choices because any technology introduction should produce some improvement in the overall business picture. It might provide a near-term revenue increase or, if done right, add value through future earning potential and greater market share.

Identification, development, and adoption of a new technology are complex matters, particularly because of the substantial risk if the technology cannot meet its objectives in a new system or application. Technologies on the immediate horizon could provide new applications and even create new markets. Consequently, system developers often are tempted to take unjustified risks. There are methodologies proven in the defense and space science fields to define and then mitigate those risks [2].

Technology for the space segment involves electronic innovations, improvements in the underlying materials and devices, as well as their conversion into a manufacturable form. Historically, spacecraft design and manufacture were labor-intensive, requiring many of the elements of basic research and custom engineering. That situation is changing due to the impact of satellite constellations with tens or even hundreds of the same spacecraft design. The evolution of direct to user services, such as DTH and hand-held mobile services, has provided the impetus to create more powerful and capable spacecraft. In addition, the need for greater capacity and service flexibility has supported the introduction of digital processing repeaters and multibeam antennas. Movement to higher frequencies such as Ka band has increased the demand for millimeter wave technology, compared to the rather slow pace of innovation during the previous decades.

Dr. Michael Daugherty of the Aerospace Corporation has provided a blueprint that identifies the likely technology introductions in spacecraft of the next decades. Starting with Dr. Daugherty's approach and recommendations, we have added other technology options that could prove important in the future. These options are identified in Tables 13.2, 13.3, and 13.4 for repeater, antenna, and bus, respectively.

13.2.1 Repeater Innovations

The technologies identified in Table 13.2 are representative of what developers can use for satellites to be launched after the year 2000. Gradual improvement in TWT and SSPA performance and reliability can be expected through industry R&D efforts. New devices and process technologies appear from year to year to add new options for making the payload more flexible. The GEO and non-GEO digital processing satellites can maintain their lead over terrestrial options only if they can demonstrate throughput performance, flexibility, and reliability for applications that users need and want. The thrust is for greater power to reduce user terminal size and cost, better switching and reconfiguration of the payload, and improvements in digital processing to increase payload capacity and services. Optical technology can be introduced to provide cross-links that do not interfere with the RF components onboard (including the physical interference of antennas and other appendages) and optical components to replace electrical signal distribution and signal routing. Taken together, those innovations improve the cost effectiveness of the satellites and the services they render.

13.2.2 Antenna System Innovations

High-gain antenna systems strongly influence the capacity and versatility of a communications satellite. Coming generations will employ highly shaped beams and sophisticated RF networks to increase bandwidth through frequency reuse techniques. Large, deployable reflectors have been in existence for years, yet satellite buyers are concerned about the risk of malfunction for apertures larger than about 6m. Programs like Thuraya and APMT demonstrate that capability at 12m. In the future, application of that size and larger is attractive, including the possibility of higher frequencies of operation.

Direct radiating arrays have long been used for L- and S-band payloads, such as Inmarsat II, ICO, and Globalstar. In 1985, a simple Ku-band planar array was flown on Morelos 1 for the receive function. That device provided the desired area coverage gain over Mexico in a much smaller physical size than would have been possible with a parabolic reflector. On the other hand, the planar array could support only 500 MHz of bandwidth in one polarization. The direct radiating arrays of individual cupped dipoles or helixes also have been

Table 13.2
Near-Term Space Segment Technological Innovations Applicable to the Repeater

Technology area	Projected improvement	Comments
RF components		
Multichip microwave integrated circuit (MMIC)	Higher levels of integration compared to conventional designs	Being introduced on a limited basis, due to number of competent suppliers
	High electron mobility power transistor (HEMPT) for receivers and power modules	Greater integration and higher power; flight experience needed
TWTA, SSPA	Increased output power and dc-to-RF efficiency	Continued improvement in performance and long life
Microwave switching	Greater use of satellite-switched TDMA and flexible routing	Improvement in packaging and low-power capability
Optical components		
Optical transceiver	Low-profile laser cross-links	Need to move from demonstration to dependable systems
Higher power laser transmitter	Provides greater data rate and margin	Enhances cross-links and has potential for space-ground links; need to provide reliable devices
Incorporation of fiber-optic technology	Reduced size of repeaters and better isolation between elements (e.g., against ESD)	Ensures acceptable power, packaging, interconnection, and on-orbit reliability
Digital components and subsystems		
High-speed complementary metal oxide semiconductor (CMOS) and DSP ICs	Greater density and speed for complex processors	Improved overall performance through higher speed, radiation hardness, greater throughput; limited number of suppliers
Advanced A/D converters	Higher sampling rates, higher resolution using InPHEMT	Higher level of integration involved with InPHEMT devices; to be addressed by industry

used extensively, but those antennas sometimes complicate the packaging of the payload and introduce significant quantities of stray RF energy. In the future, direct radiating arrays could move up in frequency to provide a multiplicity of directive beams, commandable to new coverage configurations.

Table 13.3
Near-Term Space Segment Technological Innovations Applicable to the Antenna System for Uplink and Downlink Communications

Technology area	Projected improvement	Comments
Antenna reflectors		
Larger deployed reflectors (>15 m)	More gain and frequency reuse for S- and L-band services	Need to maintain surface accuracy above 15m; prevent PIM
Improved solid reflector shaping	Better cross-polarization and pattern shaping at Ku and Ka	Normal extension of existing technology, with lighter weight
Deployment of more surfaces	More services from same satellite	Takes advantage of larger spacecraft with higher power and more capacity; ability to ensure reliable deployment and pointing
Longer focal length "deep" structures	More gain	Better use of volume around spacecraft
Direct radiating arrays		
Planar array	Receive capability demonstrated for Ku band; wideband and transmit capability desired	Improved shaping and antenna efficiency; greater potential for multibeam and reconfigurable antennas
Low-level beam forming	Included in digital processor; wideband capability desired	Processor speed and size to be determined
High-level beam forming	Direct attachment of amplifiers and phase shifters	Power-handling capability
Integrated microstrip antenna	Can be applied to integrated array elements; reduces weight and manufacturing costs through better reproducibility	Limited as to thermal control and power efficiency
Lens		
Waveguide lens	Compact configuration	To demonstrate reconfigurable and multibeam service
Dielectric lens	Potential for small size and light weight	Material to be developed and demonstrated

Lens antennas offer the potential to reduce physical size yet retain the versatility of a multihorn feed system. Systems at Ka band are hampered by the small dimensions of waveguide and horn arrays. For that reason, the lens-type antenna could provide substantial benefits.

Table 13.4
Near-Term Space Segment Technological Innovations Applicable to the Spacecraft Bus

Technology area	Projected improvement	Comments
Power subsystem		
Advanced nickel hydrogen batteries	Single pressure vessel design lowers weight, which reduces launch cost and/or allows larger payload	NiH_2 technology is mature, yielding relatively small improvements; changes from previous designs should be approached with caution.
Lithium ion batteries	Higher energy density and cycle life than NiH_2. Can be exploited to reduce weight or, for same weight, increase power	Due to immaturity of this technology, space application may wait until post-2001.
Advanced silicon solar cells	Slightly higher efficiency than current silicon technology. Less costly and more rugged than GaAs cells	Currently near the physical limit in performance; improvements could be based on lower cost and other benefits of silicon.
Advanced GaAs solar cells	Improvements in manufacturing processes to reduce cost and improve yield	Offers the potential of dominating the satellite solar cell market.
Solar collectors and reflectors	Use of lightweight reflectors provides greater power without added weight of extra cells	Deployment and long-life performance to be demonstrated.
Attitude control		
True three-axis control	Includes yaw detection for real-time full three-axis control	Must add reliable yaw sensor and techniques to maintain yaw alignment.
Acoustic and laser gyros	Allow gyros to be on, continuously safely	Drive compensation is a key aspect.
Improved ACS processing	Faster, more capable onboard computing and improved uploadable software for adaptive control	Need to ensure reliable operation in the presence of the space environment.
Propulsion systems		
Ion propulsion	Xenon ion propulsion actively in use for GEO missions. Increased thrust levels needed for transfer orbit phase	Higher thrust means larger aperture, which raises technical difficulty of achieving long life.

Table 13.4 (*continued*)

Enhanced chemical rocketry	Chemical fuels (e.g., hydrazine) attractive due to high thrust levels and dependability; increased I_{sp} dependent on electric heating or other enhancers	Long-life performance demonstrated in simple devices, but higher thrust levels remain a technical challenge for long life missions.
Thermal control		
Heat pipes	Flexible-loop heat pipes allow system to bring heat past variable geometry joints	Helps cope with launch volume constraints.
Deployable thermal radiators	Extra radiating area	Key to high-power spacecraft.
Structure		
Lightweight materials	Improvements in carbon fiber and other materials to give greater strength, thermal stability, and workability at lower weight	Improvements occur as manufacturers identify and experiment with advanced lightweight composites and other material technologies.
Mechanisms		
Memory metal mechanisms	More reliable, lower-mass actuators	Material applications engineering is key.
Replacement of pyro releases by hot knives or electrical releases	More reliable, low shock	Lower mass?
Replacement of pyro valves by shaped memory metal valves	Low shock	Lower mass?
Other		
Micro- and nanotechnology (MNT)	New sensor and tiny actuators to improve onboard monitoring and self-diagnostic capabilities	Applies only to low-power, small features and appendages (i.e., applications unknown).

13.2.3 Spacecraft Bus Innovations

Improvements on the spacecraft bus side are as important as (in many cases, more important than) those associated with the communications payload. The specific technologies identified in Table 13.4 provide varying degrees of improvement in the performance or cost of operation of spacecraft. Power system advancements are most important at GEO, where transmissions must overcome the added path length. The array itself offers the opportunity to improve the precision of alignment to the sun. For example, on Iridium, the solar panels are not maintained perpendicular to the sun, thereby requiring a substantially larger panel area. With effective three-axis motion, the size of the panel can be decreased. Battery improvements are available to all missions, but LEO spacecraft can gain more because they experience eclipse several times a day. Pointing accuracy requirements of multibeam Ka-band spacecraft are very tight, so ACS performance is going to be a key objective. Onboard ACS processing is another area that requires continual attention, so the capability of space-borne computing equipment will be expanded as the newer generations of microprocessors become space qualified. That must be approached with conservatism because of the harsh radiation environment is space.

The efficiency of propulsion systems will allow standard LVs to lift more capable spacecraft because some of the orbit-raising function can be performed by the RCS. Ion propulsion is finally in service at GEO and should become a standard feature on a majority of those spacecraft. In the future, thrust levels will increase to better support orbit raising, which would be welcome as a viable backup to LAM systems. Other propulsion technologies using chemical fuels are likely to be introduced as well. The listing in Table 13.4 is not all-inclusive because many innovations are buried in the details of spacecraft design; in fact, some of the most important are process improvements rather than physical hardware design changes.

Two opposing trends are apparent in spacecraft design and manufacture: larger, more power satellites on one end of the spectrum and smaller, low-cost spacecraft on the other. For GEO satellites, the newest designs have greater power and can carry more antennas and repeater equipment than the designs that gained popularity in the early 1990s. The objective, of course, is to be able to support the more powerful payloads demanded by GEO L-, S- and Ka-band applications. With its ability to deploy and maintain large sun-oriented solar panels, the three-axis or body-stabilized configuration continues to play a leading role in the future. Spinning satellites, being simpler in design and operation than body-stabilized ones, should find an important place in small- to medium-class satellites, particularly at C band.

The creation of non-GEO systems, notably Iridium, demonstrated that smaller, low-cost spacecraft can be manufactured more quickly and for less money per kilogram than their GEO counterparts. That appears to be possible because many identical spacecraft can be constructed on a production line. Contributing to that is the shorter expected operating lifetime of spacecraft operating in LEO. Also, the value and the risk associated with one LEO spacecraft are substantially lower than for one GEO, since it contributes substantially less to overall system capability in a constellation. GEO spacecraft also tend to be more costly because customers traditionally purchase one or two at a time, and the designs are more customized for different orbit positions and transponder mixes.

A general category of bus technology improvements can be referred to as technology streamlining, indicating that many diverse innovations often combine in a beneficial way. As new spacecraft are designed and built, numerous small weight reductions are introduced in the various subsystems. A lighter-weight material might be found for the thermal blankets, the wiring might be simplified by the use of remote fiber optics, and advanced composites containing graphite might be used in parts of the structure to add strength at the same time weight is reduced. The individual weight saving per item would be small (a few kilograms here or there), but in aggregate the reduction can represent several percentage points of total spacecraft mass. Each new generation of spacecraft bus design is significantly better than its predecessor, in terms of performance to support the payload and cost of operating in orbit. That still must be approached carefully so reliability is not sacrificed. One would have to say that with the wide variety of satellite communications missions and orbits, there is no single bus design for all. Rather, there are component and subsystem designs that can be adopted and improved. The real tasks ahead of us are to learn more about how to optimally design the bus for the mission and then to produce the complete spacecraft efficiently and dependably.

13.2.4 Launch System Development

Chapter 10 provides an overview of launch vehicles that will be available in 2000. It is difficult to make projections of major innovative technology for the next century. A significant problem is the escalating cost per kilogram of placing spacecraft mass into GTO. Fortunately, several new LV providers have entered the market, and launch costs remain steady. Surprisingly, the increased demand for launches to support the rollout of the first three non-GEO MSS programs did not exceed supply. The Space Shuttle was to have greatly reduced that cost by employing a reusable vehicle with the capability to place several spacecraft into parking orbit at the same time. However, with much of the true cost of operating the STS now accountable, the shuttle seems no more cost effective than current expendable LVs. A new reusable LV system is on the drawing boards at NASA (Figure 13.5), and perhaps such an approach could be fruitful.

Launch operations in the future might employ a space station, which is under development in the United States, Russia, and Europe. It would be

Figure 13.5   NASA's next-generation reusable launch vehicle Venture Star is on the drawing boards.

extremely costly and probably unattractive to attempt to put a space station into
GEO. However, a space station at the altitude of the parking orbit could act as
a staging area for subsequent unmanned missions. It has been suggested that
large spacecraft can be brought to the space station for final assembly and tested
in space under realistic conditions. Then a space tug or other propulsion system
would move the satellite out to the desired orbit. Assembly in space represents
an interesting possibility for communications satellites of the next century.

13.2.5 Managing Technology Risk

Designers of new generations of spacecraft as well as buyers of those products
must keep in mind the substantial risks that exist when new technology is
adopted quickly. Risk/benefit trades should be performed to ensure a proper
level of technology maturity for a particular application. According to Dr.
Daugherty, that must be an informed and conscious decision made during the
planning stages. Important but immature technology cannot be guaranteed, so
such a situation demands that one or more alternatives be identified early on to
provide what could be called an escape route (a term not used in the industry,
but one that nevertheless describes the view).

We can see many examples over the years of commercial and government satellites in which technology maturity was overestimated. The grandest example is the development of the Space Shuttle, which was put on a tight budget and followed what is termed a success-oriented schedule. The delivery of the first craft was delayed for years because many technology choices did not pan out, and alternatives had to be identified and developed. That added substantially to the cost of the program, which probably overran by a significant factor. Dr. Daugherty's advice should be heeded: "Program planners and managers must set development milestones and require that they be met before the program proceeds to the next phase. Failure to do so can cause a large system development program to be held hostage to a small technology development effort."

This should not dissuade readers from considering new technology. There are numerous situations in which a newly qualified innovation provides big gains and a competitive advantage. An early example was the gyrostat spacecraft from Hughes, which carried the first transponderized payload on Intelsat IV. That satellite showed that a space-borne repeater station could compete effectively with undersea cables and terrestrial line-of-sight microwave. More recently, Motorola took on a massive technology risk in developing the first big LEO satellite system. They put their organization to task to define and construct a system of a type not previously attempted. That puts them in a strong position to lock in an attractive market if and when it materializes.

13.3 Ground Station Technology

Satellite technology continues to undergo rapid evolution in the digital age. Many vital improvements are on the ground and now tend to drive payload requirements. The direction of technological innovation in Earth stations and networks will involve greater miniaturization and cost reduction. As a result, user terminals will gain in performance and functionality, while providing an affordable option to modern terrestrial networks. That will push VSAT and DTH applications downward toward individual users, building economy of scale not possible in many existing consumer services. On the other hand, the network, which has grown in size and diversity, must be managed effectively. That will demand advanced network management systems, such as are under development for terrestrial data communications networks that connect personal computers to servers and network backbones. The technology of digital signal processing and compression will continue to advance, making possible greater integration of voice, data, and video communications. With digital processing satellite payloads providing the ideal points for traffic routing and distribution, new applications will appear and perhaps gain dominance. This section reviews those possibilities in more detail.

13.3.1 VSAT Technology Extension

The impressive gains in the price and capability of PCs and cell phones are being transferred to the VSAT of the future. Basic to understanding the effective use of VSATs are the star and mesh architectures, first discussed in Chapter 3. The principal benefit of VSAT networks is that they can bypass the terrestrial network, particularly low-bandwidth analog local loops normally provided by the local telephone company. The first VSAT networks to appear were designed primarily for data communications and effectively applied the star architecture. With the hub station collocated with the central computer to be accessed, the star provides sufficient benefit to be competitive with multidrop private line telephone service. Telephone traffic could also pass over the link from VSAT to hub; however, it is not advisable to use another satellite hop to reach the final destination. The exception to that rule is for non-GEO satellite constellations (either with or without intersatellite links) because of the lower delay per hop. Still, there comes a point where multihop propagation delay with a hub could be unacceptable for some applications.

Greater use of the mesh topology promotes improved point-to-point service quality for GEO satellites. Point-to-point connections are particularly beneficial for high-speed data transmission, where fast response time is demanded. Mesh networks require that each VSAT have the capability to transmit sufficient uplink power to be received with another small-diameter antenna. Current VSATs operate with low power because the receiving antenna at a hub is typically considerably larger. In bent-pipe satellite payloads, the critical technology is the reliable solid state power amplifier, which is part of the VSAT outdoor equipment. Low-cost power amplifiers that could support mesh services were on the horizon at the time of this writing and should be available widely before 2000. Rain attenuation at Ka band places the additional burden that the uplink be powerful enough to provide satisfactory overall service availability. A useful feature of the VSAT would be to include automatic uplink power control to boost the level when the uplink is experiencing heavy rainfall. That has the added benefit of maximizing transponder capacity since the average power demand would be substantially less than if all carriers include adequate margin.

13.3.2 Digital Compression of Video and Voice

The state of the art of digital compression for communications has been advancing at an accelerating rate. There are three parts to the process: the mathematical algorithms used to digitize the analog signal and compress its bandwidth; the processing hardware that performs the algorithm; and the quality resulting after it has been restored to analog form. The last factor involves the perception of human subjects, that is, the quality of the signal is in the eyes and ears of the beholder. Obviously, the economics of how much capability can be obtained in a piece of processing equipment must be traded off against user acceptance of the quality of the result. New high-speed semiconductor memories and micro-processors are available to push the tradeoff in the direction of greater compres-sion for the same expense. The other benefit of compression is the reduction in transmission rate, which reduces the cost of providing the service.

In the area of video conferencing, current compression equipment oper-ating at 384 kbps generally is accepted by users, while 64 kbps quality is viewed as substandard. The equipment to produce either compressed video format is still too costly to allow widespread use, even if the cost of transmission were essentially zero. It is likely that the quality of 64 kbps compression will improve to the point where it is equal to the current 384 system (of course, this system also will improve). Equipment costs will come down significantly as the mar-ket develops and larger manufacturing volumes are possible.

The cost of transmission is also coming down, both because a lower rate is possible and because of the ability to use VSATs and advancing terrestrial transmission systems. It would not be surprising to see 64-kbps video telecon-ferencing achieve the dream of the picture telephone exhibited by AT&T at the 1964 New York World's Fair. That particular system would have required a 6-Mbps link into every establishment, 100 times the bandwidth of today's tech-nology. The ultimate video teleconferencing terminal will provide quality full-motion video at 64 kbps or less and employ a PC as its user platform.

Compression of voice signals is always the subject of investigation because of applications in mobile communications. Today, 64-kbps PCM voice at 8 bits per sample is no longer demanded, with 32 kbps ADPCM being adopted as the international standard. Further reduction has been hampered by the use of tele-phone channels for medium-speed data transmission on an analog basis. With the advent of ISDN networks, PCs will have a full 64 kbps of continuous trans-parent data transmission. Telephone channels of the future can be optimized for voice, greatly reducing the bit demand. There is already an 8-kbps transmission technology using a predictive algorithm that provides voice quality nearly indis-tinguishable from either 32 or 64 kbps. Compression down to 2,400 bps has been demonstrated at low quality but with acceptable intelligibility. Eventually, even that speed will provide adequate voice quality. Satellite networks will then have two avenues for success: more capacity for low-speed voice channels or more bandwidth (due to the use of Ka band and digital processing) to support video conferencing and other innovative applications.

13.3.3 Inexpensive TVROs for Digital Service

The importance of reducing the cost of the terminal in broadcast receiving is just as vital as it is in two-way VSAT applications. Fortunately, the simplicity of the video receive function makes cost reduction more a matter of manufacturing volume than technology innovation. The emphasis now is on using the smallest possible receiving dish to reduce cost and simplify installation. In the 1990s, services like DIRECTV and Dish TV introduced low-cost digital settop boxes with 45-cm dishes, generating what has become a true megamarket in the satellite industry. The next step is to add data communication features to provide the kind of convergence between video and computing that the prognosticators have been talking about for years. The broadband downlink can then deliver an array of services to the enhanced settop boxes.

13.4 Satellite Communication Into the Third Millennium

In 40 years, the satellite communications industry has clearly come a long way. Once seen as a technical feat and a curiosity, the geostationary communications satellite is now commonplace and indispensable in many sectors. There has been a maturation process at work: first the technology had to make economical sense, then the applications for satellite communication had to prove themselves in a competitive marketplace. Clearly, those steps have been accomplished more rapidly than anyone could have imagined.

Satellite communications is the foundation for and now defines certain industries. Cable TV in North America never could have become a $20-billion industry without satellite's reliable and low-cost delivery of programming. Transoceanic communications would still be limited to only those heavy routes that could justify the investment in undersea cable. Numerous other uses, which are yet to become industries in themselves, are establishing themselves through access to C- and Ku-band space segment. Whether it be video shopping at home, broadcasting of financial news directly to stock brokers' offices, or mobile communications by satellite with people anywhere in the world, this developmental and evolutionary process is constantly going on.

The financial community, typified by noted Wall Street firms like Bear Stearns and Goldman Sachs, are optimistic about the role of satellite communication in the coming millennium. Investors recognize the benefits of instant infrastructure and attractive user services that can be provided directly by satellite operators. This opens up business opportunities in developed regions for obtaining new revenues and for improving the quality and availability of more basic telecommunications in developing regions as well.

The technology of the satellite itself is also advancing, although the changes now tend to be less dramatic. Essentially, the applications now drive the design of the satellites to be launched in the future. It is not uncommon for a technology to be developed specifically to provide a certain capability required for an application.

Satellite communications at the end of the twentieth century and in the century following will provide many services currently available. For example, the distribution of TV programming will certainly be by way of satellites. It is the new applications, not yet introduced, that will be the most exciting, providing the base for expansion of the industry in new directions. Satellite communications will be an important part of the evolving picture of the twenty-first century.

References

[1] "Telecommunications, A Special Report," *The Wall Street Journal,* September 1997.

[2] Watts, T. W., and D. A. Freedman, *Satellite Communications—Instant Infrastructure,* Bear Stearns Equity Research, November 5, 1996.

List of Acronyms and Abbreviations

ACK	Acknowledgement of error-free reception
ACS	Attitude control subsystem
ACTS	Advanced Communications Technology Satellite
A/D	Analog to digital conversion
ADPCM	Adaptive-differential PCM
AGC	Automatic gain control
AKM	Apogee kick motor
AM	Amplitude modulation
AMF	Apogee motor firing
AMSC	American Mobile Satellite Corporation
AOR	Atlantic Ocean region
ASIC	Application-specific integrated circuit
ASK	Amplitude shift keying
ATC	Advanced Television Committee
ATM	Asynchronous transfer mode
ATS	Advanced Technology Satellite
BAe	British Aerospace
BAPTA	Bearing and power transfer assembly
BER	Bit error rate
BISDN	Broadband integrated services digital network
BPSK	Binary phase shift keying
BSB	British Satellite Broadcasting
BSS	Broadcasting satellite service
CAI	Common air interface
CATV	Community antenna television

CBR	Constant bit rate
CCIR	Commitée Consultatif Internationale de Radio
CDMA	Code division multiple access
C/I	Carrier-to-interference ratio
C/IM	Carrier-to-intermodulation ratio
C/N	Carrier-to-noise ratio
CNES	Centre National d'Etudes Spatiales
CNN	Cable News Network
COMSAT	Communications Satellite Corporation
CP	Circular polarization
CRC	Cyclic redundancy check
CS	Communications Satellite (Japan)
CTS	Communications Technology Satellite
CTS	Clear to send
D/A	Digital to analog conversion
DAB	Digital audio broadcasting
DAMA	Demand assignment multiple access
DASA	Daimler Aerospace (Germany)
DBS	Direct broadcast satellite
DCME	Digital circuit multiplication equipment
DCU	Delay communication unit
DES	Digital Encryption Standard
DFT	Discrete Fourier transform
DIO	Delivery in orbit
DMA	Defined market area
D-MAC	D-multiplex analog components
DNR	Dolby™ noise reduction
DSI	Digital speech interpolation
DSL	Digital subscriber line
DSP	Digital signal processing
DTH	Direct to home
DTV	Digital television
DVB	Digital video broadcasting
ECS	European communications satellite
EHT	Electrically heated thruster
EIRP	Effective isotropic radiated power
EMC	Electromagnetic compatability
EPC	Electrical power conditioner
ESA	European Space Agency
ESD	Electrostatic discharge
FCC	Federal Communications Commission

FDM	Frequency division multiplex
FDMA	Frequency division multiple access
FEC	Forward error correction
FET	Field effect transistor
FFT	Fast Fourier transform
FM	Frequency modulation
FSK	Frequency shift keying
FSS	Fixed satellite service
GaAs FET	Gallium arsenide field effect transistor
GEO	Geostationary Earth orbit
GMPCS	Global mobile personal communication service
GPSS	Global positioning satellite system
GSO	Geostationary satellite orbit
GTE	General Telephone and Electronics
GTO	Geostationary transfer orbit
HBO	Home Box Office
HDTV	High-definition TV
HF	High frequency
HNS	Hughes Network Systems
HPA	High-power amplifier
IBM	International Business Machines Corporation
IBS	INTELSAT Business Service
IC	Integrated circuit
IDR	Intermediate data rate
IF	Intermediate frequency
IFL	Interfacility link
IMD	Intermodulation distortion
IMUX	Input multiplexer
INTELSAT	International Telecommunications Satellite Organization
IP	Intellectual property
IRD	Integrated receiver decoder
IRR	Internal rate of return
ISDN	Integrated Services Digital Network
ISI	Intersymbol inteference
ISL	Intersatellite link
ISRO	Indian Space Research Organization
ITU	International Telecommunication Union
JC Sat	Japan communications satellite
KPA	Klystron power amplifier
KSC	Kennedy Space Center
LAM	Liquid apogee motor

LAN	Local area network
LCP	Left-hand circular polarization
LEO	Low Earth orbit
LMDS	Local multichannel distribution service
LNA	Low-noise amplifier
LNB	Low-noise block converter
LNC	Low-noise converter
LO	Local oscillator
LP	Linear polarization
LRB	Liquid rocket booster
LSA	Launch services agreement
LV	Launch vehicle
MAC	Multiplex analog components
MCPC	Multiple-channel per carrier
M&C	Monitoring and control
M-DAC	McDonnell Douglas Astronautics Company
MEO	Medium Earth orbit
MIF	Modulation improvement factor
MMDS	Multichannel microwave distribution service
MMH	Monomethylhydrazine
MPEG	Motion Picture Experts Group
MSS	Mobile satellite service
NAK	Negative acknowledgement (reception in error)
NASA	National Aeronautics and Space Administration
NASDA	National Space Development Agency (Japan)
NMS	Network management station
NOC	Network operations center
NSAB	Swedish acronym for Swedish Satellite Corporation
NTSC	National Television Standards Committee
OCC	Operations control center
O&M	Operations and maintenance
OMUX	Output multiplexer
OSI	Open Systems Interconnection
OTS	Orbital test satellite
PAL	Phase-Alteration-Line
PAM	Payload assist module
PARAMP	Parametric amplifier
PBS	Public Broadcasting Service
PBX	Private branch exchange
PCM	Pulse code modulation
PDH	Pleisiochronous digital hierarchy
PIM	Passive intermodulation

PIP	Payload integration plan
PLL	Phase locked loop
PM	Phase modulation
PN	Pseudorandom noise
POP	Point of presence
POR	Pacific Ocean region
PRN	Pseudorandom noise
PSK	Phase shift keying
PSTN	Public-switched telephone network
PTT	Post, telephone, and telegraph
QoS	Quality of service
QPSK	Quadraphase shift keying
RASCOM	Regional African Satellite Communications
RCP	Right-hand circular polarization
RCS	Reaction control system
RF	Radio frequency
RFI	Radio frequency interference
rms	Root-mean-square
RO	Receive only
RRC	Regional Radiocommunication Conference
RT	Receiver transmitter
RTS	Request to send
SAW	Surface acoustic wave
SBS	Satellite Business Systems
SCA	Subcarrier channel authorization
SCC	Satellite control center
SCC	Space Communication Corporation (Japan)
SCF	Satellite control facility
SCPC	Single channel per carrier
SDH	Synchronous digital hierarchy
SDLC	Synchronous data link control
SDMA	Space division multiple access
SECAM	Sequential couleur avec memoire
SES	Société Européene des Satéllites
SHF	Super high frequency
SNA	System network architecture
SNG	Satellite news gathering
SNMP	Simple Network Management Protocol
SONET	Synchronous optical network
SPADE	Single channel per carrier PCM multiple access demand assignment equipment

SRM	Solid rocket motor
SSB	Single sideband
SSOG	Standard System Operating Guide
SSPA	Solid state power amplifier
SS/TDMA	Satellite-switched time division multiple access
STC	Satellite Television Corporation
STS	Space Transportation System
STV	Subscription television
SUG	Satellite user group
SYNCOM	Synchronous orbit communications satellite
T&C	Telemetry and command
TDM	Time division multiplex
TDMA	Time division multiple access
TDRS	Tracking and data relay satellite
TE	Transverse electric
TED	Threshold extension demodulator
TM	Transverse magnetic
TSR	Time slot reassignment
TT&C	Tracking, telemetry, and command
TVRO	TV receive-only
TWT	Travel wave tube
TWTA	Traveling wave tube amplifier
UDMH	Unsymmetrical dimethylhydrazine
UHF	Ultra high frequency
UPC	Uplink power control
UPS	Uninterruptable power system
USSCI	United States Satellite Communications, Inc.
UT	User terminal
VBI	Vertical blanking interval
VBR	Variable bit rate
VHF	Very high frequency
VLSI	Very large scale integration
VOD	Video on demand
VSB	Vestigial sideband
VSAT	Very small aperture terminal
VSWR	Voltage standing wave ratio
WAN	Wide area network
WARC	World Administrative Radio Conference
WATS	Wide area telecommunication service
WDM	Wave division multiplex
WRC	World Radiocommunication Conference

About the Author

Bruce R. Elbert is a leading authority on satellites and telecommunications, having 35 years of working experience at Hughes, COMSAT, and the U.S. Army Signal Corps. He directed the design of several major satellite projects, including Palapa A, Indonesia's original satellite system; the Galaxy follow-on system, the largest and most successful C-band satellite TV system in the world; and the first GEO mobile satellite network to serve handheld terminals. By considering the technical, business, and operational aspects of satellite systems, Mr. Elbert has contributed to the economic success of several companies in the marketplace. In addition to an international industrial background, he is a prolific writer of books and articles, and has been featured as a speaker at numerous international conferences and seminars. Mr. Elbert is the editor of Artech's Space Technology and Applications Library, and the Technology Management and Professional Development Series. At UCLA, he teaches satellite communications, information networking, and high-tech marketing to practicing engineers. Mr. Elbert holds BEE and MSEE degrees from CCNY and the University of Maryland, respectively, and an MBA from Pepperdine University. He has two daughters, Sheri and Michelle, and resides with his wife, Cathy, on the Palos Verdes Peninsula in California.

Index

The Artech House Telecommunications Library

Vinton G. Cerf, Series Editor

For further information on these and other Artech House titles, including previously considered out-of-print books now available through our In-Print-Forever™ (IPF™) program, contact:

Artech House
685 Canton Street
Norwood, MA 02062
781-769-9750
Fax: 781-769-6334
Telex: 951-659
email: artech@artech-house.com

Artech House
Portland House, Stag Place
London SW1E 5XA England
+44 (0) 171-973-8077
Fax: +44 (0) 171-630-0166
Telex: 951-659
email: artech-uk@artech-house.com

Find us on the World Wide Web at:
www.artech-house.com